SPACE-TIME CODING: THEORY AND

This book covers the fundamental principles of space-time coding for wireless communications over multiple-input multiple-output (MIMO) channels, and sets out practical coding methods for achieving the performance improvements predicted by the theory.

Starting with background material on wireless communications and the capacity of MIMO channels, the book then reviews design criteria for space-time codes. A detailed treatment of the theory behind space-time block codes then leads on to an in-depth discussion of space-time trellis codes. The book continues with discussion of differential space-time modulation, BLAST and some other space-time processing methods. The final chapter addresses additional topics in space-time coding.

Written by one of the inventors of space-time block coding, this book is ideal for a graduate student familiar with the basics of digital communications, and for engineers implementing the theory in real systems.

The theory and practice sections can be used independently of each other. Exercises can be found at the end of the chapters.

HAMID JAFARKHANI is an associate professor in the Department of Electrical Engineering and Computer Science at the University of California, Irvine, where he is also the Deputy Director of the Center for Pervasive Communications and Computing. Before this, he worked for many years at AT&T Labs-Research where he was one of the co-inventors of space-time block coding.

SPACE-TIME CODING

THEORY AND PRACTICE

HAMID JAFARKHANI

University of California, Irvine

CAMBRIDGE
UNIVERSITY PRESS

CAMBRIDGE UNIVERSITY PRESS
Cambridge, New York, Melbourne, Madrid, Cape Town, Singapore,
São Paulo, Delhi, Dubai, Tokyo

Cambridge University Press
The Edinburgh Building, Cambridge CB2 8RU, UK

Published in the United States of America by Cambridge University Press, New York

www.cambridge.org
Information on this title: www.cambridge.org/9780521131407

First published 2005
This digitally printed version 2010

A catalogue record for this publication is available from the British Library

ISBN 978-0-521-84291-4 Hardback
ISBN 978-0-521-13140-7 Paperback

Contents

Preface

The use of multiple antennas in most future wireless communication systems seems to be inevitable. Today, the main question is how to include multiple antennas and what are the appropriate methods for specific applications. The academic interest in space-time coding and multiple-input multiple-output (MIMO) systems has been growing for the last few years. Recently, the industry has shown a lot of interest as well. It is amazing how fast the topic has emerged from a theoretical curiosity to the practice of every engineer in the field. It was just a few years ago, when I started working at AT&T Labs – Research, that many would ask "who would use more than one antenna in a real system?" Today, such skepticism is gone.

The fast growth of the interest and activities on space-time coding has resulted in a spectrum of people who actively follow the field. The range spans from mathematicians who are only curious about the interesting mathematics behind space-time coding to engineers who want to build it. There is a need for covering both the theory and practice of space-time coding in depth. This book hopes to fulfill this need. The book has been written as a textbook for first-year graduate students and as a reference for the engineers who want to learn the subject from scratch. An early version of the book has been used as a textbook to teach a course in space-time coding at the University of California, Irvine. The goal of such a course is the introduction of space-time coding to anyone with some basic knowledge of digital communications. In most cases, we start with common ideas for single-input single-output (SISO) channels and extend them to MIMO channels. Therefore, students or engineers with no knowledge of MIMO systems should be able to learn all the concepts. While graduate students might be interested in all the details and the proofs of theorems and lemmas, engineers may skip the proofs and concentrate on the results without sacrificing the continuity of the presentation.

A typical course on space-time coding may start with some background material on wireless communications and capacity of MIMO channels as covered in Chapters 1 and 2. A review of design criteria for space-time codes is covered in Chapter 3.

Chapters 4 and 5 provide a detailed treatment of the theory behind space-time block codes. A practitioner who is only interested in the structure of the codes may bypass all the proofs in these chapters and concentrate only on the examples. Chapters 6 and 7 discuss space-time trellis codes in depth. Each chapter includes discussions on the performance analysis of the codes and simulation results. For those who are more interested in the practical aspects of the topic, simulation results are sufficient and the sections on performance analysis may be skipped. The practitioners may continue with Chapter 11 and its discussion on MIMO-OFDM and Chapter 9 on receiver design. On the other hand, those who are more interested in the theory of space-time codes can follow with Chapter 8 and its treatment of differential space-time modulation. Finally, for the sake of completeness, we discuss BLAST and some other space-time processing methods in Chapters 9 and 10. Homework problems have been included at the end of each chapter.

The book includes the contribution of many researchers. I am grateful to all of them for generating excitement in the field of space-time coding. My special thanks goes to my very good friend and former colleague Professor Vahid Tarokh who introduced space-time coding to me. Also, I should thank my other colleagues at AT&T Labs – Research who initiated most of the basic concepts and ideas in space-time coding. Without the support of my department head at AT&T Labs – Research, Dr. Behzad Shahraray, I would not be able to contribute to the topic and I am thankful to him for providing the opportunity. Also, Professor Rob Calderbank has been a big supporter of the effort.

The early versions of this book have been read and reviewed by my students and others. Their comments and suggestions have improved the quality of the presentation. Especially, comments from Professor John Proakis, Professor Syed Jafar, Dr. Masoud Olfat, and Hooman Honary have resulted in many improvements. My Ph.D. students, Li Liu, Javad Kazemitabar, and Yun Zhu, have helped with the proofreading of a few chapters. Many of the presented simulation results have been double checked by Yun Zhu. I also thank the National Science Foundation (NSF) for giving me an NSF Career Award to support my research and educational goals related to this book.

Last but not least, I would like to thank my wife, Paniz, for her support and love.

Standard notation

$\lVert \cdot \rVert$	Euclidean norm
$\lVert \cdot \rVert_{\mathrm{F}}$	Frobenius norm
$\otimes$	tensor product
$*$	conjugate
$+$	Moore-Penrose generalized inverse (pseudo-inverse)
det	determinant of a matrix
E	expectation
$f_X(x)$	pdf of X
H	Hermetian
$\Im$	imaginary part
I	imaginary part
I_N	$N \times N$ identity matrix
j	$\sqrt{-1}$
K_X	covariance of vector X
$\Re$	real part
R	real part
T	transpose
Tr	trace of a matrix
Var	variance

Space-time coding notation

$\alpha_{n,m}$	path gain from transmit antenna n to receive antenna m
χ_k	a chi-square RV with $2k$ degrees of freedom
η	noise
ϕ	rotation parameter for constellations
γ	SNR
$\rho(N)$	Radon function
θ	rotation parameter for STBCs
b	number of transmit bits per channel use
C	capacity
C_{out}	outage capacity
$\mathbb{C}$	set of complex numbers
$\mathbf{C}$	$T \times N$ transmitted codeword
$\mathcal{C}$	set of super-orthogonal codes
$d_{\min}$	minimum distance
E_s	average power of transmitted symbols
f_d	Doppler shift
G_c	coding gain
G_d	diversity gain
$\mathbf{G}$	generating matrix for a STTC
$\mathcal{G}$	generator matrix for a STBC
$\mathbf{H}$	$N \times M$ channel matrix
I	number of states in a trellis
J	number of delta functions in the impulse response of a frequency selective fading channel
J	number of groups in a combined spatial multiplexing and space-time coding system
2^l	number of branches leaving each state of a trellis
K	number of transmitted symbols per block

L	number of orthogonal (data) blocks in a SOSTTC
L	size of IFFT and FFT blocks in OFDM
L-PSK	a PSK constellation with $L = 2^b$ symbols
M	number of receive antennas
N	number of transmit antennas
$\mathcal{N}$	$T \times M$ noise matrix
N_0	noise samples have a variance of $N_0/2$ per complex dimension
P	number of trellis transitions (two trellis paths differ in P transitions)
P_{out}	outage probability
Q	the memory of a trellis
r	transmission rate in bits/(s Hz)
r	received signal
$\mathbf{r}$	$T \times M$ received matrix
R	rate of a STC
$\mathbb{R}$	set of real numbers
s	transmitted signal
S_t	the state of the encoder at time t
x	indeterminant variable
$\mathbb{Z}$	set of integers

Abbreviations

ADC	Analog to Digital Converter
AGC	Automatic Gain Control
AWGN	Additive White Gaussian Noise
BER	Bit Error Rate
BLAST	Bell Labs Layered Space-Time
BPSK	Binary Phase Shift Keying
BSC	Binary Symmetric Channel
CCDF	Complementary Cumulative Distribution Function
CDF	Cumulative Distribution Function
CDMA	Code Division Multiple Access
CSI	Channel State Information
CT	Cordless Telephone
DAST	Diagonal Algebraic Space-Time
DASTBC	Diagonal Algebraic Space-Time Block Code
D-BLAST	Diagonal BLAST
DECT	Digital Cordless Telephone
DFE	Decision Feedback Equalization
DPSK	Differential Phase Shift Keying
EDGE	Enhanced Data for Global Evolution
FER	Frame Error Rate
FFT	Fast Fourier Transform
FIR	Finite Impulse Response
GSM	Global System for Mobile
IFFT	Inverse Fast Fourier Transform
iid	independent identically distributed
IMT	International Mobile Telephone
ISI	Intersymbol Interference

LAN	Local Area Network
LDSTBC	Linear Dispersion Space-Time Block Code
LOS	Line of Sight
MGF	Moment Generating Function
MIMO	Multiple-Input Multiple-Output
MISO	Multiple-Input Single-Output
MMAC	Multimedia Mobile Access Communication
MMSE	Minimum Mean Squared Error
MRC	Maximum Ratio Combining
ML	Maximum-Likelihood
MTCM	Multiple Trellis Coded Modulation
OFDM	Orthogonal Frequency Division Multiplexing
OSTBC	Orthogonal Space-Time Block Code
PAM	Pulse Amplitude Modulation
PAN	Personal Area Network
PAPR	Peak-to-Average Power Ratio
PDA	Personal Digital Assistant
PDC	Personal Digital Cellular
pdf	probability density function
PEP	Pairwise Error Probability
PHS	Personal Handyphone System
PSK	Phase Shift Keying
QAM	Quadrature Amplitude Modulation
QOSTBC	Quasi-Orthogonal Space-Time Block Code
QPSK	Quadrature Phase Shift Keying
RF	Radio Frequency
RLST	Random Layered Space-Time
RV	Random Variable
SER	Symbol Error Rate
SISO	Single-Input Single-Output
SIMO	Single-Input Multiple-Output
SM	Spatial Multiplexing
SNR	Signal to Noise Ratio
SOSTTC	Super-Orthogonal Space-Time Trellis Code
SQOSTTC	Super-Quasi-Orthogonal Space-Time Trellis Code
STBC	Space-Time Block Code
STTC	Space-Time Trellis Code
TAST	Threaded Algebraic Space-Time
TASTBC	Threaded Algebraic Space-Time Block Code

TCM	Trellis Coded Modulation
TDD	Time Division Duplexing
TDMA	Time Division Multiple Access
V-BLAST	Vertical BLAST
ZF	Zero Forcing

1

Introduction

1.1 Introduction to the book

Recent advances in wireless communication systems have increased the throughput over wireless channels and networks. At the same time, the reliability of wireless communication has been increased. As a result, customers use wireless systems more often. The main driving force behind wireless communication is the promise of portability, mobility, and accessibility. Although wired communication brings more stability, better performance, and higher reliability, it comes with the necessity of being restricted to a certain location or a bounded environment. Logically, people choose freedom versus confinement. Therefore, there is a natural tendency towards getting rid of wires if possible. The users are even ready to pay a reasonable price for such a trade-off. Such a price could be a lower quality, a higher risk of disconnection, or a lower throughput, as long as the overall performance is higher than some tolerable threshold. The main issue for wireless communication systems is to make the conversion from wired systems to wireless systems more reliable and if possible transparent. While freedom is the main driving force for users, the incredible number of challenges to achieve this goal is the main motivation for research in the field. In this chapter, we present some of these challenges. We study different wireless communication applications and the behavior of wireless channels in these applications. We provide different mathematical models to characterize the behavior of wireless channels. We also investigate the challenges that a wireless communication system faces.

Throughout the book, we provide different solutions to some of the challenges in wireless communication by using multiple antennas. The main topic of this book is how to overturn the difficulties in wireless communication by employing multiple antennas. We start with a study of the capacity increase due to the use of multiple antennas. Then, we show how to design a space-time architecture for multiple transmit antennas to improve the performance of a wireless system while keeping the transmission power intact. Most of the book discusses different space-time coding methods in detail. The detailed discussion of each method includes design, properties, encoding, decoding, performance analysis, and simulation results. We pay close attention to the complexity of encoding and decoding for each method and to different trade-offs in terms of throughput, complexity, and performance. Not only do we provide the theoretical details of each method, but also we present the details of the algorithm implementation. Our overall goal is to keep a balance between the theory and the practice of space-time coding.

1.2 Wireless applications

There are many systems in which wireless communication is applicable. Radio broadcasting is perhaps one of the earliest successful common applications. Television broadcasting and satellite communications are other important examples. However, the recent interest in wireless communication is perhaps inspired mostly by the establishment of the first generation cellular phones in the early 1980s. The first generation of mobile systems used analog transmission. The second generation of cellular communication systems, using digital transmission, were introduced in the 1990s. Both of these two systems were primarily designed to transmit speech. The success of cellular mobile systems and their appeal to the public resulted in a growing attention to wireless communications in industry and academia. Many researchers concentrated on improving the performance of wireless communication systems and expanding it to other sources of information like images, video, and data.

Also, the industry has been actively involved in establishing new standards. As a result, many new applications have been born and the performance of the old applications has been enhanced. Personal digital cellular (PDC), global system for mobile (GSM) communications, IS-54, IS-95, and IS-136 are some of the early examples of these standards. While they support data services up to 9.6 kbits/s, they are basically designed for speech. More advanced services for up to 100 kbits/s data transmission has been evolved from these standards and are called 2.5 generation. Recently, third generation mobile systems are being considered for high bit-rate services. With multimedia transmission in mind, the third generation systems are aiming towards the transmission of 144–384 kbits/s for fast moving users and up to 2.048 Mbits/s for slow moving users.

The main body of the third generation standards is known as international mobile telephone (IMT-2000). It includes the enhanced data for global evolution (EDGE) standard, which is a time division multiple access (TDMA) system and an enhancement of GSM. It also includes two standards based on wideband code division multiple access (CDMA). One is a synchronous system called CDMA2000 and the other one is an asynchronous system named WCDMA. In addition to applications demanding higher bit rates, one can use multiple services in the third-generation standards simultaneously. This means the need for improved spectral efficiency and increased flexibility to deploy new services. There are many challenges and opportunities in achieving these goals.

Of course the demand for higher bit rates does not stop with the deployment of the third-generation wireless systems. Another important application that drives the demand for high bit rates and spectral efficiency is wireless local area networks (LANs). It is widely recognized that wireless connection to the network is an inevitable part of future communication networks and systems in the emerging mobile wireless Internet. Needless to say, the design of systems with such a high spectral efficiency is a very challenging task. Perhaps the most successful standard in this area is the IEEE 802.11 class of standards. IEEE 802.11a is based on orthogonal frequency division multiplexing (OFDM) to transmit up to 54 Mbits/s of data. It transmits over the 5 GHz unlicensed frequency band. IEEE 802.11b provides up to 11 Mbits/s over the 2.45 GHz unlicensed frequency band. IEEE 802.11g uses OFDM over the 2.45 GHz unlicensed frequency band to provide a data rate of up to 54 Mbits/s. Other examples of wireless LAN standards include high performance LAN (HiperLAN) and multimedia mobile access communication (MMAC). Both HiperLAN and MMAC use OFDM. The main purpose of a wireless LAN is to provide high-speed network connections to stationary users in buildings. This is an important application of wireless communications as it provides freedom from being physically connected, portability, and flexibility to network users.

There are many other applications of wireless communications. Cordless telephone systems and wireless local loops are two important examples. Cordless telephone standards include the personal handyphone system (PHS), digital cordless telephone (DECT), and cordless telephone (CT2). Wireless personal area network (PAN) systems are utilized in applications with short distance range. IEEE 802.15 works on developing such standards. Bluetooth is a good example of how to build an ad hoc wireless network among devices that are in the vicinity of each other. The Bluetooth standard is based on frequency hop CDMA and transmits over the 2.45 GHz unlicensed frequency band. The goal of wireless PANs is to connect different portable and mobile devices such as cell phones, cordless phones, personal computers, personal digital assistants (PDAs), pagers, peripherals, and so on. The wireless PANs let these devices communicate and operate cohesively. Also, wireless PANs can replace the wire connection between different consumer electronic appliances, for example among keyboard, mouse, and computers or between television sets and cable receivers.

1.2.1 Wireless challenges

While various applications have different specifications and use different wireless technologies, most of them face similar challenges. The priority of the different challenges in wireless communications may not be the same for different applications; however, the list applies to almost all applications. Some of the challenges in wireless communications are:

- a need for high data rates;
- quality of service;
- mobility;
- portability;
- connectivity in wireless networks;
- interference from other users;
- privacy/security.

Many of the demands, for example the need for high data rates and the quality of service, are not unique to wireless communications. But, some of the challenges are specific to wireless communication systems. For example, the portability requirement results in the use of batteries and the limitation in the battery life creates a challenge for finding algorithms with low power consumptions. This requires special attention in the design of transmitters and receivers. Since the base station does not operate on batteries and does not have the same power limitations, it may be especially desirable to have asymmetric complexities in different ends.

Another example of challenges in wireless communications is the connectivity in wireless networks. The power of the received signal depends on the distance between the transmitter and the receiver. Therefore, it is important to make sure that if, because of the mobility of the nodes, their distance increases, the nodes remain connected. Also, due to the rapidly changing nature of the wireless channel, mobility brings many new challenges into the picture. Another important challenge in a wireless channel is the interference from other users or other sources of electromagnetic waves. In a wired system, the communication environment is more under control and the interference is less damaging.

While the demand for data rates and the performance of the signal processors increase exponentially, the spectrum and bandwidth are limited. The limited bandwidth of the wireless channels adds increases impairment. Increases in battery power grows slowly and there is a growing demand for smaller size terminals and handset devices. On the other hand, the users want the quality of wire-line communication and the wire-line data rates grow rapidly. Researchers face many challenges to satisfy such high expectations through the narrow pipeline of wireless channels.

The first step to solve these problems is to understand the behavior of the wireless channel. This is the main topic of the next section. We provide a brief introduction

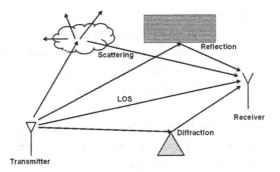

Fig. 1.1.   An example of different paths in a wireless channel.

to the topic, as it is needed in our discussion of space-time codes, and refer the interested reader to other books that concentrate on the subject [57, 103, 111, 123].

1.3 Wireless channels

One of the distinguishing characteristics of wireless channels is the fact that there are many different paths between the transmitter and the receiver. The existence of various paths results in receiving different versions of the transmitted signal at the receiver. These separate versions experience different path loss and phases. At the receiver all received signals are accumulated together creating a non-additive white Gaussian noise (AWGN) model for the wireless channels. Since an AWGN model does not describe the wireless channels, it is important to find other models that represent the channels. To portray such a model, first we study different possible paths for the received signals. Figure 1.1 demonstrates the trajectory of different paths in a typical example.

If there is a direct path between the transmitter and the receiver, it is called the line of sight (LOS). A LOS does not exist when large objects obstruct the line between the transmitter and the receiver. If LOS exists, the corresponding signal received through the LOS is usually the strongest and the dominant signal. At least, the signal from the LOS is more deterministic. While its strength and phase may change due to mobility, it is a more predictable change that is usually just a function of the distance and not many other random factors.

A LOS is not the only path that an electromagnetic wave can take from a transmitter to a receiver. An electromagnetic wave may reflect when it meets an object that is much larger than the wavelength. Through reflection from many surfaces, the wave may find its path to the receiver. Of course, such paths go through longer distances resulting in power strengths and phases other than those of the LOS path.

Another way that electromagnetic waves propagate is diffraction. Diffraction occurs when the electromagnetic wave hits a surface with irregularities like sharp edges.

Finally, scattering happens in the case where there are a large number of objects smaller than the wavelength between the transmitter and the receiver. Going through these objects, the wave scatters and many copies of the wave propagate in many different directions. There are also other phenomenona that affect the propagation of electromagnetic waves like absorbtion and refraction.

The effects of the above propagation mechanisms and their combination result in many properties of the received signal that are unique to wireless channels. These effects may reduce the power of the signal in different ways. There are two general aspects of such a power reduction that require separate treatments. One aspect is the large-scale effect which corresponds to the characterization of the signal power over large distances or the time-average behaviors of the signal. This is called attenuation or path loss and sometimes large-scale fading. The other aspect is the rapid change in the amplitude and power of the signal and this is called small-scale fading, or just fading. It relates to the characterization of the signal over short distances or short time intervals. In what follows, we explain models that explain the behavior of large-scale and small-scale fading.

1.3.1 Attenuation

Attenuation, or large-scale fading, is caused by many factors including propagation losses, antenna losses, and filter losses. The average received signal, or the large-scale fading factor, decreases logarithmically with distance. The logarithm factor, or the path gain exponent, depends on the propagation medium and the environment between the transmitter and the receiver. For example, for a free space environment, like that of satellite communications, the exponent is two. In other words, the average received power P_r is proportional to d^{-2}, where d is the distance between the transmitter and the receiver. For other propagation environments, like urban areas, the path loss exponent is usually greater than 2. In other words, if the average transmitted power is P_t, we have

$$P_r = \beta d^{-\nu} P_t, \tag{1.1}$$

where ν is the path loss exponent and β is a parameter that depends on the frequency and other factors. This is sometimes called the log-distance path loss model as the path loss and the distance have a logarithmic relationship. Calculating (1.1) at a reference distance d_0 and computing the relative loss at distance d with respect to the reference distance d_0 in decibels (dB) results in

$$L_{\text{path}} = \beta_0 + 10\nu \log_{10}\left(\frac{d}{d_0}\right), \tag{1.2}$$

where L_{path} is the path loss in dB and β_0 is the measured path loss at distance d_0 in dB. As we mentioned before, the path loss exponent, v, is a function of the environment between the transmitter and receiver. Its value is usually calculated by measuring the break signal and fitting the resulting measurements to the model. Typically, based on the empirical measurements, v is between 2 and 6. In many practical situations, the above simple model does not match with the measured data. Measurements in different locations at the same distance from the transmitter result in unequal outcomes. It has been shown empirically that many local environmental effects, such as building heights, affect the path loss. These local effects are usually random and are caused by shadowing. To model them, a Gaussian distribution around the value in (1.2) is utilized. In other words, the path loss is modeled by

$$L_{\text{path}} = \beta_0 + 10v \log_{10}\left(\frac{d}{d_0}\right) + X, \qquad (1.3)$$

where X is a zero-mean Gaussian random variable in dB with typical standard deviations ranging form 5 to 12 dB. This is called log-normal shadowing as the logarithm in dB is a normal random variable. This log-normal model is utilized in practice for the design and analysis of the system as a tool to provide the received powers. Knowing the parameters of the model, that is v, d_0, and the variance of the Gaussian, from measured data, one can generate the received power values for random locations in the system.

1.3.2 Fading

Fading, or equivalently small-scale fading, is caused by interference between two or more versions of the transmitted signal which arrive at the receiver at slightly different times. These signals, called multipath waves, combine at the receiver antenna and the corresponding matched filter and provide an effective combined signal. This resulting signal can vary widely in amplitude and phase. The rapid fluctuation of the amplitude of a radio signal over a short period of time, equivalently a short travel distance, is such that the large-scale path loss effects may be ignored. The randomness of multipath effects and fading results in the use of different statistical arguments to model the wireless channel. To understand the behavior and reasoning behind different models, we study the cause and properties of fading.

First, we study the effects of mobility on these channel models. Let us assume that the objects in the environment between the transmitter and the receiver are static and only the receiver is moving. In this case, the fading is purely a spatial phenomenon and is described completely by the distance. On the other hand, as the receiver moves through the environment, the spatial variations of the resulting signal translate into temporal variations for the receiver. In other words, due to

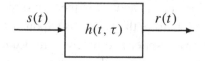

Fig. 1.2.   Modeling a multipath channel with a linear time-varying impulse response.

the mobility, there is a relationship between time and distance that creates a time-varying fading channel. Therefore, we can use time and distance interchangeably and equivalently in such a scenario. The time-varying nature of the wireless channel is also applied to the case that the surrounding objects are moving. Similarly, the resulting fluctuations in the received signal are structurally random.

As it is clear from the name, multipath fading is caused by the multiple paths that exist between the transmitter and the receiver. As we discussed before, reflection, diffraction, and scattering create several versions of the signal at the receiver. The effective combined signal is random in nature and its strength changes rapidly over a small period of time. A multipath channel can be modeled as a linear time-varying channel as depicted in Figure 1.2. The behavior of the linear time-varying impulse response depends on different parameters of the channel. For example, the speed of the mobile and surrounding objects affect the characteristic of the model. We study such behaviors in what follows.

The presence of reflecting objects and scatterers creates a constantly changing environment. Multipath propagation increases the time required for the baseband portion of the signal to reach the receiver. The resulting dissipation of the signal energy in amplitude, phase, and time may cause intersymbol interference (ISI). If the channel has a constant gain and linear phase response over a bandwidth which is greater than the bandwidth of the transmitted signal, the impulse response $h(t, \tau)$ can be approximated by a delta function at $\tau = 0$ that may have a time-varying amplitude. In other words, $h(t, \tau) = \alpha(t)\delta(\tau)$, where $\delta(\cdot)$ is the Dirac delta function. This is a narrowband channel in which the spectral characteristics of the transmitted signal are preserved at the receiver. It is called flat fading or frequency non-selective fading. An example of the impulse response for a flat fading channel is depicted in Figure 1.3. As can be seen from the figure, the narrowband nature of the channel can be checked from the time and frequency properties of the channel. In the frequency domain, the bandwidth of the signal is smaller than the bandwidth of the channel. In the time domain, the width of the channel impulse response is smaller than the symbol period. As a result, a channel might be flat for a given transmission rate, or correspondingly for a given symbol period, while the same channel is not flat for a higher transmission rate. Therefore, it is not meaningful to say a channel is flat without having some information about the transmitted signal.

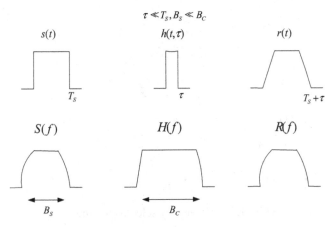

Fig. 1.3. Flat fading.

Also, we need to define the bandwidth of the channel to be able to compare it with the bandwidth of the signal. Usually the bandwidth of the channel is defined using its delay spread. To define the delay spread, let us assume that the multipath channel includes I paths and the power and delay of the ith path are p_i and τ_i, respectively. Then, the weighted average delay is

$$\overline{\tau} = \frac{\sum_{i=1}^{I} p_i \tau_i}{\sum_{i=1}^{I} p_i}. \tag{1.4}$$

The delay spread is defined as

$$\sigma_\tau = \sqrt{\overline{\tau^2} - \overline{\tau}^2}, \tag{1.5}$$

where

$$\overline{\tau^2} = \frac{\sum_{i=1}^{I} p_i \tau_i^2}{\sum_{i=1}^{I} p_i}. \tag{1.6}$$

Finally, the channel "coherence bandwidth" is approximated by

$$B_c = \frac{1}{5\sigma_\tau}. \tag{1.7}$$

As we defined earlier, in a flat fading channel, the channel coherence bandwidth B_c is much larger than the signal bandwidth B_s.

On the other hand, if the channel possesses a constant gain and linear phase over a bandwidth that is smaller than the signal bandwidth, ISI exists and the received

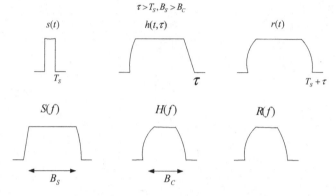

Fig. 1.4.   Frequency selective fading.

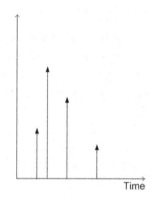

Fig. 1.5.   An approximated impulse response for a frequency selective fading.

signal is distorted. Such a wideband channel is called frequency selective fading. Figure 1.4 shows an example of the impulse response for a frequency selective fading channel. In this case, the impulse response $h(t, \tau)$ may be approximated by a number of delta functions as shown in Figure 1.5. In other words,

$$h(t, \tau) = \sum_{j=1}^{J} \alpha_j(t)\delta(\tau - \tau_j). \qquad (1.8)$$

Each delta component fades independently, that is $\alpha_j(t)$ are independent. To be more specific, for frequency selective fading, the bandwidth of the signal is larger than the coherence bandwidth of the channel. Equivalently, in the time domain, the width of the channel impulse response is larger than the symbol period. Again, the frequency selective nature of the channel depends on the transmission rate as well as the channel characteristics. In summary, based on multipath time delay, the fading channel is categorized into two types: flat and frequency selective.

   Another independent phenomenon caused by mobility is the Doppler shift in frequency. Let us assume a signal with a wavelength of λ and a mobile receiver with a velocity of v. Also, we define θ as the angle between the direction of the motion of the mobile and the direction of the arrival of the wave. In this case, the frequency change of the wave, known as Doppler shift and denoted by f_d, is given by

$$f_d = \frac{v}{\lambda} \cos \theta. \tag{1.9}$$

Since different paths have different angles, a variety of Doppler shifts corresponding to different multipath signals are observed at the receiver. In fact, the frequency change is random as the angle θ is random. The relative motion between the transmitter and the receiver results in random frequency modulation due to different Doppler shifts on each of the multipath components. Also, if the surrounding objects are moving, they create a time-varying Doppler shift on different multipath components. Such a time-varying frequency shift can be ignored if the mobile speed is much higher than that of the surrounding objects. Since the receiver observes a range of different Doppler shifts, any transmitted frequency results in a range of received frequencies. This results in a spectral broadening at the receiver. Doppler spread is a measure of such a spectral widening and is defined as the range of frequencies over which the received Doppler spectrum is not zero. If the maximum Doppler shift is f_s, the transmitted frequency, f_c, is received with components in the range $f_c - f_s$ to $f_c + f_s$. If the baseband signal bandwidth is much greater than the Doppler spread, the fading is called slow fading. In this case, the effects of Doppler spread are negligible. The channel impulse response changes at a rate much slower than the transmitted baseband signal and the channel is assumed to be static over one or several reciprocal bandwidth intervals. On the other hand, if the effects of the Doppler spread are not negligible, it is a fast-fading channel. The channel impulse response changes rapidly within the symbol duration in a fast-fading channel. In summary, based on Doppler spread, the fading channel is categorized into two types: slow and fast.

Defining the slow versus fast nature of a fading channel in terms of the signal bandwidth and Doppler spread may sound a little bit strange. Equivalently, slow- and fast-fading channels can be defined based on time domain properties. Towards this goal, first, we need to define the coherence time of a channel denoted by T_c. Two samples of a fading channel that are separated in time by less than the coherence time are highly correlated. This is a statistical measure since the definition depends on how much "correlation" is considered highly correlated. Practically, the coherence time is the duration of time in which the channel impulse response is effectively invariant. If a correlation threshold of 0.5 is chosen, the coherence time is approximated by

$$T_c = \frac{9}{16\pi f_s}, \tag{1.10}$$

where f_s is the maximum Doppler shift. If the signal duration is smaller than the coherence time, the whole signal is affected similarly by the channel and the channel is a slow fading channel. On the other hand, if the signal duration is larger than the coherence time, the channel changes are fast enough such that in practice different parts of the transmitted signal experience different channels. This is called fast fading since its main cause is the fast motion of the receiver or transmitter.

So far, we have classified fading channels based on their multipath time delay into flat and frequency selective and based on Doppler spread into slow and fast. These two phenomena are independent of each other and result in the following four types of fading channels.

- Flat Slow Fading or Frequency Non-Selective Slow Fading: When the bandwidth of the signal is smaller than the coherence bandwidth of the channel and the signal duration is smaller than the coherence time of the channel.
- Flat Fast Fading or Frequency Non-Selective Fast Fading: When the bandwidth of the signal is smaller than the coherence bandwidth of the channel and the signal duration is larger than the coherence time of the channel.
- Frequency Selective Slow Fading: When the bandwidth of the signal is larger than the coherence bandwidth of the channel and the signal duration is smaller than the coherence time of the channel.
- Frequency Selective Fast Fading: When the bandwidth of the signal is larger than the coherence bandwidth of the channel and the signal duration is larger than the coherence time of the channel.

1.4 Statistical models for fading channels

So far, we have modeled the fading channel by a linear time-varying impulse response. The impulse response was approximated by one delta function in the case of flat fading and multiple delta functions in the case of frequency selective fading. As discussed before, the nature of the multipath channel is such that the amplitude of these delta functions are random. This randomness mainly originates from the multipath and the random location of objects in the environment. Therefore, statistical models are needed to investigate the behavior of the amplitude and power of the received signal. In this section, we study some of the important models in the literature.

1.4.1 Rayleigh fading model

First, let us concentrate on the case of flat fading. The results for frequency selective channels are very similar since the amplitudes of different delta functions fade independently. We also assume that there is no LOS path between the transmitter and the receiver. Later, we will consider the case where such a LOS path exists. In a multipath channel with I multiple paths, transmitting a signal over the carrier frequency f_c results in receiving the sum of I components from different paths plus a Gaussian noise as follows:

$$r(t) = \sum_{i=1}^{I} a_i \cos(2\pi f_c t + \phi_i) + \eta(t), \qquad (1.11)$$

where a_i and ϕ_i are the amplitude and phase of the ith component, respectively, and $\eta(t)$ is the Gaussian noise. Expanding the $\cos(\cdot)$ term in (1.11) results in

$$r(t) = \cos(2\pi f_c t) \sum_{i=1}^{I} a_i \cos(\phi_i) - \sin(2\pi f_c t) \sum_{i=1}^{I} a_i \sin(\phi_i) + \eta(t). \quad (1.12)$$

It is customary in digital communications to call the first and second summations "in phase" and "quadrature" terms, respectively. The terms $A = \sum_{i=1}^{I} a_i \cos(\phi_i)$ and $B = \sum_{i=1}^{I} a_i \sin(\phi_i)$ are the summation of I random variables since the objects in the environment are randomly located. For a large value of I, as is usually the case, using the central limit theorem, the random variables A and B are independent identically distributed (iid) Gaussian random variables. The envelope of the received signal is $R = \sqrt{A^2 + B^2}$. Since A and B are iid zero-mean Gaussian random variables, the envelope follows a Rayleigh distribution. The probability density function (pdf) of a Rayleigh random variable is

$$f_R(r) = \frac{r}{\sigma^2} \exp\left(\frac{-r^2}{2\sigma^2}\right), \quad r \geq 0, \qquad (1.13)$$

where σ^2 is the variance of the random variables A and B. The received power, is an exponential random variable with a pdf:

$$f(x) = \frac{1}{2\sigma^2} \exp\left(\frac{-x}{2\sigma^2}\right), \quad x \geq 0. \qquad (1.14)$$

Note that the average power, the average of the above exponential random variable, is $E[R^2] = 2\sigma^2$ which is the sum of the variances of A and B.

The received signals in (1.11) or (1.12) represent the analog signal at the first stage of the receiver. We usually deal with the baseband digital signal after the matched filter and the sample and hold block. With a small abuse of the notation, we denote such a baseband discrete-time signal by r_t. In fact, r_t is the output of the matched filter after demodulation when the input of the receiver is $r(t)$. Similarly, s_t and η_t are the discrete-time versions of $s(t)$ and $\eta(t)$, the transmitted signal and the noise, respectively. Note that in the above analysis the transmitted signal was implicit. Then, using the above arguments, one can show that the relationship between the baseband signals is

$$r_t = \alpha s_t + \eta_t, \tag{1.15}$$

where α is a complex Gaussian random variable. In other words, the real and imaginary parts of the fade coefficient α are zero-mean Gaussian random variables. The amplitude of the fade coefficient, $|\alpha|$, is a Rayleigh random variable. The input–output relationship in (1.15) is called a fading channel model. The coefficient α is called the path gain and the additive noise component η_t is usually a Gaussian noise.

1.4.2 Ricean fading model

In a flat fading channel, if in addition to random multiple paths, a dominant stationary component exists, the Gaussian random variables A and B are not zero mean anymore. This, for example, happens when a LOS path exists between the transmitter and the receiver. In this case, the distribution of the envelope random variable, R, is a Ricean distribution with the following pdf:

$$f_R(r) = \frac{r}{\sigma^2} \exp\left(\frac{-(r^2 + D^2)}{2\sigma^2}\right) I_0\left(\frac{Dr}{\sigma^2}\right), \qquad r \geq 0, \ D \geq 0, \tag{1.16}$$

where D denotes the peak amplitude of the dominant signal and $I_0(.)$ is the modified Bessel function of the first kind and of zero-order. As expected, the Ricean distribution converges to a Rayleigh distribution when the dominant signal disappears, that is $D \to 0$.

Similarly to the case of Rayleigh fading model, the discrete-time input–output relationship in the case of a Ricean fading model is also governed by (1.15). The main difference is that the real and imaginary parts of the path gain α are Gaussian random variables with non-zero means. As a result, the distribution of the amplitude $|\alpha|$ is Ricean instead of Rayleigh.

1.4.3 Frequency selective fading models

In general, as discussed before, frequency selective fading is modeled by intersymbol interference. Therefore, the channel can be modeled by the sum of a few delta functions. In this case, the corresponding discrete-time input–output relationship is

$$r_t = \sum_{j=0}^{J-1} \alpha^j s_{t-j} + \eta_t. \tag{1.17}$$

The path gains, α^j, are independent complex Gaussian distributions and η_t represents the noise. In the case of Rayleigh fading, they are zero-mean iid complex Gaussian random variables. A special case that has been extensively utilized in the literature is the case of a two-ray Rayleigh fading model. For a two-ray Rayleigh fading model, we have

$$r_t = \alpha^0 s_t + \alpha^1 s_{t-1} + \eta_t, \tag{1.18}$$

where the real and imaginary parts of α_0 and α_1 are iid zero-mean Gaussian random variables.

1.5 Diversity

Unlike the Gaussian channel, the fading channel model in (1.15) suffers from sudden declines in the power. As we discussed before, this is due to the destructive addition of multipath signals in the propagation media. It can also be due to interference from other users. The amount of change in the received power can be sometimes more than 20 to 30 dB. The power of the thermal noise is usually not changing that much at the receiver. Therefore, the effective signal-to-noise ratio (SNR) at the receiver can go through deep fades and be dropped dramatically. Usually there is a minimum received SNR for which the receiver can reliably detect and decode the transmitted signal. If the received SNR is lower than such a threshold, a reliable recovery of the transmitted signal is impossible. This is usually called an "outage." The probability of outage can be calculated based on the statistical model that models the channel or based on the actual measurements of the channel. It is the probability of having a received power lower than the given threshold.

The main idea behind "diversity" is to provide different replicas of the transmitted signal to the receiver. If these different replicas fade independently, it is less probable to have all copies of the transmitted signal in deep fade simultaneously. Therefore, the receiver can reliably decode the transmitted signal using these received signals. This can be done, for example, by picking the signal with the highest SNR or by combining the multiple received signals. As a result, the probability of outage will be lower in the case that we receive multiple replicas of the signal using diversity. To define diversity quantitatively, we use the relationship between the received SNR, denoted by γ, and the probability of error, denoted by P_e. A tractable definition of the diversity, or diversity gain, is

$$G_d = -\lim_{\gamma \to \infty} \frac{\log(P_e)}{\log(\gamma)}, \tag{1.19}$$

where P_e is the error probability at an SNR equal to γ. In other words, diversity is the slope of the error probability curve in terms of the received SNR in a log-log scale. There are two important issues related to the concept of diversity. One is how to provide the replicas of the transmitted signal at the receiver with the lowest possible consumption of the power, bandwidth, decoding complexity and other resources. The second issue is how to use these replicas of the transmitted signal at the receiver in order to have the highest reduction in the probability of error. We study some of the methods to achieve these two goals in what follows.

1.5.1 Diversity methods

The replica of the transmitted signal can be sent through different means [99]. For example, it can be transmitted in a different time slot, a different frequency, a different polarization, or a different antenna. The goal is to send two or more copies of the signal through independent fades. Then, since it is less likely to have all the independent paths in deep fades, using appropriate combining methods, the probability of error will be lower.

When different time slots are used for diversity, it is called temporal diversity [129]. Two time intervals separated for more than the coherence time of the channel go through independent fades. Therefore, we may send copies of the transmitted signal from these separated time slots. Error-correcting codes can be utilized to reduce the amount of redundancy. In other words, sending a copy of the signal from different time slots is equivalent to using a repetition code. More efficient error-correcting codes may be used as well. If the fading is slow, that is the coherence time of the channel is large, the separation between time slots used for temporal diversity is high. In this case, the receiver suffers from a huge delay before it can start the process of decoding. The coded symbols are interleaved before sending through the channel. While interleaving increases the delay, it converts a slow fading channel to a fast fading channel that is more appropriate for temporal diversity. Temporal diversity is not bandwidth efficient because of the underlying redundancy.

Another method of diversity is frequency diversity [8]. Frequency diversity uses different carrier frequencies to achieve diversity. The signal copies are transmitted from different carrier frequencies. To achieve diversity, the carrier frequencies should be separated by more than the coherence bandwidth of the channel. In this case, different replicas of the signal experience independent fades. Similar to temporal diversity, frequency diversity suffers from bandwidth deficiency. Also the receiver needs to tune to different carrier frequencies.

One method of diversity that may not suffer from bandwidth deficiency is spatial diversity or antenna diversity [152]. Spatial diversity uses multiple antennas to achieve diversity. Multiple antennas may be used at the receiver or transmitter. If the antennas are separate enough, more than half of the wavelength, signals corresponding to different antennas fade independently. The use of multiple antennas may not be possible in small handheld devices. This is because of the fact that a minimum physical separation is needed between different antennas to achieve spatial diversity.

Spatial diversity is not the only way to use antennas for providing diversity. Angular diversity uses directional antennas to achieve diversity. Different copies of the transmitted signal are collected from different angular directions. Unlike multiple antennas, it does not need separate physical locations. Therefore, it is also good for small devices. Another diversity method is polarization diversity that uses vertically and horizontally polarized signals to achieve diversity. Because of the scattering, the arriving signal, which is not polarized, can be split into two orthogonal polarizations. If the signal goes through random reflections, its polarization state can be independent of the transmitted polarization. Unlike spatial diversity, polarization diversity does not require separate physical locations for the antennas. However, polarization diversity can only provide a diversity order of two and not more.

1.5.2 Combining methods

The multiple versions of the signals created by different diversity schemes need to be combined to improve the performance. In this section, we study different methods of combining at the receiver. We assume that multiple receive antennas are available and provide multiple replicas of the transmitted signal at the receiver. The use of multiple transmit antennas is the topic of the other chapters in this book. While we refer to multiple receive antennas, our discussion of combining methods is applicable to other forms of diversity as well. In fact, the source of diversity does not affect the method of combination with the exception of transmit antenna diversity. For example, receiving two versions of the transmitted signal through polarization diversity is the same as receiving two signals from two receive antennas for the purpose of combining. There are two main combining methods that are utilized at the receiver:

- Maximum Ratio Combining (MRC)
- Selection Combining

Figures 1.6 and 1.7 show the block diagrams of the maximum ratio combiners and the selection combiner. A hybrid scheme that combines these two main methods is also possible. In what follows, we explain the details of these combining methods.

1.5.2.1 Maximum ratio combining

We consider a system that receives M replicas of the transmitted signal through M independent paths. Let us assume r_m, $m = 1, 2, \ldots, M$, as the mth received signal

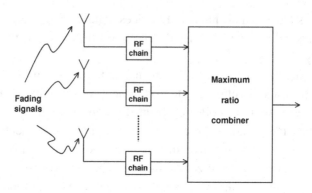

Fig. 1.6. Block diagram of maximum ratio combining.

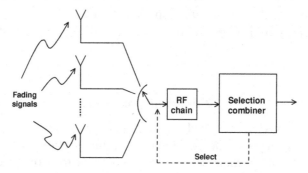

Fig. 1.7. Block diagram of selection combining.

is defined by

$$r_m = \alpha_m s + \eta_m, \tag{1.20}$$

where η_m is a white Gaussian noise sample added to the mth copy of the signal. A maximum-likelihood (ML) decoder combines these M received signals to find the most likely transmitted signal. We consider a coherent detection scheme where the receiver knows the channel path gains, α_m. Since the noise samples are independent Gaussian random variables, the received signals are also independent Gaussian random variables for the given channel path gains and transmitted signal. Therefore, the conditional joint density function of the received signals is

$$f(r_1, r_2, \ldots, r_M | s, \alpha_1, \alpha_2, \ldots, \alpha_M) = \frac{1}{(\pi N_0)^{\frac{M}{2}}} \exp\left\{ -\frac{\sum\limits_{m=1}^{M} |r_m - s\alpha_m|^2}{N_0} \right\}, \tag{1.21}$$

where $N_0/2$ is the variance of the real and imaginary parts of the complex Gaussian noise. To maximize this likelihood function, the receiver needs to find the optimal transmitted signal, $\hat{s}$, that minimizes $\sum_{m=1}^{M}|r_m - s\alpha_m|^2$. Note that with no diversity, $M = 1$, the cost function to minimize is $|r_1 - s\alpha_1|^2$ or $|r - s\alpha|^2$ for simplicity. This is equivalent to finding $\hat{s}$, among all possible transmitted signals, that is closest to $r\alpha^*$. For a constellation with equal energy symbols, for example PSK, we have

$$\hat{s} = \arg\min_s \sum_{m=1}^{M}|r_m - s\alpha_m|^2 = \arg\min_s \left[-s\sum_{m=1}^{M}\alpha_m r_m^* - s^*\sum_{m=1}^{M}\alpha_m^* r_m\right]$$

$$= \arg\min_s \left|\sum_{m=1}^{M} r_m\alpha_m^* - s\right|^2. \tag{1.22}$$

Therefore, the ML decoding is similar to that of the system with no diversity if instead of $r\alpha^*$ we use a weighted average of the received signals, $\sum_{m=1}^{M} r_m\alpha_m^*$. This is called maximum ratio combining (MRC). To summarize, MRC uses a matched filter, that is optimum receiver, for each received signal and using the optimal weights $\omega_m = \alpha_m^*$ combines the outputs of the matched filters. If the average power of the transmitted symbol is E_s, the SNR of the mth receiver is $\gamma_m = |\alpha_m|^2(E_s/N_0)$. To derive the SNR of the output of the maximum ratio combiner, first we calculate

$$\sum_{m=1}^{M} r_m\alpha_m^* = \sum_{m=1}^{M}(\alpha_m s + \eta_m)\alpha_m^* = \sum_{m=1}^{M}|\alpha_m|^2 s + \sum_{m=1}^{M}\eta_m\alpha_m^*. \tag{1.23}$$

Then, the SNR at the output of the maximum ratio combiner is

$$\gamma = \frac{\left(\sum_{m=1}^{M}|\alpha_m|^2\right)^2 E_s}{\sum_{m=1}^{M}|\alpha_m|^2 N_0} = \sum_{m=1}^{M}|\alpha_m|^2 \frac{E_s}{N_0} = \sum_{m=1}^{M}\gamma_m. \tag{1.24}$$

Therefore, the effective receive SNR of a system with diversity M is equivalent to the sum of the receive SNRs for M different paths. The importance of this M-fold increase in SNR is in the relationship between the average error probability and the average receive SNR. Let us assume that all different paths have the same average SNR, that is $E[\gamma_m] = A$. Then, using (1.24), the average SNR at the output of the maximum ratio combiner is

$$\bar{\gamma} = MA. \tag{1.25}$$

This M-fold increase in the receive average SNR results in a diversity gain of M. It can be shown that this is the maximum possible diversity gain when M copies of the signal are available in a Rayleigh fading channel.

Increasing the effective receive SNR using MRC affects the probability of error at the receiver. For a system with no diversity, the average error probability is proportional to the inverse of the SNR, A^{-1}, at high SNRs [100]. Since each of the M paths follows a Rayleigh fading distribution, the average error probability of a system with M independent Rayleigh paths is proportional to A^{-M}. As we defined before, the ratio M in the exponent of the receive SNR is called the diversity gain. Therefore, using MRC we achieve a diversity gain equal to the number of available independent paths. We provide a rigorous proof of this fact in Chapter 4.

Equal gain combining is a special case of maximum ratio combining with equal weights. In equal gain combining, co-phased signals are utilized with unit weights. The average SNR at the output of the equal gain combiner is

$$\overline{\gamma} = \left[1 + \frac{\pi}{4}(M-1)\right]A. \tag{1.26}$$

1.5.2.2 Selection combining

Using MRC, when the source of M independent signals is the receive antennas, the receiver needs to demodulate all M receive signals. In other words, M radio frequency (RF) chains are required at the receiver to provide the baseband signals. Since most RF chains are implemented by analog circuits, usually their physical size and price are high. In some applications, there is not enough room for several RF chains or their price does not justify the gain achieved by MRC. Therefore, it may be beneficial to design a combiner that uses only one RF chain.

Selection combining or antenna selection picks the signal with the highest SNR and uses it for decoding. Picking the signal is equivalent to choosing the corresponding antenna among all receive antennas. Equivalently, it is the same as selecting the best polarization in the case of polarization diversity. As before, let us assume M replicas of the transmitted signal, for example through M receive antennas. As shown before, if the fading is Rayleigh, the random variable γ_m, the SNR of the mth antenna, follows an exponential distribution. With a slight abuse of the notation, the pdf of γ_m for $m = 1, 2, \ldots, M$ is

$$f_{\gamma_m}(\gamma) = \frac{1}{E[\gamma_m]} e^{-\frac{\gamma}{E[\gamma_m]}}, \qquad \gamma > 0, \tag{1.27}$$

where $E[\gamma_m]$ is the average SNR of the mth receive signal. Let us assume that all receive signals have the same average SNR, that is $E[\gamma_m] = A$. Then, the cumulative distribution function (CDF) of the mth receive SNR is

$$F_{\gamma_m}(\gamma) = P[\gamma_m \le \gamma] = 1 - e^{-\frac{\gamma}{A}}, \qquad m = 1, 2, \ldots, M. \tag{1.28}$$

Since different receive signals are independent from each other, the probability that all of them have an SNR smaller than γ is

$$P\left[\gamma_1, \gamma_2, \ldots, \gamma_M \le \gamma\right] = \left[1 - e^{-\frac{\gamma}{A}}\right]^M. \tag{1.29}$$

On the other hand, the probability that at least one receive signal achieves an SNR greater than γ, denoted by $P_M(\gamma)$, is

$$P_M(\gamma) = 1 - \left[1 - e^{-\frac{\gamma}{A}}\right]^M. \tag{1.30}$$

The corresponding pdf is

$$f(\gamma) = \frac{M}{A}\left[1 - e^{-\frac{\gamma}{A}}\right]^{M-1} e^{-\frac{\gamma}{A}}. \tag{1.31}$$

Therefore, the average SNR at the output of the selection combiner, $\overline{\gamma}$, is

$$\overline{\gamma} = \int_0^{\infty} \gamma f(\gamma)\,d\gamma = A \sum_{m=1}^{M} \frac{1}{m}. \tag{1.32}$$

As a result, without increasing the transmission power, selection combining offers $\sum_{m=1}^{M} \frac{1}{m}$ times improvement in the average SNR. This is less than the maximum improvement ratio of M. Therefore selection combining does not provide an optimal diversity gain and as a result an optimal performance enhancement. However, its complexity is low since it only requires one RF chain. In other words, selection combining provides a trade-off between complexity and performance.

In selection combining, the receiver needs to find the strongest signal at each time instant. To avoid the monitoring of the received SNRs, one may use scanning selection combining which is a special case of selection combining. In scanning selection combining, first, the M receive signals are scanned to find the highest SNR. The corresponding signal is used until its SNR is below a predetermined threshold. Then, a new selection is done and the process is continued. In other words, a new selection is needed only if the selected signal goes through a deep fade.

Figure 1.8 compares the SNR gains of different combining methods using 1 to 10 receive antennas. As expected, MRC provides the highest gain while requiring the highest complexity. For a higher number of receive antennas, the gap between the MRC and selection combining grows. This is in a trade-off with the lower complexity of selection combining with only one RF chain. It is possible to use a number of RF chains that is neither one nor M for more than two receive antennas. Let us assume that the receiver contains J RF chains where $1 < J < M$ and $M > 2$. Then, the receiver chooses J receive signals with the highest SNR and combines them using MRC. The block diagram of such a hybrid combining method is shown in Figure 1.9. The instantaneous SNR at the output of the hybrid selection/maximal

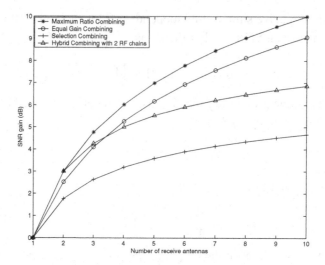

Fig. 1.8.   Comparing the gain of different combining methods.

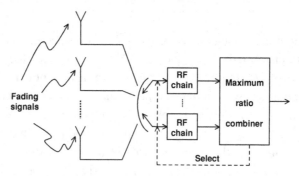

Fig. 1.9.   Block diagram of hybrid selection/maximum ratio combining.

ratio combining is

$$\gamma = \sum_{j=1}^{J} \gamma_j, \tag{1.33}$$

where γ_j is the SNR of the jth selected signal [150]. The average SNR at the output of the hybrid selection/maximal ratio combiner, $\bar{\gamma}$, is

$$\bar{\gamma} = AJ \left[1 + \sum_{m=J+1}^{M} \frac{1}{m} \right]. \tag{1.34}$$

In Figure 1.8, we also show the SNR gain for a hybrid selection/maximal ratio combiner with $J = 2$ RF chains. As depicted in the figure, using more RF chains results in a higher gain. For a small number of receive antennas, a hybrid combiner with only $J = 2$ RF chains provides most of the gain. For a higher number of receive antennas, the gap between the hybrid combiner and MRC increases and more RF chains are needed to close the gap.

1.6 Spatial multiplexing gain and its trade-off with diversity

In the last section, we mostly concentrated on diversity gains that are achieved by using multiple receive antennas. In a multiple-input multiple-output (MIMO) channel, diversity gain may be achieved by using both transmit and receive antennas. In the rest of this book, we will investigate how to use multiple transmit antennas to achieve high levels of diversity. However, multiple transmit antennas can be utilized to achieve other goals as well. For example, a higher capacity and as a result a higher transmission rate is possible by increasing the number of transmit antennas. Let us, for the sake of simplicity, assume a MIMO channel with equal number of transmit and receive antennas. Then, as we will show in Chapter 2, in a rich scattering environment the capacity increases linearly with the number of antennas without increasing the transmission power. This results in the possibility of transmitting at a higher rate, for example by using spatial multiplexing. If the number of transmit antennas is not the same as the number of receive antennas, in general one can transmit up to $\min\{N, M\}$ symbols per time slot, where N is the number of transmit antennas and M is the number of receive antennas. For example, if $M \geq N$, one can send N symbols and achieve a diversity gain of $M - N + 1$. Note that for equal number of transmit and receive antennas, the diversity gain in this case is one. On the other hand, the maximum spatial diversity while transmitting only one symbol per time slot is MN. Therefore, the advantage of a MIMO channel can be utilized in two ways: (i) to increase the diversity of the system and (ii) to increase the number of transmitted symbols. For the general case of more than one transmit antenna, $M \geq N > 1$, there is a theoretical trade-off between the number of transmit symbols and the diversity of the system. For one transmit antenna, in both cases one symbol per time slot is transmitted and the two systems coincide.

In many cases, spatial multiplexing gain refers to the fact that one can use multiple antennas to transmit at a higher rate compared to the case of one antenna. As we will show in Chapter 2, the capacity of a MIMO channel increases by raising the SNR. Since the transmission rate relates to capacity, it is reasonable to hope that the rate can be increased as the SNR increases. This argument has resulted in the following definition for spatial multiplexing gain in [159]:

$$SMG = \lim_{\gamma \to \infty} \frac{r}{\log(\gamma)}, \qquad (1.35)$$

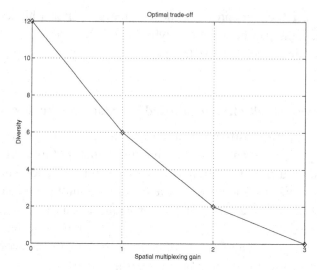

Fig. 1.10.   Optimal trade-off between spatial multiplexing gain (an indication of
rate) and diversity.

where r is the rate of the code, at the transmitter, in bits/channel use and is a
function of SNR. Note that the relationship of the above spatial multiplexing gain
to the transmission rate is similar to that of the diversity gain to the probability
of error in (1.19). The main rationale behind such a rate normalization is the fact
that SMG measures how far the rate r is from the capacity. Note that the range
of the spatial multiplexing gain is from 0 to min$\{N, M\}$. The following theorem
from [159] derives the optimal trade-off between spatial multiplexing gain, which
is an indication of the rate, and diversity, which is an indication of the probability
of error

Theorem 1.6.1 *Let us assume a code at the transmitter of a MIMO channel with N
transmit antennas and M receive antennas. For a given spatial multiplexing gain
$SMG = i$, where $i = 0, 1, \ldots, \min\{N, M\}$ is an integer, the maximum diversity
gain $G_d(i)$ is given by $G_d(i) = (N - i)(M - i)$ if the block length of the code is
greater than or equal to $N + M - 1$. The optimal trade-off curve is achieved by
connecting the points $(i, G_d(i))$ by lines.*

An example of the optimal trade-off for $N = 4$ transmit antennas and $M = 3$
receive antennas is depicted in Figure 1.10.

1.7 Closed-loop versus open-loop systems

When the transmitter does not have any information about the channel, the system
is an "open-loop" system. In this case, the receiver may estimate the channel and

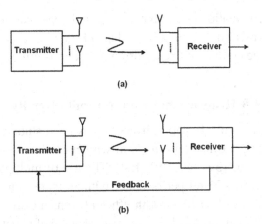

(a)

(b)

Fig. 1.11. Block diagram of (a) open-loop and (b) closed-loop multiple-input multiple-output systems.

use the channel state information (CSI) for decoding. However, the transmitter does not have access to the channel state information. On the contrary, in some systems, the receiver sends some information about the channel back to the transmitter through a feedback channel. This is called a "closed-loop" system and the transmitter can use this information to improve the performance. The block diagrams of open-loop and closed-loop systems are depicted in Figure 1.11.

In a time division duplexing (TDD) system, the radio channel is shared between the mobile and base station. In other words, the same frequency, the same channel, is utilized to transmit information in both directions using time sharing. As a result, the characteristics of the uplink channel, from mobile to base station, and the downlink channel, from base station to mobile, are similar. One possible assumption in such a TDD system is to consider the uplink and downlink channels as the reciprocal of each other. If this is a valid assumption, the channel estimation at the receiver can be utilized when transmitting back in a TDD system. Therefore, a closed-loop system can be used without sending any feedback information.

In a closed-loop system, the gain achieved by processing at the transmitter and the receiver is sometimes called "array gain." Similar to the diversity gain, the array gain results in an increase in the average receive SNR. If the transmitter knows the channel perfectly, beamforming [86] is the best solution. In fact, in such a case there is no need for spatial diversity. On the other hand, in most practical situations, there is a limited feedback information available which may not be perfect due to the quantization error or other factors. With such limited channel information, spatial diversity is still useful. The transmitted codeword can be tuned to better fit the channel based on the received information in the closed-loop system. To maintain low implementation complexity, a simple linear beamforming scheme at the transmitter is preferred. If non-perfect channel state information is available, a combined space-time coding and beamforming system is utilized [76, 87].

1.8 Historical notes on transmit diversity

Transmit diversity, a form of spatial diversity, has been studied extensively as a method of combating detrimental effects in wireless fading channels [4, 42, 51, 96, 102, 110, 135, 136, 139, 140, 153, 154]. The use of multiple transmit antennas for diversity provides better performance without increasing the bandwidth or transmission power. The first bandwidth-efficient transmit diversity scheme was proposed in [153]. It includes the delay diversity scheme of [110] as a special case. It is proved that the diversity advantage of this scheme is equal to the number of transmit antennas which is optimal [151]. Later, a multilayered space-time architecture was introduced in [43]. The scheme proposed in [43] uses spatial multiplexing to increase the data rate and not necessarily provides transmit diversity. Simple iterative decoding algorithms that have been proposed in conjunction with spatial multiplexing can achieve spatial diversity, mostly receive diversity. The criterion to achieve the maximum transmit diversity was derived in [51]. A complete study of design criteria for maximum diversity and coding gains in addition to the design of space-time trellis codes was proposed in [139]. It includes the delay diversity as a special case. Information theoretical results in [140] and [42] showed that there is a huge advantage in utilizing spatial diversity. The introduction of space-time block coding in [136] provided a theoretical framework that started an increasing interest on the subject.

The best use of multiple transmit antennas depends on the amount of channel state information that is available to the encoder and decoder. In a flat fading channel, the channel has a constant gain and linear phase response over a bandwidth which is greater than the bandwidth of the transmitted signal. Let us assume that the channel does not change rapidly during the transmission of a frame of data. The combination of the above two assumptions results in a quasi-static flat fading channel model [42]. In most practical cases, the system estimates the channel at the receiver by transmitting some known pilot signals. The receiver utilizes the corresponding received signals to estimate the channel. In such a system, a coherent detection is utilized in which the decoder uses the value of the path gains estimated at the receiver [138]. In all references that we have mentioned so far in this section, the codes are designed for the case that the receiver knows the channel. In some other cases, such an estimation of the channel is not available at the receiver or the channel changes rapidly such that the channel estimation is not useful. Then, a noncoherent detection needs to be employed. For one transmit antenna, differential detection schemes exist that neither require the knowledge of the channel nor employ pilot symbol transmission. A partial solution to generalize differential detection schemes to the case of multiple transmit antennas was proposed in [132]. This was a joint channel- and data-estimation scheme that can lead to error propagation. Noncoherent detection schemes based on unitary space-time codes were proposed in [59, 60]. The first differential decoding modulation scheme for multiple antennas that provides simple encoding and decoding based on orthogonal designs was proposed in [73, 133]. Another construction based on group codes followed in [61, 63].

1.9 Summary

- Line of Sight: A direct path between the transmitter and the receiver.
- Reflection: When the electromagnetic wave meets an object that is much larger than the wavelength.
- Diffraction: When the electromagnetic wave hits a surface with irregularities like sharp edges.
- Scattering: When the medium through which the electromagnetic wave propagates contains a large number of objects smaller than the wavelength.
- Attenuation or path loss (sometimes called large-scale fading) is due to propagation losses, filter losses, antenna losses, and so on.
- Fading is used to describe the rapid fluctuation of the amplitude of a radio signal over a short period of time or travel distance, so that large-scale path loss effects may be ignored. Fading is a time-varying phenomenon.
- Multipath: The presence of reflecting objects and scatterers creates a constantly changing environment that dissipates the signal energy in amplitude, phase, and time. A multipath channel can be modeled as a linear time-varying channel.
- Flat Slow Fading or Frequency Non-Selective Slow Fading: When the bandwidth of the signal is smaller than the coherence bandwidth of the channel and the signal duration is smaller than the coherence time of the channel.
- Flat Fast Fading or Frequency Non-Selective Fast Fading: When the bandwidth of the signal is smaller than the coherence bandwidth of the channel and the signal duration is larger than the coherence time of the channel.
- Frequency Selective Slow Fading: When the bandwidth of the signal is larger than the coherence bandwidth of the channel and the signal duration is smaller than the coherence time of the channel.
- Frequency Selective Fast Fading: When the bandwidth of the signal is larger than the coherence bandwidth of the channel and the signal duration is larger than the coherence time of the channel.
- In a Rayleigh fading model, the input–output relationship between the baseband signals is

$$r_t = \alpha s_t + \eta_t,$$

where α is a complex Gaussian random variable. In other words, the real and imaginary parts of the fade coefficient α are real zero-mean Gaussian random variables. The amplitude of the fade coefficient, $|\alpha|$, is a Rayleigh random variable.

- Diversity provides a less-attenuated replica of the transmitted signal to the receiver. Diversity is defined as

$$G_d = - \lim_{\gamma \to \infty} \frac{\log(P_e)}{\log(\gamma)},$$

where P_e is the error probability at a signal-to-noise ratio equal to γ.

- Diversity methods include temporal diversity, frequency diversity, spatial diversity, angle diversity, polarization diversity, and so on.
- The multiple versions of the signals created by different diversity schemes need to be combined to improve the performance. Two main combining methods that are utilized at the receiver are maximum ratio combining and selection combining.
- For a given spatial multiplexing gain $SMG = i$, where $i = 0, 1, \ldots, \min\{N, M\}$ is an integer, the maximum diversity gain $G_d(i)$ is given by $G_d(i) = (N - i)(M - i)$ if the block length of the code is greater than or equal to $N + M - 1$.

1.10 Problems

1 A multipath channel includes five paths with powers 0.1, 0.01, 0.02, 0.001, and 0.5 W. The corresponding delays are 0.1, 0.2, 0.3, 0.4, and 0.5 μs, respectively. If the signal bandwidth is 200 kHz, is the channel flat or frequency selective?

2 A mobile is moving at a speed of 100 km/h. The transmitted frequency is 900 MHz while the signal bandwidth is 300 kHz. Is the channel slow or fast?

3 Consider a Rayleigh random variable, R, that is constructed as the envelope of two iid zero-mean unit-variance Gaussian random variables.
   (a) What is the probability that $R < 0.1$?
   (b) If two independent random variables R_1 and R_2 have the Rayleigh distribution of R, what is the probability that $R_1 < 0.1$ and $R_2 < 0.1$?

4 Another popular model for the distribution of the envelope random variable, R, is a Nakagami distribution with the following pdf:

$$f_R(r) = \frac{2m^m r^{2m-1}}{\Gamma(m)A^m} \exp\left(\frac{-mr^2}{A}\right), \quad r \geq 0, \ m \geq 0.5,$$

where $A = E[R^2]$ and $\Gamma(\cdot)$ is the gamma function.
   (a) Show that for $m = 1$, the Nakagami distribution converges to a Rayleigh distribution.
   (b) Draw the pdf of a Nakagami distribution for $A = 1$ and different values of $m = 0.5, 1, 1.5, 2, 10$.

5 Consider a system that receives M replicas of the transmitted signal through M independent paths. The pdf of γ_m for $m = 1, 2, \ldots, M$ follows (1.27) and all receive signals have the same average SNR, $E[\gamma_m] = 1$. Draw the pdf of the SNR at the output of the selection combiner and the maximum ratio combiner for $M = 1, 2, 3, 4$.

6 For BPSK, the probability of error for a SNR of γ is $0.5 \, \text{erfc}(\sqrt{\gamma})$, where erfc is the complementary error function:

$$\text{erfc}(x) = \frac{2}{\sqrt{\pi}} \int_x^\infty \exp(-t^2) \, dt.$$

Let us assume $M = 2$ independent paths that have the same average SNR, $E[\gamma_m] = 1$ such that the pdf of γ_m for $m = 1, 2$ follows (1.27). What is the probability of error at the output of the selection combiner and the maximum ratio combiner?

2

Capacity of multiple-input multiple-output channels

2.1 Transmission model for multiple-input multiple-output channels

We consider a communication system, where N signals are transmitted from N transmitters simultaneously. For example, in a wireless communication system, at each time slot t, signals $\mathbf{C}_{t,n}$, $n = 1, 2, \ldots, N$ are transmitted simultaneously from N transmit antennas. The signals are the inputs of a multiple-input multiple-output (MIMO) channel with M outputs. Each transmitted signal goes through the wireless channel to arrive at each of the M receivers. In a wireless communication system with M receive antennas, each output of the channel is a linear superposition of the faded versions of the inputs perturbed by noise. Each pair of transmit and receive antennas provides a signal path from the transmitter to the receiver. The coefficient $\alpha_{n,m}$ is the path gain from transmit antenna n to receive antenna m. Figure 2.1 depicts a baseband discrete-time model for a flat fading MIMO channel. Based on this model, the signal $r_{t,m}$, which is received at time t at antenna m, is given by

$$r_{t,m} = \sum_{n=1}^{N} \alpha_{n,m} \mathbf{C}_{t,n} + \eta_{t,m}, \qquad (2.1)$$

where $\eta_{t,m}$ is the noise sample of the receive antenna m at time t. Based on (2.1), a replica of the transmitted signal from each transmit antenna is added to the signal of each receive antenna. Although the faded versions of different signals are mixed at each receive antenna, the existence of the M copies of the transmitted signals at the receiver creates an opportunity to provide diversity gain.

If the channel is not flat, the received signal at time t depends on the transmitted signals at times before t as well. The result is an extension to the case of one transmit and one receive antenna in (1.17). In this chapter, we only consider the case of narrowband signals for which the channel is a flat fading channel.

Another important factor in the behavior of the channel is the correlation between different path gains at different time slots. There are two general assumptions that

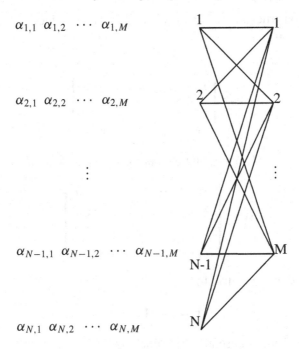

$$\alpha_{1,1}\ \alpha_{1,2}\ \cdots\ \alpha_{1,M}$$

$$\alpha_{2,1}\ \alpha_{2,2}\ \cdots\ \alpha_{2,M}$$

$$\vdots$$

$$\alpha_{N-1,1}\ \alpha_{N-1,2}\ \cdots\ \alpha_{N-1,M}$$

$$\alpha_{N,1}\ \alpha_{N,2}\ \cdots\ \alpha_{N,M}$$

Fig. 2.1.   A multiple-input multiple-output (MIMO) channel.

correspond to two practical scenarios. First, we assume a quasi-static channel, where the path gains are constant over a frame of length T' and change from frame to frame. In most cases, we assume that the path gains vary independently from one frame to another. Another assumption is to consider a correlation between the fades in adjacent time samples. One popular example of such a second-order model is the Jakes model [74].

The value of T' dictates the slow or fast nature of the fading. If a block of data is transmitted over a time frame T that is smaller than T', the fading is slow. In this case, the fades do not change during the transmission of one block of data and the values of path gains in (2.1) are constant for every frame. On the other hand, in a fast fading model, the path gains may change during the transmission of one frame of data, $T \gg T'$.

To form a more compact input-output relationship, we collect the signals that are transmitted from N transmit antennas during T time slots in a $T \times N$ matrix, $\mathbf{C}$, as follows:

$$\mathbf{C} = \begin{pmatrix} \mathbf{C}_{1,1} & \mathbf{C}_{1,2} & \cdots & \mathbf{C}_{1,N} \\ \mathbf{C}_{2,1} & \mathbf{C}_{2,2} & \cdots & \mathbf{C}_{2,N} \\ \vdots & \vdots & \ddots & \vdots \\ \mathbf{C}_{T,1} & \mathbf{C}_{T,2} & \cdots & \mathbf{C}_{T,N} \end{pmatrix}. \tag{2.2}$$

Similarly, we construct a $T \times M$ received matrix $\mathbf{r}$ that includes all received signals during T time slots:

$$\mathbf{r} = \begin{pmatrix} r_{1,1} & r_{1,2} & \cdots & r_{1,M} \\ r_{2,1} & r_{2,2} & \cdots & r_{2,M} \\ \vdots & \vdots & \ddots & \vdots \\ r_{T,1} & r_{T,2} & \cdots & r_{T,M} \end{pmatrix}. \tag{2.3}$$

Then, assuming $T < T'$, gathering the path gains in an $N \times M$ channel matrix $\mathbf{H}$

$$\mathbf{H} = \begin{pmatrix} \alpha_{1,1} & \alpha_{1,2} & \cdots & \alpha_{1,M} \\ \alpha_{2,1} & \alpha_{2,2} & \cdots & \alpha_{2,M} \\ \vdots & \vdots & \ddots & \vdots \\ \alpha_{N,1} & \alpha_{N,2} & \cdots & \alpha_{N,M} \end{pmatrix}, \tag{2.4}$$

results in the following matrix form of (2.1):

$$\mathbf{r} = \mathbf{C} \cdot \mathbf{H} + \mathcal{N}, \tag{2.5}$$

where $\mathcal{N}$ is the $T \times M$ noise matrix defined by

$$\mathcal{N} = \begin{pmatrix} \eta_{1,1} & \eta_{1,2} & \cdots & \eta_{1,M} \\ \eta_{2,1} & \eta_{2,2} & \cdots & \eta_{2,M} \\ \vdots & \vdots & \ddots & \vdots \\ \eta_{T,1} & \eta_{T,2} & \cdots & \eta_{T,M} \end{pmatrix}. \tag{2.6}$$

Different path gains may be independent from each other, that is $\alpha_{n,m}$ is independent from $\alpha_{n',m'}$ for $n \neq n'$ or $m \neq m'$. Note that the independence assumption is in the spatial domain and not necessarily in the time domain. Also, if the antennas are not far enough from each other, it is possible that some spatial correlation exists among the path gains. If the distance between two specific antennas is more than half of the wavelength, it is usually assumed that their path gains are independent of each other.

We also assume a quasi-static slow fading model such that the noise samples $\eta_{t,m}$ are independent samples of a zero-mean circularly symmetric complex Gaussian random variable. This is an additive white Gaussian noise (AWGN) assumption for a complex baseband transmission. The noise samples, channel path gains, and transmitted signals are independent from each other. Also, the bandwidth is narrow enough such that the channel is flat over a band of frequency. Such a channel is called frequency non-selective and the channel matrix is constant over the frequency band of interest. In the sequel, we use the quasi-static, non-frequency selective assumptions unless we mention otherwise.

Any of the statistical models in Section 1.4 can be utilized to model the path gains. For example, in a Rayleigh fading channel, the path gains are modeled by independent complex Gaussian random variables at each time slot. In other words, the real and imaginary parts of the path gains at each time slot are iid Gaussian random variables. Therefore, the distribution of the envelope of the path gains, $|\alpha_{n,m}|$, is Rayleigh, which is why the channel is called a Rayleigh fading channel. Also, $|\alpha_{n,m}|^2$ is a chi-square random variable with two degrees of freedom. If the real and imaginary parts have a zero mean and a variance equal to 0.5, that is $Var[\Re\{\alpha_{n,m}\}] = Var[\Im\{\alpha_{n,m}\}] = 0.5$, then $E[|\alpha_{n,m}|^2] = 1$. Note that we use $E[\cdot]$ and $Var[\cdot]$ to represent expectation and variance of a random variable, respectively. Such a normalization is possible by appropriately adjusting the transmit power.

Like single-input single-output (SISO) channels [100], the behavior of the baseband system depends on the ratio of the signal power and the noise power. Therefore, multiplying the transmission power and noise by the same factor does not affect the characteristics of the system. A fair comparison between two systems consists of the same transmission power, despite the number of transmit antennas, and the same average received signal to noise ratio (SNR). Let us denote the average power of the transmitted symbols, $\mathbf{C}_{t,n}$, by E_s. Also, let us assume that the variance of the zero-mean complex Gaussian noise is $N_0/2$ per dimension, that is $Var[\Re\{\eta_{n,m}\}] = Var[\Im\{\eta_{n,m}\}] = N_0/2$. Then, the average receive SNR is $\gamma = NE_s/N_0$. Since the performance is only a function of SNR, only the ratio of E_s/N_0 is important and not the separate values of E_s and N_0. Therefore, a normalization that removes one of these two values, that is E_s or N_0, is more compact and useful. One approach to achieve such a normalization is to normalize the average transmission power to one. For example, the average power of the transmitted symbols is normalized to $E_S = 1/N$. In this case, if the variance of the noise samples is $1/(2\gamma)$ per complex dimension, that is $N_0 = 1/\gamma$, the average power of the received signal at each receive antenna is 1 and the received SNR is γ. Note that the path gains are zero mean complex Gaussian random variables with $Var[\Re\{\alpha_{n,m}\}] = Var[\Im\{\alpha_{n,m}\}] = 0.5$.

Another approach for normalizing E_s and N_0 is to use a constellation with an average power of one for transmission symbols and unit-power noise samples. In this case, a normalization factor is considered in the input-output relationship of the MIMO channel as

$$\mathbf{r} = \sqrt{\frac{\gamma}{N}} \mathbf{C} \cdot \mathbf{H} + \mathcal{N}, \tag{2.7}$$

where again γ is the received SNR. Note that (2.5) and (2.7) describe the same system despite the difference in their forms. The main difference between the two equations is the method of normalization. Another difference is that the representation of the SNR, the main performance factor, is implicit in (2.5) and explicit in (2.7).

2.2 Capacity of MIMO channels

In this section, we study the capacity of a MIMO channel. We assume that the receiver knows the realization of the channel, that is it knows both $\mathbf{r}$ and $\mathbf{H}$. For the transmitter, we study two cases.

- The transmitter does not know the realization of the channel; however, it knows the distribution of $\mathbf{H}$. This corresponds to an open-loop system as depicted in Figure 1.11(a).
- The transmitter knows the realization of the channel. This corresponds to a closed-loop system as depicted in Figure 1.11(b).

We use a quasi-static block fading model. Under such a model, the channel path gains are fixed during a large enough block such that information theoretical results are valid. The values of path gains change from block to block based on a statistical model, for example a Rayleigh fading channel model. The resulting capacity of the channel is a random variable because the capacity is a function of the channel matrix $\mathbf{H}$. The distribution of the capacity is determined by the distribution of the channel matrix $\mathbf{H}$. The total transmitted power is kept the same for different numbers of transmit antennas to make a fair comparison. In most cases, we provide a sketch of the analysis and the interested reader is referred to [42, 140] for the details.

2.2.1 Capacity of a deterministic MIMO channel

To study the properties of the channel capacity for a random channel matrix $\mathbf{H}$, first we assume that the realization of the channel $\mathbf{H} = H$ is fixed. In other words, the channel matrix is deterministic and its value H is known at the receiver. It can be shown that for a given realization of the channel $\mathbf{H} = H$, the mutual information is maximized by a circularly symmetric complex Gaussian input codeword. Let us define K_C as the covariance of the input $\mathbf{C}$. Then, the capacity is defined as the maximum of the mutual information between the input and output given a power constraint P on the total transmission power of the input, that is $\text{Tr}(K_C) \le P$, where $\text{Tr}(K_C)$ is the trace of a matrix K_C. The explicit constraint, based on the parameters in our notation, depends on the corresponding normalization. For example, using the normalization in (2.7), the constraint is $\text{Tr}(K_C) \le N$. This is due to the fact that in this case the total transmission power of the input is N. In what follows, we use the notation in (2.7) and the constraint $\text{Tr}(K_C) \le N$.

The mutual information between input and output for the given realization is

$$I(C; r|\mathbf{H} = H) = \log_2(\det[I_M + (\gamma/N)H^H \cdot K_C \cdot H]). \qquad (2.8)$$

Then, the capacity is the maximum of the mutual information in (2.8) over all inputs satisfying $\text{Tr}(K_C) \leq P$. In other words,

$$C = \max_{\text{Tr}(K_C) \leq N} \log_2(\det[I_M + (\gamma/N)H^H \cdot K_C \cdot H]) \text{ bits/(s Hz)}. \qquad (2.9)$$

The unit bits/(s Hz) represents the fact that for a bandwidth of W, the maximum possible rate for reliable communication is CW bits/s.

The best covariance matrix for the Gaussian input that achieves the capacity depends on the realization of the channel matrix. In the case that we do not know the channel at the transmitter, that is an open-loop system, a good assumption is to distribute the input power equally among the transmit antennas. This results in a covariance matrix K_C that is a multiple of the identity matrix. Considering the constraint $\text{Tr}(K_C) \leq N$, we have $K_C = I_N$ that results in the following mutual information:

$$C_{\text{EP}} = \log_2(\det[I_M + (\gamma/N)H^H \cdot H]) \text{ bits/(s Hz)}, \qquad (2.10)$$

where the subscript "EP" stands for "equal power." The equal power allocation among different antennas is optimal for the uncorrelated iid Rayleigh fading channel. However, in the case of Ricean fading, an unequal power allocation results in a higher mutual information compared to an equal power allocation.

The capacity in (2.10) can be calculated in terms of the positive eigenvalues of the matrix $H^H \cdot H$. Let us define r as the rank of matrix H and λ_i, $i = 1, 2, \ldots, r$ as the non-zero eigenvalues of the matrix $H^H \cdot H$. Then, the eigenvalues λ_i, $i = 1, 2, \ldots, r$ are positive real numbers and the capacity in (2.10) is written as

$$C_{\text{EP}} = \sum_{i=1}^{r} \log_2[1 + (\gamma/N)\lambda_i]. \qquad (2.11)$$

When the channel is known at the transmitter, that is a closed-loop system, a nonuniform distribution of power among transmit antennas is useful. Using (2.9) and the singular value decomposition of H, one can show that in this case the capacity is

$$C = \max_{\sum_{i=1}^{r} \gamma_i = \gamma} \sum_{i=1}^{r} \log_2(1 + \gamma_i \lambda_i). \qquad (2.12)$$

The above capacity is calculated using a water-filling algorithm [27]. Note that an equal distribution of the power, $\gamma_i = \gamma/N$, in (2.12) results in the formula in (2.11). If the channel is known at the transmitter, the capacity in (2.12) provides some increase over C_{EP}. This is due to two factors: the unequal power allocation among different antennas, the diagonal elements in K_{C}, and the correlation in the optimal matrix K_{C}. It has been shown that the gain because of water-filling is mostly due to the correlation in the optimal matrix K_{C} [122]. For an equal number of transmit and receive antennas, $N = M$, the capacity increase due to an optimal water-filling disappears at high SNRs.

2.2.2 Capacity of a random MIMO channel

The capacity in (2.9) was defined for a fixed specific realization of the fading channel, $\mathbf{H} = H$, over a large block length. Every realization of the channel has some probability attached to it through the statistical model of $\mathbf{H}$. Since the channel matrix $\mathbf{H}$ is random in nature, the capacity in (2.9) is also a random variable. Let us assume an equal distribution of the input power that transforms (2.9) to (2.10), then with a small abuse of the notation, the random capacity of the channel is

$$C = \log_2\big(\det\big[I_M + (\gamma/N)\mathbf{H}^H \cdot \mathbf{H}\big]\big) \text{ bits/(s\,Hz)}, \qquad (2.13)$$

In what follows, we study the distribution, the mean, and the tail properties of the random capacity in (2.13).

In general, when $\mathbf{H}$ is a memoryless channel based on the iid Rayleigh fading model, we can calculate the distribution of the capacity. Figure 2.2 depicts the complementary cumulative distribution function (CCDF), $P(C > c)$, of the capacity (2.13) at SNR equal to $\gamma = 10$. The figure shows the complementary CDF for the four possible cases with one and two antennas at transmitter and receiver. Note that the normalization provides the same transmission power for all different cases and the comparison is fair. It is clear from Figure 2.2 that most of the time, the capacity of these four systems increase in the following order: $(N = 1, M = 1)$, $(N = 2, M = 1)$, $(N = 1, M = 2)$, and $(N = 2, M = 2)$. Normally, the SISO channel has the lowest capacity and the MIMO channel with two transmit and receive antennas has the highest capacity. Similar behavior is observed for other values of SNR.

For the case of $N \geq M$, one can derive a lower bound on the capacity in terms of chi-square random variables [42]:

$$C > \sum_{k=N-M+1}^{N} \log_2(1 + (\gamma/N) \cdot \chi_k), \qquad (2.14)$$

where χ_k is a chi-square random variable with $2k$ degrees of freedom. The probability density function (pdf) of the chi-square random variable with $2k$ degrees of

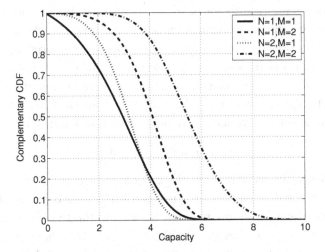

Fig. 2.2. The complementary cumulative distribution function (CCDF) of capacity for the iid Rayleigh fading model; $\gamma = 10$.

freedom is

$$f_{\chi_k}(x) = \frac{e^{-x} x^{k-1}}{[k-1]!}, \quad x > 0. \tag{2.15}$$

For the special case of $N = M$, we denote the lower bound in (2.14) by C_N

$$C_N = \sum_{k=1}^{N} \log_2(1 + (\gamma/N) \cdot \chi_k). \tag{2.16}$$

This result suggests that for large N, the lower bound C_N increases at least linearly as a function of N [42], and indicates that when the number of transmit and receive antennas are the same, the capacity of a MIMO channel increases linearly by the number of antennas. In general, the capacity increases by the minimum of the number of transmit and receive antennas. One can show that at high SNRs the capacity of a MIMO channel in (2.13) in terms of γ, the received SNR, can be described as

$$C \approx \min\{N, M\} \log_2(\gamma/N) + \sum_{k=|N-M|+1}^{\min\{N,M\}} \log_2(\chi_k), \tag{2.17}$$

where χ_k is a chi-square random variable with $2k$ degrees of freedom. Therefore, a 3 dB increase in SNR results in $\min\{N, M\}$ extra bits of capacity at high SNRs.

For the case of one transmit and one receive antenna, the capacity formula in (2.13) becomes

$$C = \log_2(1 + \gamma |\alpha|^2), \tag{2.18}$$

where $\alpha = \alpha_{1,1}$ is the fade factor of the SISO channel. In this case, the capacity coincides with the standard Shannon capacity of a Gaussian channel for a given value of α [27]. Since α is a complex Gaussian random variable, the resulting capacity in (2.18) is also a random variable that can take any non-negative value. As we discussed before, $|\alpha|^2$ is a chi-square random variable with two degrees of freedom. A chi-square random variable with two degrees of freedom is the same as an exponential random variable with unit mean as follows:

$$f_{|\alpha|^2}(x) = e^{-x}, \quad x > 0. \tag{2.19}$$

Figure 2.2 shows the distribution of the capacity. Since in this case the capacity can be any non-negative number, it is not as useful as the capacity of an AWGN channel. In other words, for any rate that we want to transmit, there is a non-zero probability that the realized channel results in a Shannon capacity smaller than the required rate. Let us define the capacity as the supremum of the rates for which there exists a code that can achieve arbitrarily small probability of error.

For the case that the channel matrix $\mathbf{H}$ is generated by an ergodic process, a non-zero capacity exists. The ergodic capacity is the maximum mutual information between the input and output if the code spans an infinite number of independent realizations of the channel matrix $\mathbf{H}$. For example, if the channel is memoryless, that is each use of the channel corresponds to an independent realization of the channel matrix $\mathbf{H}$, the ergodic capacity exists and is not zero. Note that for non-ergodic channels, the capacity may not be the same as the maximum mutual information. For example, if the channel matrix $\mathbf{H}$ is randomly chosen and remains fixed during the transmission, in general, the maximum mutual information is not the capacity. The following theorem provides the ergodic capacity of a MIMO channel when the channel is isotropic given any side information that exists at the transmitter [140].

Theorem 2.2.1 *The ergodic capacity of the MIMO channel defined in Equation (2.5) or (2.7) is:*

$$C_E = E\left[\log_2\left(\det\{I_M + (\gamma/N)\mathbf{H}^H \cdot \mathbf{H}\}\right)\right]. \tag{2.20}$$

Proof First, we note that the receiver knows the channel while the transmitter does not know the channel. Denoting the mutual information between the input and the observation of the receiver, $(\mathbf{r}, \mathbf{H})$, by $I(\mathbf{C}; (\mathbf{r}, \mathbf{H}))$, we have

$$I(\mathbf{C}; (\mathbf{r}, \mathbf{H})) = I(\mathbf{C}; \mathbf{H}) + I(\mathbf{C}; \mathbf{r}|\mathbf{H}) = I(\mathbf{C}; \mathbf{r}|\mathbf{H})$$
$$= E\left[I(\mathbf{C}; \mathbf{r}|\mathbf{H} = H)\right], \tag{2.21}$$

where the expectation is over the distribution of $\mathbf{H}$. If the channel is isotropic, the maximum of the mutual information is achieved by an independent circularly symmetric complex Gaussian random variable with $K_C = I_N$. Following the same steps that we took in Section 2.2.1 results in (2.20). $\qquad\square$

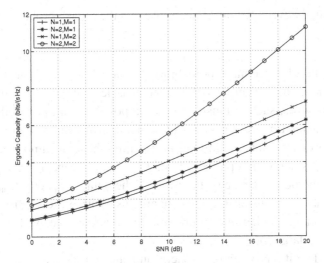

Fig. 2.3. Ergodic capacity for N transmit and M receive antennas.

Note that ergodic capacity (2.20) is in fact the mean of the capacity in (2.13). Figure 2.3 depicts the ergodic capacity plotted against SNR for different MIMO channels assuming a Rayleigh fading model. As can be seen from the figure, using multiple antennas increases the ergodic capacity. Also, the effect of multiple receive antennas in increasing the ergodic capacity is more than that of the multiple transmit antennas. The ergodic capacity of a Ricean fading channel is lower bounded by its Rayleigh component [79]. For a large number of antennas, the capacity of a Ricean fading channel converges to the capacity of the corresponding Rayleigh channel [79].

For a large number of transmit antennas and a fixed number of receive antennas, using the law of large numbers,

$$\frac{\mathbf{H}^H \cdot \mathbf{H}}{N} \longrightarrow I_M, \text{ almost surely.} \qquad (2.22)$$

As a result, the ergodic capacity is $M \log_2(1 + \gamma)$ for large N.

2.3 Outage capacity

So far we have studied the capacity of MIMO channels when the channel is ergodic. A more useful capacity concept for coding purposes is the outage capacity defined in [42]. The outage capacity C_{out} is a value that is smaller than the random variable C only with a probability P_{out} (outage probability). Similarly, one can fix the outage capacity and find the outage probability, that is the probability that the capacity random variable is smaller than the outage capacity. The following equation shows the relationship between the outage capacity and the outage probability:

$$P_{\text{out}} = P(C < C_{\text{out}}). \qquad (2.23)$$

The importance of the outage probability is that if one wants to transmit C_{out} bits/channel use, the capacity of the channel is less than C_{out} with probability P_{out}. In other words, such a transmission is impossible with probability P_{out}. For a stationary channel, if we transmit a large number of frames with a rate of C_{out} bits/channel use, the number of failures is P_{out} times the total number of frames. On the other hand, since with a probability of $1 - P_{\text{out}}$, the capacity random variable is larger than the outage capacity, the value of the outage capacity C_{out} guarantees that it is possible to transmit C_{out} bits/channel use with a probability of $1 - P_{\text{out}}$. Of course, picking a higher outage probability for a fixed received signal to noise ratio results in a larger outage capacity.

The above definitions of outage capacity and outage probability are valid for any number of transmit and receive antennas. One difficulty with a multiple antenna system is the number of independent random variables in (2.13) that affect the Shannon capacity. For a Rayleigh fading channel with N transmit antennas and M receive antennas, the Shannon capacity is a function of NM independent complex Gaussian random variables. As we discussed before, for the case of one transmit and one receive antenna, the capacity is

$$C = \log_2(1 + \gamma.\mathcal{X}), \qquad (2.24)$$

where $\mathcal{X}$ is a chi-square random variable with two degrees of freedom. In this case, to achieve one extra bit of capacity at high SNRs, one requires a 3 dB increase in signal-to-noise ratio (doubling the SNR).

For the case of one transmit antenna, that is $N = 1$, and M receive antennas, using the equality $\det[I + A \cdot B] = \det[I + B \cdot A]$, we have

$$C = \log_2\big(\det[I_M + \gamma\, \mathbf{H}^H \cdot \mathbf{H}]\big) = \log_2(1 + \gamma\, \mathbf{H} \cdot \mathbf{H}^H)$$
$$= \log_2\left(1 + \gamma \sum_{m=1}^{M} |\alpha_{1,m}|^2\right). \qquad (2.25)$$

Figure 2.2 depicts the complementary CDF of the Shannon capacity for $M = 2$ receive antennas at SNR equal to 10. A few observations regarding (2.25) for one transmit antenna are helpful. First, comparing the capacity of a channel with one receive antenna with that of a channel with M receive antennas shows that (2.25) can be derived from (2.18) by replacing $|\alpha|^2$ with $\sum_{m=1}^{M}|\alpha_{1,m}|^2$. This indicates the possibility of a diversity order of M for the case of one transmit and M receive antennas. Also, it shows that the capacity of a system that uses maximum ratio combining at the decoder is not less than that of the optimal capacity. In other

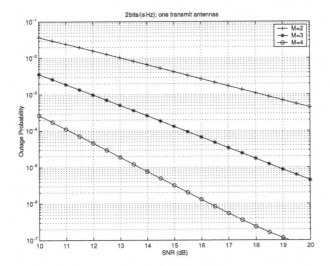

Fig. 2.4. $C_{\text{out}} = 2$ bits/(s Hz); M receive antennas, one transmit antenna.

words, one would not reduce the capacity of the system by using a maximum ratio combiner at the receiver. On the other hand, using selection combining reduces the capacity of the system to

$$C = \log_2\left(1 + \gamma \max_m |\alpha_{1,m}|^2\right). \tag{2.26}$$

Assuming independent Rayleigh fading, the capacity in (2.25) is

$$C = \log_2(1 + \gamma.\mathcal{X}_r), \tag{2.27}$$

where $\mathcal{X}_r$ is a chi-square random variable with $2M$ degrees of freedom. Since the capacity in (2.27) is only a function of one random variable, the outage probability can be calculated easily from (2.23). Simple algebraic manipulation shows that

$$P_{\text{out}} = P\left(\mathcal{X}_r < \frac{2^{C_{\text{out}}} - 1}{\gamma}\right). \tag{2.28}$$

Figure 2.4 shows the outage probability plotted against signal to noise ratio for an outage capacity of $C_{\text{out}} = 2$ bits/channel use for one transmit antenna and $M = 2, 3, 4$ receive antennas.

Similarly for a system with N transmit antennas and one receive antenna the Shannon capacity can be calculated as

$$C = \log_2(1 + (\gamma/N).\mathcal{X}_t), \tag{2.29}$$

where $\mathcal{X}_t$ is a chi-square random variable with $2N$ degrees of freedom. The

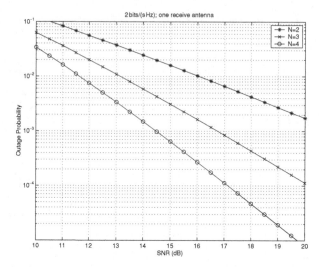

Fig. 2.5. $C_{\text{out}} = 2 \text{ bits}/(\text{s Hz})$; N transmit antennas, one receive antenna.

corresponding formula for the outage probability is

$$P_{\text{out}} = P\left(\mathcal{X}_t < N \frac{2^{C_{\text{out}}} - 1}{\gamma} \right). \tag{2.30}$$

Figure 2.5 shows the outage probability plotted against signal to noise ratio for an outage capacity $C_{\text{out}} = 2 \text{ bits}/(\text{s Hz})$ for $N = 2, 3, 4$ transmit antennas and one receive antenna. As it is clear from (2.28) and (2.30), for a given outage capacity, a system with N transmit antennas and one receive antenna requires N times more signal-to-noise ratio to provide the same outage probability as a system with $M = N$ receive antennas and one transmit antenna. This is due to the fact that the capacity formulas are derived for the same total transmission power in both cases. Mathematically, the $N \times 1$ channel and the $1 \times M$ channel provide a similar capacity for $M = N$ if the transmitted power over each path is the same. However for a fair comparison, the total transmission power needs to be divided among the N transmit antennas. Therefore, the received signal-to-noise ratio is affected by a factor of N.

While the capacity can be written as a function of one random variable for the above special cases, in general one may not be able to find such a simple formula for the capacity. Calculating the statistics of the capacity from (2.13) is always possible by empirically generating the statistics of the channel path gains $\alpha_{n,m}$ in a computer simulation. For every sample of channel matrix $\mathbf{H}$, one can calculate one sample of the capacity in (2.13).

We discussed a lower bound on the capacity in terms of chi-square random variables in (2.14). An upper bound on the capacity in terms of chi-square random variables can be derived as follows [42]. Consider a system in which the signal transmitted by every transmit antenna is received by a separate group of M receive antennas. It is clear that the capacity of such a system is more than that of the considered MIMO channel. In fact, the capacity of such a system provides an upper bound for the capacity in (2.13). The capacity for each transmit antenna and the corresponding separate group of M receive antennas is given by (2.25). Therefore, we have the following upper bound

$$C < \sum_{i=1}^{N} \log_2(1 + (\gamma/N).\chi_i), \tag{2.31}$$

where χ_i are the independent chi-square random variables with $2M$ degrees of freedom.

We note that the definition of outage capacity in (2.23) is not the only possible definition. For the sake of simplicity, let us focus on the case of the iid Rayleigh fading channel. One can define the relationship between the outage capacity and outage probability as [140]

$$P_{\text{out}} = \inf_{\text{Tr}(K_C) \leq N} P\left\{\log_2\left(\det\left[I_M + \mathbf{H}^H \cdot K_C \cdot \mathbf{H}\right]\right) < C_{\text{out}}\right\}. \tag{2.32}$$

The main difference between the two definitions is that the optimal solution in (2.32) may assign zero power to some of the transmit antennas while all antennas are used in calculating the capacity in (2.23). For example, it is clear from Figure 2.2 that for a large outage probability and one receive antenna, the capacity of a channel with one transmit antenna is more than that of a channel with two transmit antennas. Therefore, if the system has two transmit antennas and one receive antenna, it makes sense that it uses only one of the transmit antennas for very large outage probabilities. One of the diagonal elements in the optimal covariance matrix K_C in (2.32) will be zero. On the other hand, the K_C that provides the capacity C in (2.23) is the identity matrix. In fact, the capacity C in (2.13) and throughout most of Section 2.2.2 has an extra implicit constraint that the covariance matrix K_C is given. One can denote such a capacity by $C(K_C)$ and then the capacity C in (2.23) will be $C = C(I)$. Throughout the book, we use outage probabilities as a benchmark to compare the frame error probabilities of different space-time codes. Since usually all available transmit antennas are utilized by space-time codes, the system is under the constraint $K_C = I$ and the appropriate definition for our purpose is the one in (2.23).

2.4 Summary of important results

- The capacity of a MIMO channel is a function of the channel matrix $\mathbf{H}$. Considering the random nature of the channel matrix $\mathbf{H}$, the capacity of a MIMO channel subject to the input covariance matrix K_C can be considered as the following random variable:

$$C(K_C) = \log_2\left(\det\left[I_M + (\gamma/N)\mathbf{H}^H \cdot K_C \cdot \mathbf{H}\right]\right) \text{ bits/(s Hz)}.$$

- When $K_C = I$ maximizes the mutual information, for example in an iid Rayleigh fading model, the capacity is

$$C = C(I) = \log_2\big(\det\big[I_M + (\gamma/N)\mathbf{H}^H \cdot \mathbf{H}\big]\big) \text{ bits/(s Hz) or bits/channel use.}$$

- If the number of transmit and receive antennas are the same, the capacity increases at least linearly as a function of number of antennas.
- At high SNRs, a 3 dB increase in SNR results in $\min\{N, M\}$ extra bits of capacity.
- The ergodic capacity of a MIMO channel, when the channel is isotropic, given any side information that exists at the transmitter, is:

$$C_E = E\big[\log_2\big(\det\big\{I_M + (\gamma/N)\mathbf{H}^H \cdot \mathbf{H}\big\}\big)\big].$$

- The ergodic capacity is $M\log_2(1 + \gamma)$ for large N.
- The outage capacity C_{out} is a value that is smaller than the random variable C (capacity), only with a probability P_{out} (outage probability):

$$P_{out} = P(C < C_{out}).$$

2.5 Problems

1 In (2.2), the transmitted signals are collected such that the time index changes vertically and the antenna (space) index changes horizontally. Rewrite (2.5) for the case that the time index changes horizontally and the antenna (space) index changes vertically. Show that the new matrices are the transpose of the matrices in (2.5).

2 Consider a multiple-input multiple-output channel with two transmit antennas and one receive antenna. Also, assume a zero-mean unit-variance additive Gaussian noise and an average power constraint of one per transmission antenna. What is the capacity of the channel if the path gains from the first and second antennas to the receive antenna are $\alpha_1 = 1$ and $\alpha_2 = 1 + j$, respectively?

3 In problem 2, what is the capacity if the transmitter knows the channel and the average power constraint is 2 over the sum of the transmission powers from both antennas?

4 Calculate the complementary cumulative distribution function (CDF) of the capacity in (2.13) for the iid Rayleigh fading model at SNR equal to $\gamma = 20\,\text{dB}$ in the following cases:
- one transmit antenna and one receive antenna;
- two transmit antennas and one receive antenna;
- one transmit antenna and two receive antennas;
- two transmit antennas and two receive antennas.

Draw a figure containing the results.

5 Consider a MIMO channel with three transmit antennas and one receive antenna. For an outage capacity $C_{out} = 1\,\text{bit/(s Hz)}$, what is the outage probability at SNR equal to $\gamma = 10$?

3

Space-time code design criteria

3.1 Background

A code is mapping from the input bits to the transmitted symbols. As discussed in Chapter 2, we assume that symbols are transmitted simultaneously from different antennas. In this chapter, we study the performance of different codes by deriving some bounds on them. Then, we use the bounds to provide some guidance to design codes with "good" performance. Such guidance is called the design criterion. Most of the analyses in this chapter are asymptotic analysis. Therefore, different asymptotic assumptions may result in different code criteria. We concentrate on a quasi-static Rayleigh fading wireless channel and some of the important design criteria that result in achieving maximum diversity and good performance at high SNRs.

A good code follows a design criterion that adds some notion of optimality to the code. In fact, the goal of defining a design criterion is to have a guideline for designing good codes. For example let us consider transmission over a binary symmetric channel using a linear binary block channel code. The bit error rate of the system depends on the Hamming distances of the codeword pairs. Defining the set of all possible codeword pairs and the corresponding set of Hamming distances, we denote the minimum Hamming distance by $d_{\min}$. It can be shown that a code with minimum Hamming distance $d_{\min}$ can correct all error patterns of weight less than or equal to $\lfloor (d_{\min} - 1)/2 \rfloor$, where $\lfloor x \rfloor$ is the largest integer less than or equal to x. Therefore, for a given redundancy, a "good" code has a high minimum Hamming distance. The design criterion for such a code is to maximize the minimum possible Hamming distance among the codeword pairs. To compare two codes with similar redundancies, the one with higher Hamming distance is preferable. Similarly for an additive white Gaussian noise (AWGN) channel, a good design criterion is to maximize the minimum Euclidean distance among all possible codeword pairs.

In what follows, we study the design criterion for space-time codes. We derive design criteria that guarantee the maximum possible diversity gain and coding gain at high SNRs. Also, we consider maximizing the mutual information between the input and output of the system as a design criterion.

3.2 Rank and determinant criteria

In order to come up with a design criterion, first we need to quantify the effects of mistaking two codewords with each other. In the case of a space-time code, a codeword is a $T \times N$ matrix given by (2.2). Let us assume that we transmit a codeword $\mathbf{C}^1$

$$
\mathbf{C}^1 = \begin{pmatrix}
C^1_{1,1} & C^1_{1,2} & \cdots & C^1_{1,N} \\
C^1_{2,1} & C^1_{2,2} & \cdots & C^1_{2,N} \\
\vdots & \vdots & \ddots & \vdots \\
C^1_{T,1} & C^1_{T,2} & \cdots & C^1_{T,N}
\end{pmatrix}.
\tag{3.1}
$$

An error occurs if the decoder mistakenly decides that we have transmitted another codeword, for example $\mathbf{C}^2$

$$
\mathbf{C}^2 = \begin{pmatrix}
C^2_{1,1} & C^2_{1,2} & \cdots & C^2_{1,N} \\
C^2_{2,1} & C^2_{2,2} & \cdots & C^2_{2,N} \\
\vdots & \vdots & \ddots & \vdots \\
C^2_{T,1} & C^2_{T,2} & \cdots & C^2_{T,N}
\end{pmatrix}.
\tag{3.2}
$$

If the codebook, the set of all codewords, only contains $\mathbf{C}^1$ and $\mathbf{C}^2$, we denote the pairwise error probability of transmitting $\mathbf{C}^1$ and detecting it as $\mathbf{C}^2$ by $P(\mathbf{C}^1 \to \mathbf{C}^2)$. Note that in general when the codebook contains I codewords, using the union bound, the probability of error when we transmit $\mathbf{C}^1$ is upper bounded by

$$
P(\text{error}|\mathbf{C}^1 \text{is sent}) \leq \sum_{i=2}^{I} P(\mathbf{C}^1 \to \mathbf{C}^i).
\tag{3.3}
$$

The overall bound on error probability can be calculated by using (3.3).

In what follows we calculate the pairwise error probability $P(\mathbf{C}^1 \to \mathbf{C}^2)$ and use it to define the design criteria. In order to calculate the pairwise error probability, first we assume a fixed known channel matrix $\mathbf{H}$ and then calculate the average error by computing the expected value over the distribution of the $\mathbf{H}$.

Without loss of generality, we use (2.5) to represent the input–output relationship of the channel and the corresponding normalization from Section 2.1. The average symbol transmission power from each antenna is $E_S = 1/N$ and the variance of a noise sample is $E[|\eta_{t,m}|^2] = N_0 = 1/\gamma$. We consider the distribution of the received signals for a known codeword $\mathbf{C}$ and channel matrix $\mathbf{H}$, that is $f(\mathbf{r}|\mathbf{C}, \mathbf{H})$. Note that the linear combination of independent Gaussian random variables is a Gaussian random variable. Since we assume a Gaussian noise $\mathcal{N}$ with independent components, for a fixed $\mathbf{C}$ and $\mathbf{H}$, the received vector $\mathbf{r}$ is also a multivariate, multidimensional, Gaussian random variable. Therefore,

$$
\begin{aligned}
f(\mathbf{r}|\mathbf{C}, \mathbf{H}) &= \frac{1}{(\pi N_0)^{\frac{M \times M}{2}}} \exp \left\{ \frac{-\text{Tr}[(\mathbf{r} - \mathbf{C} \cdot \mathbf{H})^H (\mathbf{r} - \mathbf{C} \cdot \mathbf{H})]}{N_0} \right\} \\
&= \left(\frac{\gamma}{\pi} \right)^{\frac{M \times M}{2}} \exp \left\{ -\gamma \text{Tr}[(\mathbf{r} - \mathbf{C} \cdot \mathbf{H})^H (\mathbf{r} - \mathbf{C} \cdot \mathbf{H})] \right\}.
\end{aligned}
\tag{3.4}
$$

The Frobenius norm of matrix A is denoted by $||A||_F$ and is defined as

$$
||A||_F = \sqrt{\text{Tr}(A^H \cdot A)} = \sqrt{\text{Tr}(A \cdot A^H)}.
\tag{3.5}
$$

One can rewrite (3.4) in terms of a Frobenius norm as follows:

$$
\begin{aligned}
f(\mathbf{r}|\mathbf{C}, \mathbf{H}) &= \left(\frac{\gamma}{\pi} \right)^{\frac{M \times M}{2}} \exp \left\{ -\gamma ||\mathbf{r} - \mathbf{C} \cdot \mathbf{H}||_F^2 \right\} \\
&= \left(\frac{\gamma}{\pi} \right)^{\frac{M \times M}{2}} \exp \left\{ -\gamma \sum_{t=1}^{T} \sum_{m=1}^{M} |(\mathbf{r} - \mathbf{C} \cdot \mathbf{H})_{t,m}|^2 \right\},
\end{aligned}
\tag{3.6}
$$

where $||\mathbf{r} - \mathbf{C} \cdot \mathbf{H}||_F^2 = \sum_{t=1}^{T} \sum_{m=1}^{M} |(\mathbf{r} - \mathbf{C} \cdot \mathbf{H})_{t,m}|^2$ is derived from the definition of the Frobenius norm in (3.5). Maximum-likelihood (ML) decoding decides in favor of a codeword that maximizes $f(\mathbf{r}|\mathbf{C}, \mathbf{H})$. Let us assume that we transmit $\mathbf{C}^1$, the received vector is $\mathbf{r}^1 = \mathbf{C}^1 \cdot \mathbf{H} + \mathcal{N}^1$ and given the channel matrix $\mathbf{H}$, the pairwise error probability is calculated by

$$
P(\mathbf{C}^1 \rightarrow \mathbf{C}^2 | \mathbf{H}) = P(||\mathbf{r}^1 - \mathbf{C}^1 \cdot \mathbf{H}||_F^2 - ||\mathbf{r}^1 - \mathbf{C}^2 \cdot \mathbf{H}||_F^2 > 0 | \mathbf{H}).
\tag{3.7}
$$

We rewrite (3.7) to calculate the pairwise error probability as follows:

$$
\begin{aligned}
&P(\mathbf{C}^1 \rightarrow \mathbf{C}^2 | \mathbf{H}) \\
&= P \left\{ \text{Tr}\left[(\mathbf{r}^1 - \mathbf{C}^1 \cdot \mathbf{H})^H \cdot (\mathbf{r}^1 - \mathbf{C}^1 \cdot \mathbf{H}) - (\mathbf{r}^1 - \mathbf{C}^2 \cdot \mathbf{H})^H \cdot (\mathbf{r}^1 - \mathbf{C}^2 \cdot \mathbf{H}) \right] > 0 | \mathbf{H} \right\} \\
&= P \left(\text{Tr}\left\{ [(\mathbf{C}^1 - \mathbf{C}^2) \cdot \mathbf{H} + \mathcal{N}^1]^H \cdot [(\mathbf{C}^1 - \mathbf{C}^2) \cdot \mathbf{H} + \mathcal{N}^1] - \mathcal{N}^{1^H} \mathcal{N}^1 \right\} < 0 | \mathbf{H} \right) \\
&= P \left(\text{Tr}\left\{ \mathbf{H}^H \cdot (\mathbf{C}^1 - \mathbf{C}^2)^H \cdot (\mathbf{C}^1 - \mathbf{C}^2) \cdot \mathbf{H} \right\} - X < 0 | \mathbf{H} \right) \\
&= P \left(||(\mathbf{C}^1 - \mathbf{C}^2) \cdot \mathbf{H}||_F^2 < X | \mathbf{H} \right) \\
&= P \left(X > ||(\mathbf{C}^2 - \mathbf{C}^1) \cdot \mathbf{H}||_F^2 | \mathbf{H} \right),
\end{aligned}
\tag{3.8}
$$

where given $\mathbf{H}$, $X = \text{Tr}\{\mathcal{N}^{1^H} \cdot (\mathbf{C}^2 - \mathbf{C}^1) \cdot \mathbf{H} + \mathbf{H}^H \cdot (\mathbf{C}^2 - \mathbf{C}^1)^H \cdot \mathcal{N}^1\}$ is a zero mean Gaussian random variable with variance $2N_0||(\mathbf{C}^2 - \mathbf{C}^1) \cdot \mathbf{H}||_F^2 = (2/\gamma)||(\mathbf{C}^2 - \mathbf{C}^1) \cdot \mathbf{H}||_F^2$. Therefore, one can calculate the pairwise error probability using the Q function.

$$P\left(\mathbf{C}^1 \to \mathbf{C}^2|\mathbf{H}\right)$$

$$= Q\left(\frac{||(\mathbf{C}^2 - \mathbf{C}^1) \cdot \mathbf{H}||_F^2}{\sqrt{(2/\gamma)||(\mathbf{C}^2 - \mathbf{C}^1) \cdot \mathbf{H}||_F^2}}\right) = Q\left(\sqrt{\frac{\gamma}{2}}||(\mathbf{C}^2 - \mathbf{C}^1) \cdot \mathbf{H}||_F\right), \quad (3.9)$$

where

$$Q(x) = \frac{1}{\sqrt{2\pi}} \int_x^{\infty} e^{\frac{-y^2}{2}} \, dy. \tag{3.10}$$

Therefore, it remains to calculate $||(\mathbf{C}^2 - \mathbf{C}^1) \cdot \mathbf{H}||_F^2$ to derive the conditional pairwise error probability. Let us define the error (difference) matrix $D(\mathbf{C}^1, \mathbf{C}^2) = \mathbf{C}^2 - \mathbf{C}^1$. In what follows, we write the pairwise error probability in terms of the eigenvalues of a matrix $\mathbf{A}(\mathbf{C}^1, \mathbf{C}^2) = D(\mathbf{C}^1, \mathbf{C}^2)^H \cdot D(\mathbf{C}^1, \mathbf{C}^2) = (\mathbf{C}^2 - \mathbf{C}^1)^H \cdot (\mathbf{C}^2 - \mathbf{C}^1)$. Since $D(\mathbf{C}^1, \mathbf{C}^2)$ is a square root of $\mathbf{A}(\mathbf{C}^1, \mathbf{C}^2)$, the eigenvalues of $\mathbf{A}(\mathbf{C}^1, \mathbf{C}^2)$ denoted by λ_n, $n = 1, 2, \ldots, N$ are nonnegative real numbers, $\lambda_n \geq 0$. Using the singular value decomposition theorem [62], we have

$$\mathbf{A}(\mathbf{C}^1, \mathbf{C}^2) = V^H \cdot \Lambda \cdot V, \tag{3.11}$$

where $\Lambda = diag(\lambda_1, \lambda_2, \cdots, \lambda_N)$. Therefore,

$$||(\mathbf{C}^2 - \mathbf{C}^1) \cdot \mathbf{H}||_F^2 = \mathrm{Tr}\left[\mathbf{H}^H \cdot \mathbf{A}(\mathbf{C}^1, \mathbf{C}^2) \cdot \mathbf{H}\right]$$

$$= \mathrm{Tr}\left[\mathbf{H}^H \cdot V^H \cdot \Lambda \cdot V \cdot \mathbf{H}\right]. \tag{3.12}$$

Since the elements of $\mathbf{H}$ are independent Gaussian random variables, the elements of $V \cdot \mathbf{H}$ are also Gaussian. We denote the (n, m)th element of $V \cdot \mathbf{H}$ by $\beta_{n,m}$. Therefore,

$$||(\mathbf{C}^2 - \mathbf{C}^1) \cdot \mathbf{H}||_F^2$$

$$= \mathrm{Tr}\left[\begin{pmatrix} \beta_{1,1}^* & \beta_{2,1}^* & \cdots & \beta_{N,1}^* \\ \beta_{1,2}^* & \beta_{2,2}^* & \cdots & \beta_{N,2}^* \\ \vdots & \vdots & \ddots & \vdots \\ \beta_{1,M}^* & \beta_{2,M}^* & \cdots & \beta_{N,M}^* \end{pmatrix} \cdot \begin{pmatrix} \lambda_1 & 0 & \cdots & 0 \\ 0 & \lambda_2 & \cdots & 0 \\ \vdots & \vdots & \ddots & \vdots \\ 0 & 0 & \cdots & \lambda_N \end{pmatrix} \cdot \begin{pmatrix} \beta_{1,1} & \beta_{1,2} & \cdots & \beta_{1,M} \\ \beta_{2,1} & \beta_{2,2} & \cdots & \beta_{2,M} \\ \vdots & \vdots & \ddots & \vdots \\ \beta_{N,1} & \beta_{N,2} & \cdots & \beta_{N,M} \end{pmatrix}\right]$$

$$= \mathrm{Tr}\left[\begin{pmatrix} \sum_{n=1}^N \lambda_n|\beta_{n,1}|^2 & & \cdots & \\ & \sum_{n=1}^N \lambda_n|\beta_{n,2}|^2 & \cdots & \\ & & & \\ & & \cdots & \sum_{n=1}^N \lambda_n|\beta_{n,M}|^2 \end{pmatrix}\right]. \tag{3.13}$$

Calculating the trace in the last equation results in

$$||(\mathbf{C}^2 - \mathbf{C}^1) \cdot \mathbf{H}||_F^2 = \sum_{m=1}^{M} \sum_{n=1}^{N} \lambda_n |\beta_{n,m}|^2. \tag{3.14}$$

Applying (3.14) in (3.9) results in

$$P(\mathbf{C}^1 \rightarrow \mathbf{C}^2 | \mathbf{H}) = Q\left(\sqrt{\frac{\gamma}{2} \sum_{m=1}^{M} \sum_{n=1}^{N} \lambda_n |\beta_{n,m}|^2}\right). \tag{3.15}$$

A good upper bound on the Q function is $Q(x) \leq \frac{1}{2} e^{\frac{-x^2}{2}}$. Therefore, we can calculate an upper bound on the conditional pairwise error probability as follows:

$$P(\mathbf{C}^1 \rightarrow \mathbf{C}^2 | \mathbf{H}) \leq \frac{1}{2} \exp\left(-\frac{\gamma}{4} \sum_{m=1}^{M} \sum_{n=1}^{N} \lambda_n |\beta_{n,m}|^2\right). \tag{3.16}$$

Note that since $\beta_{n,m}$ are Gaussian, their magnitudes, $|\beta_{n,m}|$, are Rayleigh with the probability density function

$$f(|\beta_{n,m}|) = 2|\beta_{n,m}| \exp(-|\beta_{n,m}|^2). \tag{3.17}$$

Using the distributions of $|\beta_{n,m}|$, we can calculate the expected value of the pairwise error probability,

$$P(\mathbf{C}^1 \rightarrow \mathbf{C}^2) = E[P(\mathbf{C}^1 \rightarrow \mathbf{C}^2 | \mathbf{H})] \leq \frac{1}{\prod_{n=1}^{N}[1 + (\gamma \lambda_n/4)]^M}. \tag{3.18}$$

If matrix $\mathbf{A}(\mathbf{C}^1, \mathbf{C}^2)$ is full rank, none of its eigenvalues is zero. On the other hand, if its rank is $r < N$, without loss of generality, we have $\lambda_1 \geq \lambda_2 \geq \cdots \geq \lambda_r > 0$ and $\lambda_{r+1} = \cdots = \lambda_N = 0$. At high SNRs, one can neglect the one in the denominator of the inequality (3.18) and write the following upper bound based on the nonzero eigenvalues:

$$P(\mathbf{C}^1 \rightarrow \mathbf{C}^2) \leq \frac{4^r M}{(\prod_{n=1}^{r} \lambda_n)^M \gamma^{rM}}. \tag{3.19}$$

Without loss of generality, let us assume that the worst case is transmitting $\mathbf{C}^1$ and decoding it as $\mathbf{C}^2$. We can define the diversity gain G_d and the coding gain G_c using the right-hand side of (3.19). Using $(G_c \gamma)^{-G_d}$ to represent the right-hand side of (3.19), the diversity of the code is equal to rM. In other words, the diversity is equal to the rank of matrix $\mathbf{A}(\mathbf{C}^1, \mathbf{C}^2)$ or equivalently the rank of the difference matrix $D(\mathbf{C}^1, \mathbf{C}^2)$ multiplied by the number of receive antennas. Similarly, the coding gain relates to the product of the nonzero eigenvalues of the matrix $\mathbf{A}(\mathbf{C}^1, \mathbf{C}^2)$. A full diversity of MN is possible if the matrix $\mathbf{A}(\mathbf{C}^1, \mathbf{C}^2)$ is full rank. In this case, the coding gain relates to the products of eigenvalues $\prod_{n=1}^{N} \lambda_n$ or equivalently the determinant of matrix $\mathbf{A}(\mathbf{C}^1, \mathbf{C}^2)$. We define the coding gain distance (CGD) between codewords $\mathbf{C}^1$ and $\mathbf{C}^2$ as $\mathrm{CGD}(\mathbf{C}^1, \mathbf{C}^2) = \det(A(\mathbf{C}^1, \mathbf{C}^2))$.[1] Therefore, a good design criterion to guarantee full diversity is to make sure that for all possible codewords $\mathbf{C}^i$ and $\mathbf{C}^j$, $i \neq j$, the matrix $\mathbf{A}(\mathbf{C}^i, \mathbf{C}^j)$ is full rank [51, 139]. Then, to increase the coding gain for a full diversity code, an additional good design criterion is to maximize the minimum determinant of matrices $\mathbf{A}(\mathbf{C}^i, \mathbf{C}^j)$ for all $i \neq j$.

The above two criteria for designing space-time codes are called rank and determinant criteria. For any two codewords $\mathbf{C}^i \neq \mathbf{C}^j$, the rank criterion suggests that the error matrix $D(\mathbf{C}^i, \mathbf{C}^j) = \mathbf{C}^j - \mathbf{C}^i$ has to be full rank for all $i \neq j$ in order to obtain full diversity NM. The determinant criterion says that the minimum determinant of $\mathbf{A}(\mathbf{C}^i, \mathbf{C}^j) = D(\mathbf{C}^i, \mathbf{C}^j)^H D(\mathbf{C}^i, \mathbf{C}^j)$ among all $i \neq j$ has to be large to obtain high coding gains.

3.3 Trace criterion

In this section, first we recalculate (3.9) by rearranging the norm $||(\mathbf{C}^2 - \mathbf{C}^1) \cdot \mathbf{H}||_F$ as follows

$$
\begin{aligned}
||(\mathbf{C}^2 - \mathbf{C}^1) \cdot \mathbf{H}||_F^2 &= \mathrm{Tr}[D(\mathbf{C}^1, \mathbf{C}^2) \cdot \mathbf{H} \cdot \{D(\mathbf{C}^1, \mathbf{C}^2) \cdot \mathbf{H}\}^H] \\
&= \mathrm{Tr}[D(\mathbf{C}^1, \mathbf{C}^2) \cdot \mathbf{H} \cdot \mathbf{H}^H \cdot D(\mathbf{C}^1, \mathbf{C}^2)^H].
\end{aligned} \tag{3.20}
$$

Note that (3.20) is for a given channel matrix $\mathbf{H}$. We calculate the expected value of (3.20) over the distribution of the channel matrix. Since the elements of $\mathbf{H}$ are independent Gaussian random variables, we have

$$
E[\mathbf{H} \cdot \mathbf{H}^H] = M I_N, \tag{3.21}
$$

where I_N is the $N \times N$ identity matrix. Therefore, we can calculate

$$
\begin{aligned}
E[||(\mathbf{C}^2 - \mathbf{C}^1) \cdot \mathbf{H}||_F^2] &= M \, \mathrm{Tr}[D(\mathbf{C}^1, \mathbf{C}^2) \cdot D(\mathbf{C}^1, \mathbf{C}^2)^H] \\
&= M \, \mathrm{Tr}[\mathbf{A}(\mathbf{C}^1, \mathbf{C}^2)] \\
&= M ||D(\mathbf{C}^1, \mathbf{C}^2)||_F^2.
\end{aligned} \tag{3.22}
$$

Note that $||D(\mathbf{C}^1, \mathbf{C}^2)||_F$ is a metric on the codebook of the space-time code. In other words, (i) it is symmetric $||D(\mathbf{C}^1, \mathbf{C}^2)||_F = ||D(\mathbf{C}^2, \mathbf{C}^1)||_F$, (ii) it is zero if and only if $\mathbf{C}^1 = \mathbf{C}^2$, and (iii) the triangle inequality $||D(\mathbf{C}^1, \mathbf{C}^2)||_F \leq ||D(\mathbf{C}^1, \mathbf{C}^3)||_F + ||D(\mathbf{C}^3, \mathbf{C}^2)||_F$ holds. Therefore, it can be used as a distance measure between space-time codewords the same way that Euclidean distance is used

---

[1] In general, if we have a matrix with rank less than N, the distance can be defined as the product of the non-zero eigenvalues of $\mathbf{A}(\mathbf{C}^1, \mathbf{C}^2)$.

for the codewords of trellis coded modulation schemes for SISO channels. This distance measure is of course independent of the channel matrix.

Another approach to further study the behaviors of the norm $||(C^2 - C^1) \cdot H||_F$ is to consider a large number of receive antennas. In this case, we have

$$\lim_{M \to \infty} \frac{H \cdot H^H}{M} = I_N. \tag{3.23}$$

Therefore, it can be shown that

$$\lim_{M \to \infty} \frac{||(C^2 - C^1) \cdot H||_F^2}{M} = ||D(C^1, C^2)||_F^2. \tag{3.24}$$

Since the right-hand side of (3.24) is not a function of the channel matrix H, using (3.9) for a large number of receive antennas, we have

$$P(C^1 \to C^2) \approx P(C^1 \to C^2|H) \approx Q\left(\sqrt{M\gamma/2}\,||D(C^1, C^2)||_F\right). \tag{3.25}$$

Following the approach in Section 3.2, we use the upper bound $Q(x) \le \frac{1}{2}e^{\frac{-x^2}{2}}$ to derive the following upper bound for a large number of receive antennas

$$P(C^1 \to C^2) \le \frac{1}{2}\exp\left(-M\,||D(C^1, C^2)||_F^2\,\frac{\gamma}{4}\right). \tag{3.26}$$

Another approach to derive a similar bound is to use the central limit theorem to calculate the norm in (3.14) [21]. The distribution of $|\beta_{n,m}|^2$ is chi-square with unit mean and variance for all values of n and m. Thus, the central limit theorem shows that $\sum_{m=1}^M \sum_{n=1}^N \lambda_n |\beta_{n,m}|^2 = \sum_{m=1}^M \sum_{n=1}^r \lambda_n |\beta_{n,m}|^2$ is a Gaussian random variable for large rM, where r is the rank of matrix $D(C^1, C^2)$. Note that $\lambda_n = 0$ for $n > r$ and the mean and variance of the Gaussian random variable are $M \sum_{n=1}^r \lambda_n$ and $M \sum_{n=1}^r \lambda_n^2$, respectively. Integrating over the Gaussian distribution of $Y = \sum_{m=1}^M \sum_{n=1}^r \lambda_n |\beta_{n,m}|^2$ in Inequality (3.16) results in

$$P(C^1 \to C^2) \le \int_{y=0}^{\infty} \frac{1}{2}\exp\left(\frac{-\gamma\,y}{4}\right) f_Y(y)\,dy. \tag{3.27}$$

Calculating the integral in (3.27) and using the upper bound $Q(x) \le \frac{1}{2}e^{\frac{-x^2}{2}}$ provides the following upper bound which is very similar to that of (3.26):

$$P(C^1 \to C^2) \le \frac{1}{4}\exp\left(-M\,||D(C^1, C^2)||_F^2\,\frac{\gamma}{4}\right). \tag{3.28}$$

As can be seen from bounds (3.26) and (3.28), the pairwise error probability is related to the metric $||D(C^1, C^2)||_F$. It is argued in [65] and [22] that a good design criterion is to maximize the minimum distance $||D(C^i, C^j)||_F$ among all possible $i \ne j$. This is called "trace criterion" because $||D(C^1, C^2)||_F^2 = \mathrm{Tr}[A(C^1, C^2)]$. As

Equivalent channel

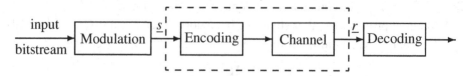

Fig. 3.1.   Block diagram of a space-time coding system.

shown above, a good property of such a design criterion is that $||D(\mathbf{C}^1, \mathbf{C}^2)||_F$ is a metric on the codebook of the space-time code. Therefore, it provides all the good properties of a distance measure.

3.4 Maximum mutual information criterion

Another design criterion that suggests a completely different design philosophy is the maximum mutual information criterion. This criterion selects the code parameters to maximize the mutual information between the transmitted and received signals. Let us denote the vector of data symbols by $\underline{s}$. The codeword $\mathbf{C}$ is defined as a function of the symbol vector $\underline{s}$ using the structure of the code. As shown in Figure 3.1, the combination effect of the encoder and the channel on the data symbols can be considered as an "equivalent channel" H. Then, the input–output relationship between the data symbols and the received signals $\underline{r}$ is

$$\underline{r} = \underline{s} \cdot H + \underline{N}, \qquad (3.29)$$

where $\underline{N}$ is the additive Gaussian noise of the channel. The goal is to design the code such that the mutual information between input $\underline{s}$ and output $\underline{r}$ is maximized. Such a design criterion is mostly beneficial if one wants to maximize the throughput spatial multiplexing gain. In fact, the main argument behind such a design criterion is that for the goal of maximizing the throughput, the code should not limit the capacity of the MIMO channel. Therefore, the "best" code is the one for which the mutual information between input $\underline{s}$ and output $\underline{r}$ is equal to the capacity of the original channel. Of course, like any other capacity argument, such an optimization is subjected to some power constraints on the input. As we will discuss in Chapters 9 and 10, there may be many codes with such a property and the maximum mutual information criterion does not provide a unique design for a given structure. However, it is possible to combine it with other design criteria as we will discuss in Chapter 10.

3.5 Summary of important results

- Rank criterion: The error matrix $D(\mathbf{C}^i, \mathbf{C}^j) = \mathbf{C}^j - \mathbf{C}^i$ has to be full rank for all $i \neq j$ in order to obtain full diversity NM.
- Determinant criterion: The minimum determinant of

$$\mathbf{A}(\mathbf{C}^i, \mathbf{C}^j) = D(\mathbf{C}^i, \mathbf{C}^j)^H D(\mathbf{C}^i, \mathbf{C}^j),$$

among all $i \neq j$, the minimum CGD of the code, has to be large to obtain high coding gains.
- Trace criterion: The minimum trace of $\mathbf{A}(\mathbf{C}^i, \mathbf{C}^j) = D(\mathbf{C}^i, \mathbf{C}^j)^H D(\mathbf{C}^i, \mathbf{C}^j)$ among all $i \neq j$ has to be large to obtain high coding gains.
- Maximum mutual information criterion: The mutual information between the transmitted and received signals has to be large to obtain high throughput.

3.6 Problems

To calculate the pairwise error probability, one can utilize the singular value decomposition theorem for $D(\mathbf{C}^1, \mathbf{C}^2)$ instead of $\mathbf{A}(\mathbf{C}^1, \mathbf{C}^2)$ [65]. Without loss of generality, we assume $T \geq N$. There are unitary matrices U and W, respectively $T \times T$ and $N \times N$, such that

$$D(\mathbf{C}^1, \mathbf{C}^2) = U^H \cdot \Sigma \cdot W,$$

where Σ is a $T \times N$ matrix with elements $\sigma_{1,1} \geq \sigma_{2,2} \geq \cdots \geq \sigma_{r,r} > \sigma_{r+1,r+1} = \cdots = \sigma_{N,N} = 0$ and $\sigma_{t,n} = 0$, $t \neq n$. Note that r is the rank of $D(\mathbf{C}^1, \mathbf{C}^2)$ which is the same as the rank of $\mathbf{A}(\mathbf{C}^1, \mathbf{C}^2)$ and $\sigma_{n,n} = \sqrt{\lambda_n}$, $n = 1, 2, \ldots, N$.

1 Rewriting (3.8) in terms of U, Σ, and W show that

$$P(\mathbf{C}^1 \to \mathbf{C}^2|\mathbf{H}) = P(X > \mathrm{Tr}[\mathbf{H}^H \cdot W^H \cdot \Sigma^H \cdot \Sigma \cdot W \cdot \mathbf{H}]).$$

2 Defining $\zeta = \Sigma \cdot W \cdot \mathbf{H}$ and $\eta = U \cdot \mathcal{N}^1$, show that

$$P(\mathbf{C}^1 \to \mathbf{C}^2|\mathbf{H}) = P(\mathrm{Tr}[\eta^H \cdot \zeta + \zeta^H \cdot \eta] > ||\zeta||_F^2). \tag{1}$$

3 Using $||\zeta||_F = ||(\mathbf{C}^2 - \mathbf{C}^1) \cdot \mathbf{H}||_F$ write the conditional pairwise error probability as

$$P(\mathbf{C}^1 \to \mathbf{C}^2|\mathbf{H}) = Q\left(\frac{||\zeta||_F}{\sqrt{2N_0}}\right). \tag{2}$$

Equations (1) and (2) calculate the conditional pairwise error probability for a given channel matrix $\mathbf{H}$. To calculate the pairwise error probability, one needs to compute the expected value of the conditional probability. Using Jensen's inequality and following the argument in [65], with a small abuse of the notation, we have

$$E\big[P(\mathbf{C}^1 \to \mathbf{C}^2|\mathbf{H})\big] \geq Q\left(\sqrt{\frac{E[||\zeta||_F^2]}{2N_0}}\right).$$

4 Since the elements of $\mathbf{H}$ are independent zero mean complex Gaussian random variables, ζ is also a zero mean Gaussian random variable. Show that $E[||\zeta||_F^2] = ||D(\mathbf{C}^1, \mathbf{C}^2)||_F^2$ for $M = 1$ receive antenna and using the normalization $1/N_0 = \gamma$ show that

$$E\big[P(\mathbf{C}^1 \to \mathbf{C}^2|\mathbf{H})\big] \geq Q\left(\sqrt{\gamma||D(\mathbf{C}^1, \mathbf{C}^2)||_F^2/2}\right).$$

4

Orthogonal space-time block codes

4.1 Introduction

In this section, we study the design of space-time block codes (STBCs) to transmit information over a multiple antenna wireless communication system. We assume that fading is quasi-static and flat as explained in Chapter 2. We consider a wireless communication system where the transmitter contains N transmit antennas and the decoder contains M receive antennas. We follow our notations in Equation (2.1) for the input–output relation of the MIMO channel. The goal of space-time coding is to achieve the maximum diversity of NM, the maximum coding gain, and the highest possible throughput. In addition, the decoding complexity is very important. In a typical wireless communication system the mobile transceiver has a limited available power through a battery and should be a small physical device. To improve the battery life, low complexity encoding and decoding is very crucial. On the other hand, the base station is not as restricted in terms of power and physical size. One can put multiple independent antennas in a base station. Therefore, in many practical situations, a very low complexity system with multiple transmit antennas is desirable. Space-time block coding is a scheme to provide these properties. Despite the name, a STBC can be considered as a modulation scheme for multiple transmit antennas that provide full diversity and very low complexity encoding and decoding.

4.2 Alamouti code

We start our discussion of space-time block coding with a simple example. Let us assume a system with $N = 2$ transmit antennas and one receive antenna, employing Alamouti code as in Figure 4.1 [4, 136]. To transmit b bits/cycle, we use a modulation scheme that maps every b bits to one symbol from a constellation with 2^b symbols. The constellation can be any real or complex constellation, for example

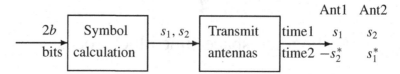

Fig. 4.1.   Transmitter block diagram for Alamouti code.

PAM, PSK, QAM, and so on. First, the transmitter picks two symbols from the constellation using a block of $2b$ bits. If s_1 and s_2 are the selected symbols for a block of $2b$ bits, the transmitter sends s_1 from antenna one and s_2 from antenna two at time one. Then at time two, it transmits $-s_2^*$ and s_1^* from antennas one and two, respectively. Therefore, the transmitted codeword is

$$\mathbf{C} = \begin{pmatrix} s_1 & s_2 \\ -s_2^* & s_1^* \end{pmatrix}. \tag{4.1}$$

To check if the code provides full diversity, we need to calculate the rank of all possible difference matrices $D(\mathbf{C}, \mathbf{C}')$ and show that it is equal to two for every $\mathbf{C}' \neq \mathbf{C}$. Let us consider a different pair of symbols (s_1', s_2') and the corresponding codeword

$$\mathbf{C}' = \begin{pmatrix} s_1' & s_2' \\ -s_2'^* & s_1'^* \end{pmatrix}. \tag{4.2}$$

The difference matrix $D(\mathbf{C}, \mathbf{C}')$ is given by

$$D(\mathbf{C}, \mathbf{C}') = \begin{pmatrix} s_1' - s_1 & s_2' - s_2 \\ s_2^* - s_2'^* & s_1'^* - s_1^* \end{pmatrix}. \tag{4.3}$$

The determinant of the difference matrix $\det[D(\mathbf{C}, \mathbf{C}')] = |s_1' - s_1|^2 + |s_2' - s_2|^2$ is zero if and only if $s_1' = s_1$ and $s_2' = s_2$. Therefore, $D(\mathbf{C}, \mathbf{C}')$ is always full rank when $\mathbf{C}' \neq \mathbf{C}$ and the Alamouti code satisfies the determinant criterion. It provides a diversity of $2M$ for M receive antennas and therefore is a full diversity code. Note that the code transmits one symbol, b bits, per time slot. This is the maximum number of possible symbols for a full diversity code [139]. Also, it can be shown that for one receive antenna, the maximum mutual information criterion in Section 3.4 is satisfied. Note that the maximum mutual information criterion is not valid for more than one receive antenna [55, 105].

Let us assume that the path gains from transmit antennas one and two to the receive antenna are α_1 and α_2, respectively. Then, based on our model in Equation (2.1), the decoder receives signals r_1 and r_2 at times one and two, respectively, such that

$$\begin{cases} r_1 = \alpha_1 s_1 + \alpha_2 s_2 + \eta_1 \\ r_2 = -\alpha_1 s_2^* + \alpha_2 s_1^* + \eta_2. \end{cases} \tag{4.4}$$

For a coherent detection scheme where the receiver knows the channel path gains α_1 and α_2, the maximum-likelihood detection amounts to minimizing the decision metric

$$|r_1 - \alpha_1 s_1 - \alpha_2 s_2|^2 + |r_2 + \alpha_1 s_2^* - \alpha_2 s_1^*|^2, \tag{4.5}$$

over all possible values of s_1 and s_2. Such a decoding scheme requires a full search over all possible pairs (s_1, s_2) and in general its complexity grows exponentially by the number of transmit antennas. Expanding the cost function (4.5), one can ignore the common term $|r_1|^2 + |r_2|^2$. Then, the cost function (4.5) decomposes into two parts, one of which,

$$|s_1|^2 \sum_{n=1}^{2} |\alpha_n|^2 - [r_1 \alpha_1^* s_1^* + r_1^* \alpha_1 s_1 + r_2 \alpha_2^* s_1 + r_2^* \alpha_2 s_1^*], \tag{4.6}$$

is only a function of s_1, and the other one

$$|s_2|^2 \sum_{n=1}^{2} |\alpha_n|^2 - [r_1 \alpha_2^* s_2^* + r_1^* \alpha_2 s_2 - r_2 \alpha_1^* s_2 - r_2^* \alpha_1 s_2^*], \tag{4.7}$$

is only a function of s_2. Therefore, instead of minimizing the cost function (4.5) over all possible values of (s_1, s_2), one can simultaneously minimize the cost functions (4.6) and (4.7) over all possible values of s_1 and s_2, respectively. As a result the decoding complexity of the code increases linearly, instead of exponentially, by the number of transmit antennas. In addition, if all the constellation symbols have equal energies, for example PSK, the terms $|s_1|^2 \sum_{n=1}^{2} |\alpha_n|^2$ and $|s_2|^2 \sum_{n=1}^{2} |\alpha_n|^2$ can be ignored. As a result, the maximum-likelihood decoding can be further simplified for equal energy constellations. In fact, the receiver should minimize

$$|s_1 - r_1 \alpha_1^* - r_2^* \alpha_2|^2 \tag{4.8}$$

to decode s_1 and minimize

$$|s_2 - r_1 \alpha_2^* + r_2^* \alpha_1|^2 \tag{4.9}$$

to decode s_2. Therefore, the decoding consists of first calculating

$$\begin{cases} \tilde{s}_1 = r_1 \alpha_1^* + r_2^* \alpha_2 \\ \tilde{s}_2 = r_1 \alpha_2^* - r_2^* \alpha_1. \end{cases} \tag{4.10}$$

Then, to decode s_1, the receiver finds the closest symbol to $\tilde{s}_1$ in the constellation. Similarly, the decoding of s_2 consists of finding the closest symbol to $\tilde{s}_2$ in the

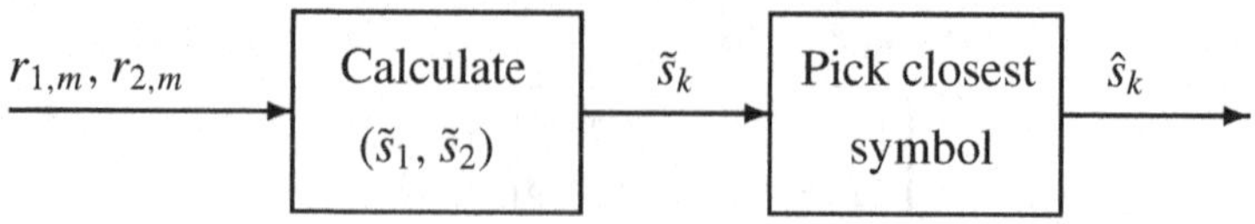

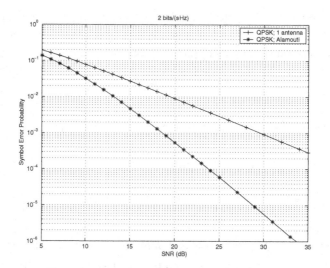

$$\tilde{s}_1 = \sum_{m=1}^{M} [r_{1,m}\alpha_{1,m}^* + r_{2,m}^*\alpha_{2,m}]$$

$$\tilde{s}_2 = \sum_{m=1}^{M} [r_{1,m}\alpha_{2,m}^* - r_{2,m}^*\alpha_{1,m}]$$

Fig. 4.2.   Receiver block diagram for Alamouti code with MRC.

Fig. 4.3.   Performance of the Alamouti code.

constellation. Figure 4.2 shows a block diagram of the decoder with M receive antennas. Note that we use maximum ratio combining for the maximum-likelihood decoding with more than one receive antenna. In this case all of the above formulas are valid when every cost function is the sum of the corresponding cost functions for each receive antenna. The details are discussed in Section 4.3.

The performance of the Alamouti code over a quasi-static flat Rayleigh fading channel is provided in Figure 4.3. The quasi-static assumption is true if the channel does not change over a frame of $T = 2$ symbols. Figure 4.3 shows the SNR versus symbol error probability plot for a system using a QPSK constellation and one receive antenna. As can be seen, the performance of the Alamouti code with two transmit antennas is much better than that of the system with one transmit antenna. At a symbol error probability of 10^{-3}, the Alamouti code provides more than 11 dB improvement. More importantly, due to the higher diversity gain of the Alamouti code, the gap increases for higher SNR values. In fact, the error rate decreases inversely with γ^2, where γ is the received SNR. In other words, the diversity gain of the code is two. This provides dual diversity that is the same as the diversity of a system with one transmit antenna and two receive antennas using maximum ratio combining.

The Alamouti code provides two important properties.

- **Simple decoding:** Each symbol is decoded separately using only linear processing.
- **Maximum diversity:** The code satisfies the rank criterion and therefore provides the maximum possible diversity.

These are very desirable properties that can be achieved for two transmit antennas. An important question is: Is it possible to design similar codes for a greater number of transmit antennas? We study this question in what follows.

4.3 Maximum-likelihood decoding and maximum ratio combining

Before discussing the design of codes for more than two transmit antennas, we study the structure of maximum-likelihood (ML) decoding. ML decoding amounts to finding the codeword $\hat{\mathbf{C}}$ that maximizes the density function in (3.4). Equivalently, ML decoding finds the codeword $\hat{\mathbf{C}}$ that solves the following minimization problem:

$$\hat{\mathbf{C}} = \underset{\mathbf{C}}{\operatorname{argmin}}\left\{\operatorname{Tr}\left[(\mathbf{r} - \mathbf{C} \cdot \mathbf{H})^H \cdot (\mathbf{r} - \mathbf{C} \cdot \mathbf{H})\right]\right\}. \tag{4.11}$$

Expanding the cost function in (4.11) and considering the fact that $\mathbf{r}^H \cdot \mathbf{r}$ is independent of the transmitted codeword and the choice of the codeword, we obtain

$$\begin{aligned}
\hat{\mathbf{C}} &= \underset{\mathbf{C}}{\operatorname{argmin}}\left\{\operatorname{Tr}\left[\mathbf{r}^H \cdot \mathbf{r} + \mathbf{H}^H \cdot \mathbf{C}^H \cdot \mathbf{C} \cdot \mathbf{H} - \mathbf{H}^H \cdot \mathbf{C}^H \cdot \mathbf{r} - \mathbf{r}^H \cdot \mathbf{C} \cdot \mathbf{H}\right]\right\} \\
&= \underset{\mathbf{C}}{\operatorname{argmin}}\left\{\operatorname{Tr}\left[\mathbf{H}^H \cdot \mathbf{C}^H \cdot \mathbf{C} \cdot \mathbf{H}\right] - 2\Re\left(\operatorname{Tr}\left[\mathbf{H}^H \cdot \mathbf{C}^H \cdot \mathbf{r}\right]\right)\right\}.
\end{aligned} \tag{4.12}$$

Now, we show the relationship between the cost function for multiple receive antennas and that of one receive antenna. Let us define $\mathbf{H}_m$ as the mth column of $\mathbf{H}$. Representing $\mathbf{H}$ in terms of its columns results in

$$\operatorname{Tr}\left[\mathbf{H}^H \cdot \mathbf{C}^H \cdot \mathbf{C} \cdot \mathbf{H}\right] = \operatorname{Tr}\left[\begin{pmatrix} \mathbf{H}_1^H \\ \mathbf{H}_2^H \\ \vdots \\ \mathbf{H}_M^H \end{pmatrix} \cdot \mathbf{C}^H \cdot \mathbf{C} \cdot (\mathbf{H}_1 \, \mathbf{H}_2 \, \cdots \, \mathbf{H}_M)\right]$$

$$= \sum_{m=1}^{M} \mathbf{H}_m^H \cdot \mathbf{C}^H \cdot \mathbf{C} \cdot \mathbf{H}_m. \tag{4.13}$$

Similarly, defining $\mathbf{r}_m$ as the mth column of $\mathbf{r}$ results in

$$
\text{Tr}[\mathbf{H}^H \cdot \mathbf{C}^H \cdot \mathbf{r}] = \text{Tr}\left[\begin{pmatrix} \mathbf{H}_1^H \\ \mathbf{H}_2^H \\ \vdots \\ \mathbf{H}_M^H \end{pmatrix} \cdot \mathbf{C}^H \cdot (\mathbf{r}_1\, \mathbf{r}_2\, \cdots\, \mathbf{r}_M) \right]
$$

$$
= \sum_{m=1}^{M} \mathbf{H}_m^H \cdot \mathbf{C}^H \cdot \mathbf{r}_m. \tag{4.14}
$$

Therefore, using (4.13) and (4.14) in (4.12), we can calculate the cost function of the minimization problem in terms of the received signals and path gains for M receive antennas as follows:

$$
\hat{\mathbf{C}} = \operatorname*{argmin}_{\mathbf{C}} \left\{ \sum_{m=1}^{M} \left[\mathbf{H}_m^H \cdot \mathbf{C}^H \cdot \mathbf{C} \cdot \mathbf{H}_m - 2\Re\big(\mathbf{H}_m^H \cdot \mathbf{C}^H \cdot \mathbf{r}_m\big) \right] \right\}. \tag{4.15}
$$

Note that if we only have the receive antenna m, $M = 1$, the corresponding minimization cost function is

$$
\mathbf{H}_m^H \cdot \mathbf{C}^H \cdot \mathbf{C} \cdot \mathbf{H}_m - 2\Re\big\{ \mathbf{H}_m^H \cdot \mathbf{C}^H \cdot \mathbf{r}_m \big\}. \tag{4.16}
$$

Therefore, we can write the cost function for only one receive antenna and add the correct summation in front of it to achieve the ML decoding formulas for the general case of M receive antennas. We call this maximum ratio combining (MRC). As a result, using ML decoding, it suffices to consider only one receive antenna. Then, the general ML decoding formulas can be achieved using MRC. Therefore, when discussing the decoding, without loss of generality, we consider only one receive antenna in most cases.

4.4 Real orthogonal designs

To design space-time codes that provide the properties of the Alamouti code for more than two transmit antennas, first we should understand why these codes behave in this way. To ease such an understanding, we rewrite (4.4) in a matrix form as follows:

$$
(r_1,\, r_2^*) = (s_1,\, s_2)\Omega + (\eta_1,\, \eta_2^*), \tag{4.17}
$$

where

$$
\Omega = \Omega(\alpha_1, \alpha_2) = \begin{pmatrix} \alpha_1 & \alpha_2^* \\ \alpha_2 & -\alpha_1^* \end{pmatrix}. \tag{4.18}
$$

Multiplying both sides of (4.17) by Ω^H gives

$$
(\tilde{s}_1,\, \tilde{s}_2) = (r_1,\, r_2^*)\Omega^H = (|\alpha_1|^2 + |\alpha_2|^2)(s_1,\, s_2) + \mathcal{N}, \tag{4.19}
$$

where $\mathcal{N}$ is also a Gaussian noise. Equation (4.19) consists of two separate equations for decoding the two transmitted symbols. This separation is the main reason for the first property of simple ML decoding. The second property comes from the factor $|\alpha_1|^2 + |\alpha_2|^2$ in the right-hand side of (4.19). Although we have mathematically proved that the Alamouti code provides full diversity, an explanation based on the term $|\alpha_1|^2 + |\alpha_2|^2$ is also possible. When there is only one transmit antenna available, the power of the signal is affected by a factor of $|\alpha|^2$ due to the path gain. In a deep faded environment $|\alpha|^2$ is very small and the noise dominates the signal. On the other hand, (4.19) shows that using the Alamouti code $|\alpha_1|^2 + |\alpha_2|^2$ should be small for a noise dominated channel. To have a small $|\alpha_1|^2 + |\alpha_1|^2$, both $|\alpha_1|^2$ and $|\alpha_1|^2$ must be small. This is less likely since we assume that α_1 and α_2 are independent.

Derivation of (4.19) shows that the following equation is the main reason for both properties:

$$\Omega \cdot \Omega^H = (|\alpha_1|^2 + |\alpha_1|^2)I_2, \tag{4.20}$$

where I_2 is a 2×2 identity matrix. If we show the structure of the Alamouti code by the following generator matrix

$$\mathcal{G} = \begin{pmatrix} x_1 & x_2 \\ -x_2^* & x_1^* \end{pmatrix}, \tag{4.21}$$

(4.20) is a result of the orthogonality of the columns of $\mathcal{G}$ and the following property:

$$\mathcal{G}^H \mathcal{G} = (|x_1|^2 + |x_2|^2)I_2. \tag{4.22}$$

Note that the generator matrix of a space-time code has similarities and differences with the conventional concept of the generating matrix for linear codes in the classic coding theory. Both of them represent the redundancy of the corresponding codes. For example, in the Alamouti code of (4.21), we transmit two symbols while in general a 2×2 matrix can transmit four symbols. On the other hand, the redundancy in linear block codes is used to correct errors and corresponds to coding gain, while the redundancy in space-time block codes is utilized to achieve diversity gain.

So far we have defined the generator matrix $\mathcal{G}$ and showed that since the generator matrix satisfies (4.22) for every possible indeterminant pair (x_1, x_2), it provides full diversity and simple ML decoding. The next natural step is to investigate the possibility of similar generator matrices for more than two transmit antennas. For real numbers, matrices satisfying (4.22) are called orthogonal designs. A complete study of real orthogonal designs was provided by Radon and Hurwitz early in the twentieth century [49, 101]. Our framework above lets us use their study to design real space-time block codes, as we discuss in this section.

First, we provide a review of Radon–Hurwitz theory.

Table 4.1. *Values of function* $\rho(N)$

a	$N = 2^a$	$\rho(N)$
1	2	2
2	4	4
3	8	8
4	16	9
5	32	10
6	64	12
7	128	16
8	256	17

Definition 4.4.1 *A set of* $N \times N$ *real matrices* $\{B_1, B_2, \ldots, B_L\}$ *is a size* L *Hurwitz–Radon family of matrices if*

$$
\begin{aligned}
B_l^T B_l &= I_N, & l &= 1, 2, \cdots, L \\
B_l^T &= -B_l, & l &= 1, 2, \cdots, L \\
B_l B_{l'} &= -B_{l'} B_l, & 1 &\leq l < l' \leq L.
\end{aligned}
\tag{4.23}
$$

As we see in what follows, for a given matrix dimension, a bigger size Hurwitz–Radon family is more desirable. The following theorem due to Radon determines the maximum size of a Hurwitz–Radon family [49].

Theorem 4.4.1 *Radon Theorem: We write any positive integer number* N *as* $N = 2^a b$, *where* $a = 4c + d$, $c \geq 0$, $0 \leq d < 4$ *and* b *is an odd number. Any* $N \times N$ *Hurwitz–Radon family contains less than* $\rho(N) = 8c + 2^d \leq N$ *matrices.*

Table 4.1 tabulates the value of $\rho(N)$ when N is a power of two. As we show in the sequel, these values, that is $N = 2^a$, play an important role in the theory of orthogonal designs. Based on the above theorem, the maximum possible size of a Hurwitz–Radon family is $\rho(N) - 1$. It is important to know when such a maximum size is possible. First, we remind ourselves of a few definitions from matrix theory.

Definition 4.4.2 *An integer matrix is defined as a matrix that all of its elements are* $-1, 0,$ *or* 1.

Definition 4.4.3 *The tensor product* $A \otimes B$ *of matrices* $A = [a_{ij}]$, *with size* $I \times J$, *and* B *is defined as*

$$
A \otimes B = \begin{pmatrix}
a_{11} B & a_{12} B & \cdots & a_{1J} B \\
a_{21} B & a_{22} B & \cdots & a_{2J} B \\
\vdots & \vdots & \ddots & \vdots \\
a_{I1} B & a_{I2} B & \cdots & a_{IJ} B
\end{pmatrix}.
\tag{4.24}
$$

The following lemma provides the maximum size of a Hurwitz–Radon family of integer matrices for any number N.

Lemma 4.4.1 *A Hurwitz–Radon family of size $\rho(N)-1$ whose members are integer matrices exists for any positive integer N.*

Proof First, let us note that by definition $\rho(N) = \rho(2^a)$ if $N = 2^a b$ and b is an odd number. Let us assume $\{A_1, A_2, \ldots, A_L\}$ be a family of $2^a \times 2^a$ Hurwitz–Radon integer matrices and its size be $L = \rho(2^a) - 1 = \rho(N) - 1$. Then, $\{A_1 \otimes I_b, A_2 \otimes I_b, \ldots, A_L \otimes I_b\}$ is a family of $N \times N$ Hurwitz–Radon integer matrices of size $L = \rho(N) - 1$. Therefore, it suffices to prove the lemma for $N = 2^a$. We prove the lemma by using induction. For $a = 1$, we have $N = 2$, $c = 0$, $d = 1$, and $\rho(N) = \rho(2) = 2$. Therefore, a Hurwitz–Radon family of size $\rho(2) - 1 = 1$ has only one member, for example

$$R = \begin{pmatrix} 0 & 1 \\ -1 & 0 \end{pmatrix}. \tag{4.25}$$

Also, for $a = 2$, we have $\rho(N) = N = 4$ and a Hurwitz–Radon family of size $\rho(4) - 1 = 3$ is $\{Q \otimes R, R \otimes I_2, P \otimes R\}$, where

$$P = \begin{pmatrix} 0 & 1 \\ 1 & 0 \end{pmatrix}, \quad Q = \begin{pmatrix} 1 & 0 \\ 0 & -1 \end{pmatrix}. \tag{4.26}$$

In addition, for $a = 3$, we have $\rho(N) = N = 8$ and a Hurwitz–Radon family of size $\rho(8) - 1 = 7$ is $\{Q \otimes Q \otimes R, I_2 \otimes R \otimes I_2, I_2 \otimes P \otimes R, R \otimes Q \otimes I_2, P \otimes Q \otimes R, R \otimes P \otimes Q, R \otimes P \otimes P\}$. Now, we define the following sequence of numbers to use induction

$$\begin{aligned} N_1 &= 2^{4s+3}, \\ N_2 &= 2^{4(s+1)}, \\ N_3 &= 2^{4(s+1)+1}, \\ N_4 &= 2^{4(s+1)+2}, \\ N_5 &= 2^{4(s+1)+3}, \end{aligned} \tag{4.27}$$

for which the corresponding $\rho(N)$ numbers are

$$\begin{aligned} \rho(N_2) &= \rho(N_1) + 1, \\ \rho(N_3) &= \rho(N_1) + 2, \\ \rho(N_4) &= \rho(N_1) + 4, \\ \rho(N_5) &= \rho(N_1) + 8. \end{aligned} \tag{4.28}$$

Since an example for $N = 2^3$ is constructed, to prove the lemma using induction, we need to show that if the lemma is correct for N_1, it is also correct for N_2, N_3, N_4, and N_5. To go from N_1 to N_2, we use the fact that if $\{A_1, A_2, \ldots, A_L\}$ is a family of $N \times N$ Hurwitz–Radon integer matrices, then

$$\{R \otimes I_N\} \cup \{Q \otimes A_l, \; l = 1, 2, \ldots, L\} \tag{4.29}$$

is a family of $2N \times 2N$ Hurwitz–Radon integer matrices with size $L + 1$. Note that $N_2 = 2N_1$ and $\rho(N_2) = \rho(N_1) + 1$. To go from N_1 to N_3, we use the fact that if, in addition, $\{B_1, B_2, \ldots, B_K\}$ is a family of $M \times M$ Hurwitz–Radon integer matrices, then

$$\begin{aligned} \{P \otimes I_M \otimes A_l, \; l &= 1, 2, \ldots, L\} \\ &\cup \{Q \otimes B_k \otimes I_N, \; k = 1, 2, \ldots, K\} \cup \{R \otimes I_{NM}\} \end{aligned} \tag{4.30}$$

is a family of $2NM \times 2NM$ Hurwitz–Radon integer matrices with size $L + K + 1$. Note that we go from N_1 to N_3 by using $M = N_1$ and $N = 2$ in (4.30). We also go from N_1 to N_4 by using $M = N_1$ and $N = 4$ in (4.30). The last transition from N_1 to N_5 is achieved similarly by using $M = N_1$ and $N = 8$ in (4.30). $\square$

Before we define orthogonal designs, we mention that a Hurwitz–Radon family of size $N - 1$ exists if and only if $N = 2, 4, 8$. This is easily shown from Theorem 4.4.1 and the fact that $\rho(N) = N$ if and only if $N = 2, 4, 8$.

Definition 4.4.4 *A real orthogonal design of size N is an $N \times N$ orthogonal matrix $\mathcal{G}_N$ with real entries $x_1, -x_1, x_2, -x_2, \ldots, x_N, -x_N$ such that*

$$\mathcal{G}_N^T \mathcal{G}_N = (x_1^2 + x_2^2 + \cdots + x_N^2) I_N. \tag{4.31}$$

Using Radon Theorem 4.4.1, we prove the following theorem on the existence of real orthogonal designs [136].

Theorem 4.4.2 *A real orthogonal design exists if and only if $N = 2, 4, 8$.*

Proof Let us rewrite $\mathcal{G}_N = \sum_{n=1}^{N} x_n A_n$, where A_n is an integer matrix. Then,

$$\mathcal{G}_N^T \mathcal{G}_N = \sum_{n=1}^{N} \sum_{n'=1}^{N} x_n x_{n'} A_{n'}^T A_n. \tag{4.32}$$

To have (4.31) and (4.32) simultaneously and for all possible indeterminate variables $x_1, x_2, \ldots, x_N$, we should have

$$\begin{aligned} A_{n'}^T A_n + A_n^T A_{n'} &= 0, \quad 1 \le n < n' \le N \\ A_n^T A_n = A_n A_n^T &= I_N, \quad n = 1, 2, \ldots, N. \end{aligned} \tag{4.33}$$

By defining $B_n = A_1^T A_n$ for $n = 1, 2, \ldots, N$, we have $B_1 = I_N$ and the following

Hurwitz–Radon family of size $N - 1$ is needed:

$$\begin{aligned}
B_n^T B_n &= I_N & n &= 2, \ldots, N \\
B_n^T &= -B_n & n &= 2, \ldots, N \\
B_n B_{n'} &= -B_{n'} B_n & 1 &\leq n < n' \leq N.
\end{aligned} \tag{4.34}$$

Using Radon Theorem 4.4.1, we have $\rho(N) = N - 1$ and therefore $N = 2, 4, 8$. $\square$

In fact, the proof of the theorem is a constructive proof and we provide the three existing orthogonal designs. Note that the designs are not unique. In the constructive proof we have one degree of freedom which is the choice of A_1 such that $A_1^T A_1 = I$. We usually pick $A_1 = I$ which results in $A_n = B_n$ for $n = 1, 2, \ldots, N$. However, for any orthogonal design $\mathcal{G}_N = \sum_{n=1}^{N} x_n A_n$ and any unitary matrix $U (U^T U = I)$, the matrix $U\mathcal{G}_N = \sum_{n=1}^{N} x_n U A_n$ is also an orthogonal design. This is easily shown as follows

$$(U\mathcal{G}_N)^T U\mathcal{G}_N = \mathcal{G}_N^T U^T U\mathcal{G}_N = \mathcal{G}_N^T \mathcal{G}_N = (x_1^2 + x_2^2 + \cdots + x_N^2)I_N. \tag{4.35}$$

A 2×2 design is achieved by using a Hurwitz–Radon family of size one, that is R in (4.25), as follows

$$\mathcal{G}_2 = x_1 I_2 + x_2 R = \begin{pmatrix} x_1 & x_2 \\ -x_2 & x_1 \end{pmatrix}. \tag{4.36}$$

A 4×4 design that can be identified by using a Hurwitz–Radon family of size three. The following Hurwitz–Radon family was constructed in the proof of Lemma 4.4.1

$$B_2 = \begin{pmatrix} 0 & 1 & 0 & 0 \\ -1 & 0 & 0 & 0 \\ 0 & 0 & 0 & -1 \\ 0 & 0 & 1 & 0 \end{pmatrix},$$

$$B_3 = \begin{pmatrix} 0 & 0 & 1 & 0 \\ 0 & 0 & 0 & 1 \\ -1 & 0 & 0 & 0 \\ 0 & -1 & 0 & 0 \end{pmatrix}, \tag{4.37}$$

$$B_4 = \begin{pmatrix} 0 & 0 & 0 & 1 \\ 0 & 0 & -1 & 0 \\ 0 & 1 & 0 & 0 \\ -1 & 0 & 0 & 0 \end{pmatrix}.$$

Therefore, a 4×4 orthogonal design is derived as follows:

$$\mathcal{G}_4 = x_1 I_4 + x_2 B_2 + x_3 B_3 + x_4 B_4 = \begin{pmatrix} x_1 & x_2 & x_3 & x_4 \\ -x_2 & x_1 & -x_4 & x_3 \\ -x_3 & x_4 & x_1 & -x_2 \\ -x_4 & -x_3 & x_2 & x_1 \end{pmatrix}. \quad (4.38)$$

Similarly a Hurwitz–Radon family of size seven is

$$B_2 = \begin{pmatrix} 0 & 1 & 0 & 0 & 0 & 0 & 0 & 0 \\ -1 & 0 & 0 & 0 & 0 & 0 & 0 & 0 \\ 0 & 0 & 0 & -1 & 0 & 0 & 0 & 0 \\ 0 & 0 & 1 & 0 & 0 & 0 & 0 & 0 \\ 0 & 0 & 0 & 0 & 0 & -1 & 0 & 0 \\ 0 & 0 & 0 & 0 & 1 & 0 & 0 & 0 \\ 0 & 0 & 0 & 0 & 0 & 0 & 0 & 1 \\ 0 & 0 & 0 & 0 & 0 & 0 & -1 & 0 \end{pmatrix}, \quad B_3 = \begin{pmatrix} 0 & 0 & 1 & 0 & 0 & 0 & 0 & 0 \\ 0 & 0 & 0 & 1 & 0 & 0 & 0 & 0 \\ -1 & 0 & 0 & 0 & 0 & 0 & 0 & 0 \\ 0 & -1 & 0 & 0 & 0 & 0 & 0 & 0 \\ 0 & 0 & 0 & 0 & 0 & 0 & 1 & 0 \\ 0 & 0 & 0 & 0 & 0 & 0 & 0 & 1 \\ 0 & 0 & 0 & 0 & -1 & 0 & 0 & 0 \\ 0 & 0 & 0 & 0 & 0 & -1 & 0 & 0 \end{pmatrix},$$

$$B_4 = \begin{pmatrix} 0 & 0 & 0 & 1 & 0 & 0 & 0 & 0 \\ 0 & 0 & -1 & 0 & 0 & 0 & 0 & 0 \\ 0 & 1 & 0 & 0 & 0 & 0 & 0 & 0 \\ -1 & 0 & 0 & 0 & 0 & 0 & 0 & 0 \\ 0 & 0 & 0 & 0 & 0 & 0 & 0 & 1 \\ 0 & 0 & 0 & 0 & 0 & 0 & -1 & 0 \\ 0 & 0 & 0 & 0 & 0 & 1 & 0 & 0 \\ 0 & 0 & 0 & 0 & -1 & 0 & 0 & 0 \end{pmatrix}, \quad B_5 = \begin{pmatrix} 0 & 0 & 0 & 0 & 1 & 0 & 0 & 0 \\ 0 & 0 & 0 & 0 & 0 & 1 & 0 & 0 \\ 0 & 0 & 0 & 0 & 0 & 0 & -1 & 0 \\ 0 & 0 & 0 & 0 & 0 & 0 & 0 & -1 \\ -1 & 0 & 0 & 0 & 0 & 0 & 0 & 0 \\ 0 & -1 & 0 & 0 & 0 & 0 & 0 & 0 \\ 0 & 0 & 1 & 0 & 0 & 0 & 0 & 0 \\ 0 & 0 & 0 & 1 & 0 & 0 & 0 & 0 \end{pmatrix},$$

$$B_6 = \begin{pmatrix} 0 & 0 & 0 & 0 & 0 & 1 & 0 & 0 \\ 0 & 0 & 0 & 0 & -1 & 0 & 0 & 0 \\ 0 & 0 & 0 & 0 & 0 & 0 & 0 & -1 \\ 0 & 0 & 0 & 0 & 0 & 0 & 1 & 0 \\ 0 & 1 & 0 & 0 & 0 & 0 & 0 & 0 \\ -1 & 0 & 0 & 0 & 0 & 0 & 0 & 0 \\ 0 & 0 & 0 & -1 & 0 & 0 & 0 & 0 \\ 0 & 0 & 1 & 0 & 0 & 0 & 0 & 0 \end{pmatrix}, \quad B_7 = \begin{pmatrix} 0 & 0 & 0 & 0 & 0 & 0 & 1 & 0 \\ 0 & 0 & 0 & 0 & 0 & 0 & 0 & -1 \\ 0 & 0 & 0 & 0 & 1 & 0 & 0 & 0 \\ 0 & 0 & 0 & 0 & 0 & -1 & 0 & 0 \\ 0 & 0 & -1 & 0 & 0 & 0 & 0 & 0 \\ 0 & 0 & 0 & 1 & 0 & 0 & 0 & 0 \\ -1 & 0 & 0 & 0 & 0 & 0 & 0 & 0 \\ 0 & 1 & 0 & 0 & 0 & 0 & 0 & 0 \end{pmatrix},$$

$$B_8 = \begin{pmatrix} 0 & 0 & 0 & 0 & 0 & 0 & 0 & 1 \\ 0 & 0 & 0 & 0 & 0 & 0 & 1 & 0 \\ 0 & 0 & 0 & 0 & 0 & 1 & 0 & 0 \\ 0 & 0 & 0 & 0 & 1 & 0 & 0 & 0 \\ 0 & 0 & 0 & -1 & 0 & 0 & 0 & 0 \\ 0 & 0 & -1 & 0 & 0 & 0 & 0 & 0 \\ 0 & -1 & 0 & 0 & 0 & 0 & 0 & 0 \\ -1 & 0 & 0 & 0 & 0 & 0 & 0 & 0 \end{pmatrix}. \quad (4.39)$$

The corresponding 8×8 orthogonal design is given below:

$$\mathcal{G}_8 = \begin{pmatrix} x_1 & x_2 & x_3 & x_4 & x_5 & x_6 & x_7 & x_8 \\ -x_2 & x_1 & -x_4 & x_3 & -x_6 & x_5 & x_8 & -x_7 \\ -x_3 & x_4 & x_1 & -x_2 & x_7 & x_8 & -x_5 & -x_6 \\ -x_4 & -x_3 & x_2 & x_1 & x_8 & -x_7 & x_6 & -x_5 \\ -x_5 & x_6 & -x_7 & -x_8 & x_1 & -x_2 & x_3 & x_4 \\ -x_6 & -x_5 & -x_8 & x_7 & x_2 & x_1 & -x_4 & x_3 \\ -x_7 & -x_8 & x_5 & -x_6 & -x_3 & x_4 & x_1 & x_2 \\ -x_8 & x_7 & x_6 & x_5 & -x_4 & -x_3 & -x_2 & x_1 \end{pmatrix}. \quad (4.40)$$

A real space-time block code can be constructed using the above orthogonal designs. The encoder of such a code first selects N symbols $s_1, s_2, \ldots, s_N$ from a real constellation, for example PAM, using Nb input bits. To encode such an input block, the indeterminate variables $x_1, x_2, \ldots, x_N$ in the orthogonal design are replaced by the N symbols $s_1, s_2, \ldots, s_N$ and the (t, n)th element of the resulting matrix is transmitted at time t from antenna n. Note that N different symbols are transmitted simultaneously from N transmit antennas at each time slot t, $t = 1, 2, \ldots, N$. The resulting real space-time block code transmits one symbol per time slot. The following theorem shows that the resulting code provides full diversity,

Theorem 4.4.3 *The diversity order of the above space-time block codes is NM for M receive antennas.*

Proof To prove that the codes give full diversity, we need to show that if two different codewords $\mathcal{G}(s_1, s_2, \ldots, s_N) \neq \mathcal{G}(s_1', s_2', \ldots, s_N')$ are transmitted, the difference matrix $\mathcal{G}(s_1, s_2, \ldots, s_N) - \mathcal{G}(s_1', s_2', \ldots, s_N')$ is full rank. From the linearity of the code we have $\mathcal{G}(s_1, s_2, \ldots, s_N) - \mathcal{G}(s_1', s_2', \ldots, s_N') = \mathcal{G}(s_1 - s_1', s_2 - s_2', \ldots, s_N - s_N')$. To show that this is a nonsingular matrix, we use the fact that

$$
\det\left[\mathcal{G}\left(s_1 - s_1', s_2 - s_2', \ldots, s_N - s_N'\right)^T \cdot \mathcal{G}\left(s_1 - s_1', s_2 - s_2', \ldots, s_N - s_N'\right)\right]
$$
$$
= \sum_{n=1}^{N} \left(s_n - s_n'\right)^2. \tag{4.41}
$$

Therefore, the difference matrix is singular if and only if $s_n = s_n'$, $\forall n = 1, 2, \ldots, N$. $\qquad\square$

We also show that the symbols can be decoded separately. Without loss of generality, let us concentrate on the case of one receive antenna for the sake of simplicity. For more than one receive antenna, we can use maximum ratio combining to come up with similar formulas. Replacing x_n with s_n in the orthogonal designs to transmit them over the fading channel and using the input–output relationship in (2.5) results in

$$
\mathbf{r}^T = \mathbf{H}^T \cdot \mathbf{C}^T + \mathcal{N}^T
$$
$$
= \mathbf{H}^T \sum_{n=1}^{N} s_n B_n^T + \mathcal{N}^T
$$
$$
= \sum_{n=1}^{N} s_n \Omega_n + \mathcal{N}^T \tag{4.42}
$$
$$
= (s_1, s_2, \ldots, s_N) \cdot \Omega + \mathcal{N}^T,
$$

where $\Omega_n = \mathbf{H}^T B_n^T$ is the nth row of an $N \times N$ matrix $\Omega = \Omega(\alpha_1, \alpha_2, \ldots, \alpha_N)$. To be more precise, Ω is built by putting $\Omega_n = \mathbf{H}^T B_n^T$ in its nth row. To decode the symbols, one can multiply both sides of (4.42) by Ω^T to arrive at

$$\mathbf{r}^T \cdot \Omega^T = (s_1, s_2, \ldots, s_N) \cdot \Omega \cdot \Omega^T + \mathcal{N}^T \cdot \Omega^T. \tag{4.43}$$

To show that the real transmitted symbols can be decoded separately, we need to show that

$$\Omega \cdot \Omega^T = \left(\sum_{n=1}^{N} \alpha_n^2 \right) I_N. \tag{4.44}$$

Once (4.44) is proved, (4.42) can be written as

$$\mathbf{r}^T \cdot \Omega^T = \left(\sum_{n=1}^{N} \alpha_n^2 \right) (s_1, s_2, \ldots, s_N) + \mathcal{N}^T \cdot \Omega^T. \tag{4.45}$$

To prove (4.44), we consider the (i, j)th element of $\Omega \cdot \Omega^T$ which is $\Omega_i \Omega_j^T = \mathbf{H}^T B_i^T B_j \mathbf{H}$. Note that this is a scalar number and thus equal to its transpose. For $i = j$, we have $\Omega_i \Omega_i^T = \mathbf{H}^T B_i^T B_i \mathbf{H} = \mathbf{H}^T \mathbf{H} = \sum_{n=1}^{N} \alpha_n^2$. For $i \neq j$, we calculate the transpose of $\mathbf{H}^T B_i^T B_j \mathbf{H}$ to show that

$$\mathbf{H}^T B_i^T B_j \mathbf{H} = \left[\mathbf{H}^T B_i^T B_j \mathbf{H} \right]^T = \mathbf{H}^T B_j^T B_i \mathbf{H} = - \left[\mathbf{H}^T B_i^T B_j \mathbf{H} \right]. \tag{4.46}$$

The only number which is equal to its negative is zero. Therefore, $\Omega_i \Omega_j^T = \mathbf{H}^T B_i^T B_j \mathbf{H} = 0$ and (4.44) is proved.

Since the matrix Ω plays an important role in the decoding algorithm of space-time block codes, we calculate it for different examples. For the 2×2 orthogonal design in (4.36), Ω is

$$\Omega(\alpha_1, \alpha_2) = \begin{pmatrix} \alpha_1 & \alpha_2 \\ \alpha_2 & -\alpha_1 \end{pmatrix}. \tag{4.47}$$

For the 4×4 orthogonal design in (4.38) Ω is a 4×4 matrix defined by

$$\Omega(\alpha_1, \alpha_2, \alpha_3, \alpha_4) = \begin{pmatrix} \alpha_1 & \alpha_2 & \alpha_3 & \alpha_4 \\ \alpha_2 & -\alpha_1 & -\alpha_4 & \alpha_3 \\ \alpha_3 & \alpha_4 & -\alpha_1 & -\alpha_2 \\ \alpha_4 & -\alpha_3 & \alpha_2 & -\alpha_1 \end{pmatrix}. \tag{4.48}$$

And, finally, for the 8×8 orthogonal design in (4.40) Ω is a 8×8 matrix defined by

$$\Omega(\alpha_1, \alpha_2, \alpha_3, \alpha_4, \alpha_5, \alpha_6, \alpha_7, \alpha_8) =$$

$$\begin{pmatrix}
\alpha_1 & \alpha_2 & \alpha_3 & \alpha_4 & \alpha_5 & \alpha_6 & \alpha_7 & \alpha_8 \\
\alpha_2 & -\alpha_1 & -\alpha_4 & \alpha_3 & -\alpha_6 & \alpha_5 & \alpha_8 & -\alpha_7 \\
\alpha_3 & \alpha_4 & -\alpha_1 & -\alpha_2 & \alpha_7 & \alpha_8 & -\alpha_5 & -\alpha_6 \\
\alpha_4 & -\alpha_3 & \alpha_2 & -\alpha_1 & \alpha_8 & -\alpha_7 & \alpha_6 & -\alpha_5 \\
\alpha_5 & \alpha_6 & -\alpha_7 & -\alpha_8 & -\alpha_1 & -\alpha_2 & \alpha_3 & \alpha_4 \\
\alpha_6 & -\alpha_5 & -\alpha_8 & \alpha_7 & \alpha_2 & -\alpha_1 & -\alpha_4 & \alpha_3 \\
\alpha_7 & -\alpha_8 & \alpha_5 & -\alpha_6 & -\alpha_3 & \alpha_4 & -\alpha_1 & \alpha_2 \\
\alpha_8 & \alpha_7 & \alpha_6 & \alpha_5 & -\alpha_4 & -\alpha_3 & -\alpha_2 & -\alpha_1
\end{pmatrix} \quad (4.49)$$

One can directly calculate and show that $\Omega \cdot \Omega^T = (\sum_{n=1}^{N} \alpha_n^2) I_N$, $N = 2, 4, 8$, for Ω in (4.47), (4.48), and (4.49).

Note that orthogonal designs are not unique for a given N. In fact, the above orthogonal designs are different from the original orthogonal designs in [136]. We picked the above method to construct orthogonal designs because of a simpler presentation. However, different examples of orthogonal designs provide similar properties. Another method to construct an orthogonal design using the Hurwitz–Radon family of matrices is to denote the $N - 1$ members of the $N \times N$ integer Hurwitz–Radon family by $A_2, \ldots, A_N$ for $N = 2, 4, 8$ and $A_1 = I_N$. Then, denoting $X = (x_1, x_2, \ldots, x_N)$, we design the generator matrix $\mathcal{G}$ by choosing $A_n X^T$ as its nth column. Following this approach results in the following orthogonal designs for $N = 2, 4, 8$:

$$\mathcal{G}_2 = \begin{pmatrix} x_1 & x_2 \\ x_2 & -x_1 \end{pmatrix}, \quad (4.50)$$

$$\mathcal{G}_4 = \begin{pmatrix}
x_1 & x_2 & x_3 & x_4 \\
x_2 & -x_1 & x_4 & -x_3 \\
x_3 & -x_4 & -x_1 & x_2 \\
x_4 & x_3 & -x_2 & -x_1
\end{pmatrix}, \quad (4.51)$$

$$\mathcal{G}_8 = \begin{pmatrix}
x_1 & x_2 & x_3 & x_4 & x_5 & x_6 & x_7 & x_8 \\
x_2 & -x_1 & x_4 & -x_3 & x_6 & -x_5 & -x_8 & x_7 \\
x_3 & -x_4 & -x_1 & x_2 & -x_7 & -x_8 & x_5 & x_6 \\
x_4 & x_3 & -x_2 & -x_1 & -x_8 & x_7 & -x_6 & x_5 \\
x_5 & -x_6 & x_7 & x_8 & -x_1 & x_2 & -x_3 & -x_4 \\
x_6 & x_5 & x_8 & -x_7 & -x_2 & -x_1 & x_4 & -x_3 \\
x_7 & x_8 & -x_5 & x_6 & x_3 & -x_4 & -x_1 & -x_2 \\
x_8 & -x_7 & -x_6 & -x_5 & x_4 & x_3 & x_2 & -x_1
\end{pmatrix}. \quad (4.52)$$

Note that in this case the matrix Ω in (4.42) is defined differently. In fact, one can show that the (k, t)th element of Ω is defined as follows

$$\Omega_{k,t} = \begin{cases} \alpha_n & \text{if and only if } \mathcal{G}_{t,n} = x_k \\ -\alpha_n & \text{if and only if } \mathcal{G}_{t,n} = -x_k. \end{cases} \quad (4.53)$$

Due to the structure of the orthogonal designs, each symbol x_k appears only once in every column. Therefore the above definition determines Ω without any ambiguity. As a reference, we provide Ω for the orthogonal design of (4.52):

$$\Omega(\alpha_1, \alpha_2, \alpha_3, \alpha_4, \alpha_5, \alpha_6, \alpha_7, \alpha_8) =$$

$$\begin{pmatrix}
\alpha_1 & -\alpha_2 & -\alpha_3 & -\alpha_4 & -\alpha_5 & -\alpha_6 & -\alpha_7 & -\alpha_8 \\
\alpha_2 & \alpha_1 & \alpha_4 & -\alpha_3 & \alpha_6 & -\alpha_5 & -\alpha_8 & \alpha_7 \\
\alpha_3 & -\alpha_4 & \alpha_1 & \alpha_2 & -\alpha_7 & -\alpha_8 & \alpha_5 & \alpha_6 \\
\alpha_4 & \alpha_3 & -\alpha_2 & \alpha_1 & -\alpha_8 & \alpha_7 & -\alpha_6 & \alpha_5 \\
\alpha_5 & -\alpha_6 & \alpha_7 & \alpha_8 & \alpha_1 & \alpha_2 & -\alpha_3 & -\alpha_4 \\
\alpha_6 & \alpha_5 & \alpha_8 & -\alpha_7 & -\alpha_2 & \alpha_1 & \alpha_4 & -\alpha_3 \\
\alpha_7 & \alpha_8 & -\alpha_5 & \alpha_6 & \alpha_3 & -\alpha_4 & \alpha_1 & -\alpha_2 \\
\alpha_8 & -\alpha_7 & -\alpha_6 & -\alpha_5 & \alpha_4 & \alpha_3 & \alpha_2 & \alpha_1
\end{pmatrix} \tag{4.54}$$

So far we have shown that the above real space-time block codes have all the nice properties of the Alamouti code. Unfortunately, based on the Radon Theorem, this is only possible for $N = 2, 4, 8$ transmit antennas. Note that the Radon Theorem is valid for square matrices. In what follows we generalize orthogonal designs to non-square matrices and show how to design space-time block codes for real constellations that can send one symbol per time slot for any number of transmit antennas, while providing full diversity and simple separate decoding.

4.5 Generalized real orthogonal designs

In this section, we generalize orthogonal designs to non-square real matrices. We start the discussion by the definition of a generalized orthogonal design from [136].

Definition 4.5.1 *A generalized real orthogonal design is a $T \times N$ matrix $\mathcal{G}$ with real entries $x_1, -x_1, x_2, -x_2, \dots, x_K, -x_K$ such that*

$$\mathcal{G}^T \mathcal{G} = \kappa(x_1^2 + x_2^2 + \cdots + x_K^2) I_N, \tag{4.55}$$

where I_N is the $N \times N$ identity matrix and κ is a constant.

A real space-time block code is defined by using $\mathcal{G}$ as a transmission matrix. Let us assume a constellation with 2^b symbols. For each block of Kb bits, the encoder first picks K symbols $(s_1, s_2, \dots, s_K)$ from the constellation. Then, x_k is replaced by s_k in $\mathcal{G}$ to arrive at $\mathbf{C} = \mathcal{G}(s_1, s_2, \dots, s_K)$. At time $t = 1, 2, \dots, T$, the (t, n)th element of $\mathbf{C}$, $\mathbf{C}_{t,n}$, is transmitted from antenna n for $n = 1, 2, \dots, N$. There are three parameters in Definition 4.5.1 that define different properties of the corresponding codes. The number of transmit antennas is N. For each block, K symbols are transmitted over T time slots. Therefore, we define the rate of the code as $R = K/T$. Space-time block codes from orthogonal designs in Section 4.4 are special cases of generalized orthogonal designs for which $T = K = N$ and therefore $R = 1$. For the sake of brevity, when it is clear from the text, we may drop the word "generalized" in our sentences. We call an orthogonal design with rate $R = 1$ a full-rate orthogonal design. One can show that the generalized real orthogonal designs provide full diversity and separate decoding of the symbols. The proof is very similar to the proof in Section 4.4.

So far we have shown how to design space-time block codes using orthogonal designs defined in Definition 4.5.1. We need to show when generalized orthogonal designs exist and provide examples that can be used for different number of transmit antennas. The goal is to find transmission matrices that provide rates as high as possible while providing full diversity and allowing the ML decoding of the symbols to be performed separately. Particularly we are interested in designing full rate codes. The following theorem shows that a full rate real space-time block code exists for any given number of transmit antennas.

Theorem 4.5.1 *For any number of transmit antennas, N, there exists a full rate, $R = 1$, real space-time block code with a block size $T = \min\left[2^{4c+d}\right]$, where the minimization is over all possible integer values of c and d in the set $\left\{c \geq 0, d \geq 0 \mid 8c + 2^d \geq N\right\}$.*

Proof For a given number N, we pick the smallest number T such that $N \leq \rho(T)$. Let us define $X = (x_1, x_2, \ldots, x_T)$ as the row vector of T variables. Based on the constructive proof of Lemma 4.4.1, there exists a Hurwitz–Radon family of $T \times T$ integer matrices with size $\rho(T) - 1$. We denote the members of this family by $A_2, A_3, \ldots, A_{\rho(T)}$. Adding $A_1 = I_T$ to this family, we construct a generator matrix $\mathcal{G}$ such that its columns are the vectors $A_n X^T$ for $n = 1, 2, \ldots, N$, that is

$$\mathcal{G} = \left(A_1 X^T \; A_2 X^T \; \cdots \; A_N X^T\right) = (g_1 g_2 \; \cdots \; g_N). \tag{4.56}$$

Note that $\mathcal{G}$ is a $T \times N$ matrix with T indeterminate variables. The (i, j)th element of $\mathcal{G}^T \mathcal{G}$ is $g_i^T g_j = X A_i^T A_j X^T$. For $i = j$, we have $X A_i^T A_i X^T = X X^T = \sum_{t=1}^T x_t^2$. For $i \neq j$, since $X A_i^T A_j X^T$ is a number, we have

$$X A_i^T A_j X^T = \left[X A_i^T A_j X^T\right]^T = X A_j^T A_i X^T = -\left[X A_i^T A_j X^T\right] = 0. \tag{4.57}$$

Therefore, we have $\mathcal{G}^T \mathcal{G} = \left(\sum_{t=1}^T x_t^2\right) I_N$ and $\mathcal{G}$ defines a full rate, $R = 1$, real space-time block code. It remains to show that $T = \min[2^{4c+d}]$ over the set defined in the theorem. First, we show that T is a power of two. Suppose $T = 2^a b$ where $b > 1$ is an odd number. We have $N \leq \rho(T) = \rho(2^a)$ and $2^a < T$. This is in contradiction with the fact that T is the smallest integer such that $N \leq \rho(T)$. The set on which the minimization is applied derives from the definition of $\rho(2^a)$. Note that one cannot construct an orthogonal design using a smaller T, for example $T' < T$. To show this, let us assume that $\mathcal{G}$ is a $T' \times N$ orthogonal design with T' indeterminate variables. Since the columns of $\mathcal{G}$ are linear combinations of the elements of $X = (x_1, x_2, \ldots, x_{T'})$, the nth column of $\mathcal{G}$ can be written as $B_n X^T$ where B_n are $T' \times T'$

matrices. The orthogonality of the columns of $\mathcal{G}$ results in

$$
\begin{aligned}
B_n^T B_n &= I_{T'}, && n = 1, \ldots, N \\
B_n^T B_{n'} &= -B_{n'}^T B_n, && 1 \le n < n' \le N.
\end{aligned}
\tag{4.58}
$$

Defining $A_n = B_1^T B_n$ for $n = 2, 3, \ldots, N$ results in a $T' \times T'$ family of Hurwitz–Radon matrices with $N - 1$ members. Therefore, $N - 1 \le \rho(T') - 1$ and since by definition T is the smallest of such numbers, we have $T' \ge T$ and $T' < T$ is impossible. $\square$

Since the proof is constructive, for any given number of transmit antennas, Theorem 4.5.1 shows how to design a real orthogonal STBC (OSTBC) that provides full rate and full diversity in addition to a simple ML decoding. Also, the corresponding block size T is the minimum possible. This is an important parameter that dictates the delay of the code. In fact, an orthogonal design with minimum block length T and the corresponding STBC is called "delay-optimal" [136]. Theorem 4.5.1 shows that the block lengths for delay-optimal codes are powers of two. To emphasize the constructive nature of our proofs, we provide the following example.

Example 4.5.1 *In this example, we would like to design a full-rate OSTBC for $N = 9$ antennas. First, using Table 4.1 and the proof of Theorem 4.5.1, we have $T = K = 16$. Therefore, we consider $X = (x_1, x_2, \ldots, x_{16})$. The 16×16 family of Hurwitz–Radon contains $N - 1 = 8$ members. Adding the identity matrix $B_1 = I_{16}$, we use nine 16×16 matrices $B_1, B_2, \ldots, B_9$ to construct a 16×9 real orthogonal design. The nth column of $\mathcal{G}$ is $B_n X^T$ for $n = 1, 2, \ldots, 9$. Therefore, we have*

$$
\mathcal{G}_{16\times9} =
\begin{pmatrix}
x_1 & x_2 & x_3 & x_4 & x_5 & x_6 & x_7 & x_8 & x_9 \\
x_2 & -x_1 & x_4 & -x_3 & x_6 & -x_5 & -x_8 & x_7 & x_{10} \\
x_3 & -x_4 & -x_1 & x_2 & -x_7 & -x_8 & x_5 & x_6 & x_{11} \\
x_4 & x_3 & -x_2 & -x_1 & -x_8 & x_7 & -x_6 & x_5 & x_{12} \\
x_5 & -x_6 & x_7 & x_8 & -x_1 & x_2 & -x_3 & -x_4 & x_{13} \\
x_6 & x_5 & x_8 & -x_7 & -x_2 & -x_1 & x_4 & -x_3 & x_{14} \\
x_7 & x_8 & -x_5 & x_6 & x_3 & -x_4 & -x_1 & -x_2 & x_{15} \\
x_8 & -x_7 & -x_6 & -x_5 & x_4 & x_3 & x_2 & -x_1 & x_{16} \\
x_9 & -x_{10} & -x_{11} & -x_{12} & -x_{13} & -x_{14} & -x_{15} & -x_{16} & -x_1 \\
x_{10} & x_9 & -x_{12} & x_{11} & -x_{14} & x_{13} & x_{16} & -x_{15} & -x_2 \\
x_{11} & x_{12} & x_9 & -x_{10} & x_{15} & x_{16} & -x_{13} & -x_{14} & -x_3 \\
x_{12} & -x_{11} & x_{10} & x_9 & x_{16} & -x_{15} & x_{14} & -x_{13} & -x_4 \\
x_{13} & x_{14} & -x_{15} & -x_{16} & x_9 & -x_{10} & x_{11} & x_{12} & -x_5 \\
x_{14} & -x_{13} & -x_{16} & x_{15} & x_{10} & x_9 & -x_{12} & x_{11} & -x_6 \\
x_{15} & -x_{16} & x_{13} & -x_{14} & -x_{11} & x_{12} & x_9 & x_{10} & -x_7 \\
x_{16} & x_{15} & x_{14} & x_{13} & -x_{12} & -x_{11} & -x_{10} & x_9 & -x_8
\end{pmatrix}.
\tag{4.59}
$$

The corresponding matrix Ω is defined by

$$\Omega(\alpha_1, \alpha_2, \alpha_3, \alpha_4, \alpha_5, \alpha_6, \alpha_7, \alpha_8, \alpha_9) =$$

$$
\begin{pmatrix}
\alpha_1 & -\alpha_2 & -\alpha_3 & -\alpha_4 & -\alpha_5 & -\alpha_6 & -\alpha_7 & -\alpha_8 & -\alpha_9 & 0 & 0 & 0 & 0 & 0 & 0 & 0 \\
\alpha_2 & \alpha_1 & \alpha_4 & -\alpha_3 & \alpha_6 & -\alpha_5 & -\alpha_8 & \alpha_7 & 0 & -\alpha_9 & 0 & 0 & 0 & 0 & 0 & 0 \\
\alpha_3 & -\alpha_4 & \alpha_1 & \alpha_2 & -\alpha_7 & -\alpha_8 & \alpha_5 & \alpha_6 & 0 & 0 & -\alpha_9 & 0 & 0 & 0 & 0 & 0 \\
\alpha_4 & \alpha_3 & -\alpha_2 & \alpha_1 & -\alpha_8 & \alpha_7 & -\alpha_6 & \alpha_5 & 0 & 0 & 0 & -\alpha_9 & 0 & 0 & 0 & 0 \\
\alpha_5 & -\alpha_6 & \alpha_7 & \alpha_8 & \alpha_1 & \alpha_2 & -\alpha_3 & -\alpha_4 & 0 & 0 & 0 & 0 & -\alpha_9 & 0 & 0 & 0 \\
\alpha_6 & \alpha_5 & \alpha_8 & -\alpha_7 & -\alpha_2 & \alpha_1 & \alpha_4 & -\alpha_3 & 0 & 0 & 0 & 0 & 0 & -\alpha_9 & 0 & 0 \\
\alpha_7 & \alpha_8 & -\alpha_5 & \alpha_6 & \alpha_3 & -\alpha_4 & \alpha_1 & -\alpha_2 & 0 & 0 & 0 & 0 & 0 & 0 & -\alpha_9 & 0 \\
\alpha_8 & -\alpha_7 & -\alpha_6 & -\alpha_5 & \alpha_4 & \alpha_3 & \alpha_2 & \alpha_1 & 0 & 0 & 0 & 0 & 0 & 0 & 0 & -\alpha_9 \\
\alpha_9 & 0 & 0 & 0 & 0 & 0 & 0 & 0 & \alpha_1 & \alpha_2 & \alpha_3 & \alpha_4 & \alpha_5 & \alpha_6 & \alpha_7 & \alpha_8 \\
0 & \alpha_9 & 0 & 0 & 0 & 0 & 0 & 0 & -\alpha_2 & \alpha_1 & -\alpha_4 & \alpha_3 & -\alpha_6 & \alpha_5 & \alpha_8 & -\alpha_7 \\
0 & 0 & \alpha_9 & 0 & 0 & 0 & 0 & 0 & -\alpha_3 & \alpha_4 & \alpha_1 & -\alpha_2 & \alpha_7 & \alpha_8 & -\alpha_5 & -\alpha_6 \\
0 & 0 & 0 & \alpha_9 & 0 & 0 & 0 & 0 & -\alpha_4 & -\alpha_3 & \alpha_2 & \alpha_1 & \alpha_8 & -\alpha_7 & \alpha_6 & -\alpha_5 \\
0 & 0 & 0 & 0 & \alpha_9 & 0 & 0 & 0 & -\alpha_5 & \alpha_6 & -\alpha_7 & -\alpha_8 & \alpha_1 & -\alpha_2 & \alpha_3 & \alpha_4 \\
0 & 0 & 0 & 0 & 0 & \alpha_9 & 0 & 0 & -\alpha_6 & -\alpha_5 & -\alpha_8 & \alpha_7 & \alpha_2 & \alpha_1 & -\alpha_4 & \alpha_3 \\
0 & 0 & 0 & 0 & 0 & 0 & \alpha_9 & 0 & -\alpha_7 & -\alpha_8 & \alpha_5 & -\alpha_6 & -\alpha_3 & \alpha_4 & \alpha_1 & \alpha_2 \\
0 & 0 & 0 & 0 & 0 & 0 & 0 & \alpha_9 & -\alpha_8 & \alpha_7 & \alpha_6 & \alpha_5 & -\alpha_4 & -\alpha_3 & -\alpha_2 & \alpha_1
\end{pmatrix}
\qquad (4.60)
$$

Removing a column of an orthogonal design results in another orthogonal design that can be used to design a STBC with one less transmit antenna. We call such a column removal process "shortening." If the original orthogonal design is delay-optimal, the shortened orthogonal design is also delay-optimal. We use this property to design the following orthogonal designs and the corresponding STBCs.

Example 4.5.2 *For $N = 3$ transmit antennas, the following 4×3 orthogonal design, derived from (4.38), is full rate and delay-optimal.*

$$
\mathcal{G} = \begin{pmatrix}
x_1 & x_2 & x_3 \\
-x_2 & x_1 & -x_4 \\
-x_3 & x_4 & x_1 \\
-x_4 & -x_3 & x_2
\end{pmatrix}.
\qquad (4.61)
$$

Example 4.5.3 *For $N = 7$ transmit antennas, the following 8×7 orthogonal design, derived from (4.40), is full rate and delay-optimal.*

$$
\mathcal{G} = \begin{pmatrix}
x_1 & x_2 & x_3 & x_4 & x_5 & x_6 & x_7 \\
-x_2 & x_1 & -x_4 & x_3 & -x_6 & x_5 & x_8 \\
-x_3 & x_4 & x_1 & -x_2 & x_7 & x_8 & -x_5 \\
-x_4 & -x_3 & x_2 & x_1 & x_8 & -x_7 & x_6 \\
-x_5 & x_6 & -x_7 & -x_8 & x_1 & -x_2 & x_3 \\
-x_6 & -x_5 & -x_8 & x_7 & x_2 & x_1 & -x_4 \\
-x_7 & -x_8 & x_5 & -x_6 & -x_3 & x_4 & x_1 \\
-x_8 & x_7 & x_6 & x_5 & -x_4 & -x_3 & -x_2
\end{pmatrix}.
\qquad (4.62)
$$

Example 4.5.4 *Similarly, the following orthogonal designs are full rate and delay-optimal for* $N = 5$ *and* $N = 6$ *transmit antennas, respectively.*

$$
\mathcal{G} = \begin{pmatrix}
x_1 & x_2 & x_3 & x_4 & x_5 \\
-x_2 & x_1 & -x_4 & x_3 & -x_6 \\
-x_3 & x_4 & x_1 & -x_2 & x_7 \\
-x_4 & -x_3 & x_2 & x_1 & x_8 \\
-x_5 & x_6 & -x_7 & -x_8 & x_1 \\
-x_6 & -x_5 & -x_8 & x_7 & x_2 \\
-x_7 & -x_8 & x_5 & -x_6 & -x_3 \\
-x_8 & x_7 & x_6 & x_5 & -x_4
\end{pmatrix},
\tag{4.63}
$$

$$
\mathcal{G} = \begin{pmatrix}
x_1 & x_2 & x_3 & x_4 & x_5 & x_6 \\
-x_2 & x_1 & -x_4 & x_3 & -x_6 & x_5 \\
-x_3 & x_4 & x_1 & -x_2 & x_7 & x_8 \\
-x_4 & -x_3 & x_2 & x_1 & x_8 & -x_7 \\
-x_5 & x_6 & -x_7 & -x_8 & x_1 & -x_2 \\
-x_6 & -x_5 & -x_8 & x_7 & x_2 & x_1 \\
-x_7 & -x_8 & x_5 & -x_6 & -x_3 & x_4 \\
-x_8 & x_7 & x_6 & x_5 & -x_4 & -x_3
\end{pmatrix}.
\tag{4.64}
$$

So far we have provided examples of full-rate real orthogonal designs and the corresponding STBCs. Following the first design method in Section 4.4, where an orthogonal design was defined as $\mathcal{G} = \sum_{k=1}^{K} x_k A_k$, we can define a different class of orthogonal designs. We define this new class through an example.

Example 4.5.5 *Let us consider the nine* 16×16 *matrices* $B_1, B_2, \ldots, B_9$ *in Example 4.5.1. We design an OSTBC to transmit* $K = 9$ *real symbols in* $T = 16$ *time slots from* $N = 16$ *transmit antennas by* $\mathcal{G} = \sum_{k=1}^{9} x_k B_k$. *The rate of such an orthogonal design is* $R = K/T = 9/16$. *Similar to the codes in Section 4.4, it is easy to show that the constructed STBC provides all the nice properties of real orthogonal STBCs at lower rates. Note that removing the columns of* $\mathcal{G}$ *results in OSTBCs with the same rate for a smaller number of transmit antennas. One can design rate* $R = 9/16$ *codes for* $N = 1, 2, \ldots, 16$ *antennas by removing columns of* $\mathcal{G}$.

Using a similar method, one can design square orthogonal STBCs to transmit $K = \rho(T)$ real symbols for any T that is a power of two. The rate of such a code is $R = K/T = \rho(T)/T$ and can be easily calculated from Table 4.1. These orthogonal codes are designed for $N = T$ transmit antennas or less. Based on the Radon Theorem, first we start with a Hurwitz–Radon family of matrices and the identity matrix to come up with $K = \rho(T)$ matrices $B_1, B_2, \ldots, B_K$. Then, we construct the code as $\mathcal{G} = \sum_{k=1}^{K} x_k B_k$. The generator matrix $\mathcal{G}$ is an OSTBC for

$N = T$ transmit antennas. If a smaller number of transmit antenna are required, we remove some of the columns of $\mathcal{G}$.

We emphasize that the above two design strategies result in codes with different rates from orthogonal designs. There is another way of presenting the full-rate codes designed by the first method. Since the elements of the non-square generator matrices are linear combinations of the indeterminate variables $x_1, x_2, \ldots, x_K$, they can be written as

$$\mathcal{G} = \sum_{k=1}^{K} x_k E_k, \tag{4.65}$$

where matrices E_k are $T \times N$ real matrices. We rewrite the orthogonality of generator matrices in (4.55) by using (4.65) as follows

$$
\begin{aligned}
\mathcal{G}^T \mathcal{G} &= \left(\sum_{k=1}^{K} x_k E_k^T \right) \sum_{k'=1}^{K} (x_{k'} E_{k'}) \\
&= \sum_{k=1}^{K} \sum_{k'=1}^{K} (E_k^T E_{k'}) x_k x_{k'} \\
&= \sum_{k=1}^{K} x_k^2 I_N.
\end{aligned}
\tag{4.66}
$$

Since (4.66) should be true for any choice of indeterminate variables, we must have

$$
\begin{cases}
E_k^T E_{k'} + E_{k'}^T E_k = 0_N & k \neq k' \\
E_k^T E_k = I_N & k = 1, 2, \ldots, K
\end{cases}
\tag{4.67}
$$

where 0_N is an $N \times N$ zero matrix. Equations (4.67) hold for a set of non-square matrices and are very similar to (4.23) for a Hurwitz–Radon family of matrices. Therefore, (4.67) and their existence through the constructive proof of Theorem 4.5.1 are referred to as the generalization of Radon–Hurwitz theory to non-square matrices [136].

The importance of (4.65) is its use to derive a simple ML decoding algorithm. Similar to the case of orthogonal designs, we show how to separately decode the symbols of a STBC from generalized orthogonal designs. The input–output relationship in (2.5) for transmitting $(s_1, s_2, \ldots, s_K)$ over a fading channel results in

$$
\begin{aligned}
\mathbf{r}^T &= \mathbf{H}^T \cdot \mathbf{C}^T + \mathcal{N}^T \\
&= \mathbf{H}^T \sum_{k=1}^{K} s_k E_k^T + \mathcal{N}^T \\
&= \sum_{k=1}^{K} s_k \Omega_k + \mathcal{N}^T \\
&= (s_1, s_2, \ldots, s_K) \cdot \Omega + \mathcal{N}^T,
\end{aligned}
\tag{4.68}
$$

where $\Omega_k = \mathbf{H}^T E_k^T$ is the k^T row of a $K \times T$ matrix Ω and contains T elements. These elements are a function of N path gains $\alpha_1, \alpha_2, \ldots, \alpha_N$. Note that

$$\Omega \cdot \Omega^T = \left(\sum_{n=1}^{N} \alpha_n^2 \right) I_K. \tag{4.69}$$

To show that (4.69) is true, we consider the (k, k')th element of $\Omega \cdot \Omega^T$ which is $\Omega_k \Omega_{k'}^T = \mathbf{H}^T E_k^T E_{k'} \mathbf{H}$. For the diagonal elements in (4.69), that is $k = k'$, we have $\Omega_k \Omega_k^T = \mathbf{H}^T E_k^T E_k \mathbf{H} = \mathbf{H}^T \mathbf{H} = \sum_{n=1}^{N} \alpha_n^2$. For $k \neq k'$, we calculate the transpose of $\mathbf{H}^T E_k^T E_{k'} \mathbf{H}$ which is a number to show that

$$\begin{aligned} \mathbf{H}^T E_k^T E_{k'} \mathbf{H} &= \left[\mathbf{H}^T E_k^T E_{k'} \mathbf{H} \right]^T \\ &= \mathbf{H}^T E_{k'}^T E_k \mathbf{H} \\ &= -\left[\mathbf{H}^T E_k^T E_{k'} \mathbf{H} \right]. \end{aligned} \tag{4.70}$$

Therefore, $\Omega_k \Omega_{k'}^T = \mathbf{H}^T E_k^T E_{k'} \mathbf{H} = 0$ and (4.69) is proved.

To decode the symbols, one can multiply both sides of (4.68) by Ω^T to arrive at

$$\mathbf{r}^T \cdot \Omega^T = \left(\sum_{n=1}^{N} \alpha_n^2 \right) (s_1, s_2, \ldots, s_K) + \mathcal{N}^T \cdot \Omega^T. \tag{4.71}$$

We conclude this section by reminding that orthogonal designs are not unique. In fact, multiplying any orthogonal design $\mathcal{G}$ by a matrix U with the property $U^T \cdot U = I$ results in another orthogonal design $\mathcal{G}' = U \cdot \mathcal{G}$. This is easy to see from $\mathcal{G}'^T \cdot \mathcal{G}' = \mathcal{G}^T \cdot U^T \cdot U \cdot \mathcal{G} = \mathcal{G}^T \cdot \mathcal{G}$.

4.6 Complex orthogonal designs

So far we have studied real STBCs that provide full diversity and simple ML decoding for any number of transmit antennas. The proposed structures only work for real constellations. Naturally, it is interesting to extend these schemes to complex signal constellations.

Definition 4.6.1 *A complex orthogonal design of size N is an $N \times N$ orthogonal matrix $\mathcal{G}_N$ with complex entries $x_1, -x_1, x_2, -x_2, \ldots, x_N, -x_N$, their conjugates $x_1^*, -x_1^*, x_2^*, -x_2^*, \ldots, x_N^*, -x_N^*$, and multiples of these indeterminate variables by $j = \sqrt{-1}$ or $-j$ such that*

$$\mathcal{G}_N^H \cdot \mathcal{G}_N = \left(|x_1|^2 + |x_2|^2 + \cdots + |x_N|^2 \right) I_N. \tag{4.72}$$

In this section, we study the existence of complex orthogonal designs. The following construction shows the relationship between real and complex orthogonal designs.

Construction: Given a complex orthogonal design of size N, we replace each complex variable $x_n = \Re\{x_n\} + \Im\{x_n\}j$, $1 \le n \le N$ by the 2×2 real matrix

$$\begin{pmatrix} \Re\{x_n\} & \Im\{x_n\} \\ -\Im\{x_n\} & \Re\{x_n\} \end{pmatrix}. \tag{4.73}$$

The resulting $2N \times 2N$ matrix is a real orthogonal design of size $2N$.

Note that the representation in (4.73) results in representing x_n^* by

$$\begin{pmatrix} \Re\{x_n\} & -\Im\{x_n\} \\ \Im\{x_n\} & \Re\{x_n\} \end{pmatrix}. \tag{4.74}$$

Since real orthogonal designs only exist for $N = 2, 4, 8$, the above construction shows that a complex orthogonal design can only exist for $N = 2$ or $N = 4$. For $N = 2$, the following matrix used in the construction of Alamouti code in Section 4.2 is a complex orthogonal design

$$\mathcal{G}_2 = \begin{pmatrix} x_1 & x_2 \\ -x_2^* & x_1^* \end{pmatrix}. \tag{4.75}$$

The following theorem shows that (4.75) is the only possible complex orthogonal design.

Theorem 4.6.1 *A complex orthogonal design exists if and only if $N = 2$.*

Proof So far we have shown that a complex orthogonal design can only exist for $N = 2, 4$. We have an example for $N = 2$. To prove the theorem we only need to show that for $N = 4$ a complex orthogonal design does not exist. The lengthy proof is provided in [136]. □

4.7 Generalized complex orthogonal designs

Similar to the case of real orthogonal designs, we generalize the theory of complex orthogonal designs to non-square matrices. We start with the definition of a generalized complex orthogonal design.

Definition 4.7.1 *A generalized complex orthogonal design is a $T \times N$ matrix $\mathcal{G}$ with entries that are linear combinations of the indeterminate variables $x_1, x_2, \ldots, x_K$ and their conjugates such that*

$$\mathcal{G}^H \cdot \mathcal{G} = \kappa \left(|x_1|^2 + |x_2|^2 + \cdots + |x_K|^2 \right) I_N, \tag{4.76}$$

where I_N is the $N \times N$ identity matrix and κ is a constant.

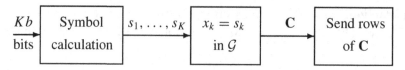

Fig. 4.4.   Encoder block diagram for orthogonal space-time block codes.

Note that $\kappa = 1$ is possible by an appropriate normalization of $\mathcal{G}$ elements. Also, multiplying a generalized complex orthogonal design by a unitary matrix results in another generalized complex orthogonal design. In other words, if $\mathcal{G}$ is a generalized complex orthogonal design and U is unitary, that is $U^H \cdot U = I$, then $\mathcal{G}' = U \cdot \mathcal{G}$ is also a generalized complex orthogonal design. This can be shown using the fact that $\mathcal{G}'^H \cdot \mathcal{G}' = \mathcal{G}^H \cdot U^H \cdot U \cdot \mathcal{G} = \mathcal{G}^H \cdot \mathcal{G}$. Similarly, $\mathcal{G}' = \mathcal{G} \cdot U$ is a generalized complex orthogonal design as well since

$$
\begin{aligned}
\mathcal{G}'^H \cdot \mathcal{G}' &= U^H \cdot \mathcal{G}^H \cdot \mathcal{G} \cdot U \\
&= \kappa \sum_{k=1}^{K} |x_k|^2 U^H \cdot I_N \cdot U \\
&= \kappa \sum_{k=1}^{K} |x_k|^2 I_N.
\end{aligned}
\tag{4.77}
$$

A STBC for any complex constellation can be constructed using a generalized complex orthogonal design. The number of transmission antennas is N. We assume that transmission at the baseband employs a signal constellation with 2^b elements. At time slot 1, Kb bits arrive at the encoder and select constellation signals $s_1, s_2, \ldots, s_K$. Setting $x_k = s_k$ for $k = 1, 2, \ldots, K$ in $\mathcal{G}$, we arrive at a matrix $\mathbf{C} = \mathcal{G}(s_1, s_2, \ldots, s_K)$ whose entries are linear combinations of $s_1, s_2, \ldots, s_K$ and their conjugates. So, while $\mathcal{G}$ contains indeterminate variables $x_1, x_2, \ldots, x_K$, matrix $\mathbf{C}$ contains specific constellation symbols that are transmitted from N antennas for each Kb bits as follows. If $\mathbf{C}_{t,n}$ represents the element in the tth row and the nth column of $\mathbf{C}$, the entries $\mathbf{C}_{t,n}$, $n = 1, 2, \ldots, N$ are transmitted simultaneously from transmit antennas $1, 2, \ldots, N$ at each time slot $t = 1, 2, \ldots, T$. So, the nth column of $\mathbf{C}$ represents the transmitted symbols from the nth antenna and the tth row of $\mathbf{C}$ represents the transmitted symbols at time slot t. Note that $\mathbf{C}$ is basically defined using $\mathcal{G}$. Therefore, the orthogonality of the columns of $\mathcal{G}$ allows a simple ML decoding scheme, which will be explained in what follows. Since T time slots are used to transmit K symbols, we define the rate of the code to be $R = K/T$. The block diagram of the encoder is shown at Figure 4.4. The following theorem shows that the STBCs designed from generalized complex orthogonal designs provide full diversity and separate decoding.

Theorem 4.7.1 *A complex space-time block code designed from a $T \times N$ generalized complex orthogonal design provides a diversity of NM for M receive antennas and a separate maximum-likelihood decoding of its symbols.*

Proof First, we show that STBCs from generalized complex orthogonal designs provide full diversity. We need to show that the difference matrix $\mathcal{G}(s_1, s_2, \ldots, s_K) - \mathcal{G}(s_1', s_2', \ldots, s_K')$ is full rank (nonsingular) for any two distinct set of inputs $(s_1, s_2, \ldots, s_K) \neq (s_1', s_2', \ldots, s_K')$. Since $\mathcal{G}(s_1, s_2, \ldots, s_K)$ elements are linear combinations of $s_1, s_2, \ldots, s_K$, we have $\mathcal{G}(s_1, s_2, \ldots, s_K) - \mathcal{G}(s_1', s_2', \ldots, s_K') = \mathcal{G}(s_1 - s_1', s_2 - s_2', \ldots, s_K - s_K')$. Therefore, it is adequate to show that $\mathcal{G}(s_1 - s_1', s_2 - s_2', \ldots, s_K - s_K')$ is nonsingular. Equivalently, we can show that $\det[\mathcal{G}^H(s_1 - s_1', s_2 - s_2', \ldots, s_K - s_K') \cdot \mathcal{G}(s_1 - s_1', s_2 - s_2', \ldots, s_K - s_K')]$ is not zero for any two distinct set of inputs. From the definition of generalized complex orthogonal designs in (4.76), we have

$$\det\left[\mathcal{G}^H(s_1 - s_1', s_2 - s_2', \ldots, s_K - s_K') \cdot \mathcal{G}(s_1 - s_1', s_2 - s_2', \ldots, s_K - s_K')\right]$$
$$= \kappa \sum_{k=1}^{K} |s_k - s_k'|^2. \tag{4.78}$$

Therefore, the above determinant is zero if and only if $(s_1, s_2, \ldots, s_K) = (s_1', s_2', \ldots, s_K')$, which means the difference matrix is full rank.

Second, we show that the symbols of STBCs from generalized complex orthogonal designs can be decoded separately. This is achieved due to the orthogonality of the columns of $\mathcal{G}$. The proof is similar to the proof for real orthogonal designs. The maximum-likelihood decoding amounts to finding the codeword $\hat{\mathbf{C}}$ that maximizes the density function in (3.4) or equivalently minimizes the cost function in (4.12). Therefore, the most likely transmitted signals are

$$\underset{s_1, s_2, \ldots, s_K}{\text{argmin}} \left\{ \kappa \sum_{k=1}^{K} |s_k|^2 \text{Tr}\left[\mathbf{H}^H \cdot \mathbf{H}\right] - \text{Tr}\left[\mathbf{H}^H \cdot \mathbf{C}^H \cdot \mathbf{r} - \mathbf{r}^H \cdot \mathbf{C} \cdot \mathbf{H}\right] \right\}. \tag{4.79}$$

The argument of the minimization in (4.79) is a linear combination of s_k and $|s_k|^2$ and can be written as $\sum_{k=1}^{K} f_k(s_k)$ where $f_k(s_k)$, $k = 1, 2, \ldots, K$ is only a function of s_k. As a result, the minimization in (4.79) is equivalent to K independent minimizations $\min_{s_k} f_k(s_k)$, $k = 1, 2, \ldots, K$. Therefore, transmitted symbols can be decoded separately. □

The formulas for $f_k(s_k)$, $k = 1, 2, \ldots, K$ depend on the structure of the corresponding STBC. Specific decoding formulas for many of the existing STBCs from orthogonal designs are provided in the sequel.

Note that for $M = 1$ receive antenna, the minimization cost function in (4.79) is

$$\kappa \sum_{n=1}^{N} |\alpha_{n,1}|^2 \sum_{k=1}^{K} |s_k|^2 - 2\Re\left\{ \sum_{n=1}^{N} \sum_{t=1}^{T} \alpha_{n,1}^* C_{t,n}^* r_{t,1} \right\}. \tag{4.80}$$

Also, the general ML decoding cost function for M receive antennas in (4.79) can be rewritten as

$$\sum_{m=1}^{M}\left[\kappa\sum_{n=1}^{N}|\alpha_{n,m}|^{2}\sum_{k=1}^{K}|s_{k}|^{2}-2\Re\left\{\sum_{n=1}^{N}\sum_{t=1}^{T}\alpha_{n,m}^{*}\mathbf{C}_{t,n}^{*}r_{t,m}\right\}\right]. \tag{4.81}$$

Therefore, as discussed in Section 4.3, MRC can be used to derive the general cost function for M receive antennas from that of the one receive antenna. In other words, the formulas for $f_k(s_k)$, $k = 1, 2, \ldots, K$ for $M > 1$ receive antennas can be calculated from the corresponding formulas for one receive antenna. First, we write the cost function for different receive antennas assuming only that receive antenna exists. Then, we compute the final cost function by adding all of the M calculated intermediate cost functions. More precisely, we calculate (4.80), replace index 1 with m and put a summation $\sum_{m=1}^{M}$ in front of it to arrive at the minimization cost function in (4.81).

So far we have shown that STBCs designed from generalized complex orthogonal designs provide both desirable properties of the Alamouti code. We call these codes orthogonal space-time block codes. One important question is the existence of orthogonal STBCs or equivalently the existence of the corresponding generalized complex orthogonal designs at a given rate. In what follows, we discuss different examples of orthogonal STBCs in addition to the existence of high rate codes. First, we show that rate half, $R = 0.5$, codes exist for any number of transmit antennas and present how to design them. Then, we provide the highest rate available orthogonal STBCs for different number of transmit antennas.

Construction: Let us assume a rate R real orthogonal design with a $T \times N$ transmission matrix $\mathcal{G}$. Note that $K = RT$ symbols are transmitted in each block of T time slots. The conjugate of $\mathcal{G}$ is a $T \times N$ matrix denoted by $\mathcal{G}^*$ that is derived by replacing x_k with x_k^* in $\mathcal{G}$. We design a $2T \times N$ complex orthogonal design by concatenating $\mathcal{G}$ and $\mathcal{G}^*$ as follows

$$\mathcal{G}_c = \begin{pmatrix} \mathcal{G} \\ \mathcal{G}^* \end{pmatrix} = \begin{pmatrix} \sum_{k=1}^{K} x_k E_k \\ \sum_{k=1}^{K} x_k^* E_k \end{pmatrix}. \tag{4.82}$$

To show that $\mathcal{G}_c$ is a complex orthogonal design, we calculate $\mathcal{G}_c^H \cdot \mathcal{G}_c$ in terms of real matrices E_k.

$$\mathcal{G}_c^H \cdot \mathcal{G}_c = \sum_{k=1}^{K}\sum_{k'=1}^{K}\left(x_k^* x_{k'} E_k^T E_{k'} + x_k x_{k'}^* E_k^T E_{k'}\right). \tag{4.83}$$

Applying (4.67) to (4.83) results in

$$\mathcal{G}_c^H \cdot \mathcal{G}_c = 2 \sum_{k=1}^{K} |x_k|^2 I_N. \tag{4.84}$$

Using the above construction, one can design a rate $R/2$ generalized complex orthogonal design from any rate R generalized real orthogonal designs. Based on Theorem 4.7.1, (4.84) shows that the resulting generalized complex orthogonal design provides full diversity and simple ML decoding.

To derive another method of a simple ML decoding algorithm, we denote the $2T \times 1$ received vector by $\mathbf{r}$ and the $2T \times N$ codeword by $\mathbf{C}$ for one receive antenna. While we provide the formulas for one receive antenna, for the sake of simplicity, similar formulas are valid for more receive antennas using maximum ratio combining. We use the $T \times N$ matrix $\mathbf{C}_R$ to denote the first T rows of $\mathbf{C}$. Note that while $\mathbf{C}_R$ corresponds to the real orthogonal design $\mathcal{G}$, its elements are not necessarily real numbers because it is derived by replacing x_k with complex constellation elements s_k in $\mathcal{G}$. We write the received vector in terms of transmitted signals, channel path gains and noise as

$$\mathbf{r} = \mathbf{C} \cdot \mathbf{H} + \mathcal{N} = \begin{pmatrix} \mathbf{C}_R \\ \mathbf{C}_R^* \end{pmatrix} \cdot \mathbf{H} + \mathcal{N} = \begin{pmatrix} \mathbf{C}_R \cdot \mathbf{H} \\ \mathbf{C}_R^* \cdot \mathbf{H} \end{pmatrix} + \begin{pmatrix} \mathcal{N}_1 \\ \mathcal{N}_2 \end{pmatrix}. \tag{4.85}$$

Since the path gains are complex numbers, matrix $\mathbf{H}^*$, the conjugate of $\mathbf{H} = (\alpha_1, \alpha_2, \dots, \alpha_N)$, is achieved by replacing α_n with α_n^*. We define a $2T \times 1$ vector $\mathbf{r}'$ as

$$\mathbf{r}' = \left(r_1, r_2, \dots, r_T, r_{T+1}^*, r_{T+2}^*, \dots, r_{2T}^* \right)^T$$
$$= \begin{pmatrix} \mathbf{C}_R \cdot \mathbf{H} \\ \mathbf{C}_R \cdot \mathbf{H}^* \end{pmatrix} + \begin{pmatrix} \mathcal{N}_1 \\ \mathcal{N}_2^* \end{pmatrix}. \tag{4.86}$$

Its transpose can be written as the concatenation of two row vectors.

$$\mathbf{r}'^T = \left[(\alpha_1, \alpha_2, \dots, \alpha_N) \cdot \mathbf{C}_R^T, \ (\alpha_1^*, \alpha_2^*, \dots, \alpha_N^*) \cdot \mathbf{C}_R^T \right] + (\mathcal{N}_1^T, \mathcal{N}_2^H). \tag{4.87}$$

Note that in Section 4.5 we showed that there exists a $K \times T$ matrix $\Omega_R(\alpha_1, \alpha_2, \dots, \alpha_N)$ where

$$\mathbf{H}^T \cdot \mathbf{C}_R^T = (s_1, s_2, \dots, s_K) \cdot \Omega_R. \tag{4.88}$$

For complex path gains, using an approach similar to that of Section 4.5, one can show

$$\Omega_R \cdot \Omega_R^H = \sum_{n=1}^{N} |\alpha_n|^2 I_K. \tag{4.89}$$

To write $\mathbf{r}'$ in terms of transmitted symbols $s_1, s_2, \dots, s_K$, we define the following $K \times 2T$ matrix Ω by concatenating two $K \times T$ matrices.

$$\Omega = \left[\Omega_R(\alpha_1, \alpha_2, \dots, \alpha_N), \Omega_R(\alpha_1^*, \alpha_2^*, \dots, \alpha_N^*) \right]. \tag{4.90}$$

Then using (4.88) in (4.87) and applying (4.89) and (4.90) results in

$$\mathbf{r}'^T = (s_1, s_2, \ldots, s_K) \cdot \Omega + (\mathcal{N}_1^T, \mathcal{N}_2^H). \tag{4.91}$$

It is easy to verify that

$$\Omega \cdot \Omega^H = 2 \sum_{n=1}^{N} |\alpha_n|^2 I_K. \tag{4.92}$$

Therefore, multiplying both sides of (4.91) by Ω^H results in the following formula that provides a simple algorithm to separately decode the transmitted symbols:

$$\mathbf{r}'^T \cdot \Omega^H = 2 \sum_{n=1}^{N} |\alpha_n|^2 (s_1, s_2, \ldots, s_K) + (\mathcal{N}_1^T, \mathcal{N}_2^H) \cdot \Omega^H. \tag{4.93}$$

Since we have shown in Section 4.5 that a full rate, $R = 1$, generalized real orthogonal design exists for any number of transmit antennas, we conclude that a rate half, $R = 0.5$, generalized complex orthogonal design exists for any number of transmit antennas. An example for four transmit antennas is provided below.

Example 4.7.1 *In this example, we consider a STBC that utilizes four transmit antennas and which is defined by:*

$$\mathcal{G}_{448} = \begin{pmatrix} x_1 & x_2 & x_3 & x_4 \\ -x_2 & x_1 & -x_4 & x_3 \\ -x_3 & x_4 & x_1 & -x_2 \\ -x_4 & -x_3 & x_2 & x_1 \\ x_1^* & x_2^* & x_3^* & x_4^* \\ -x_2^* & x_1^* & -x_4^* & x_3^* \\ -x_3^* & x_4^* & x_1^* & -x_2^* \\ -x_4^* & -x_3^* & x_2^* & x_1^* \end{pmatrix}. \tag{4.94}$$

In most of the examples in this chapter, we use $\mathcal{G}_{NKT}$ to represent an OSTBC for N transmit antennas delivering K symbols over T time slots. The code in (4.94) transmits $K = 4$ symbols at $T = 8$ time slots and the rate of the code is $R = 0.5$. The code takes 4b bits at each block of length $T = 8$ to select constellation symbols s_1, s_2, s_3, s_4. Then, replacing $x_k = s_k$ for $k = 1, 2, 3, 4$ in $\mathcal{G}_{448}$ results in the transmitted codeword $\mathbf{C}$. The tth row of $\mathbf{C}$ is transmitted from antennas $n = 1, 2, 3, 4$ at time t.

The decoder uses the following matrix Ω for separately decoding symbols s_1, s_2, s_3, s_4.

$$\Omega = \begin{pmatrix} \alpha_1 & \alpha_2 & \alpha_3 & \alpha_4 & \alpha_1^* & \alpha_2^* & \alpha_3^* & \alpha_4^* \\ \alpha_2 & -\alpha_1 & -\alpha_4 & \alpha_3 & \alpha_2^* & -\alpha_1^* & -\alpha_4^* & \alpha_3^* \\ \alpha_3 & \alpha_4 & -\alpha_1 & -\alpha_2 & \alpha_3^* & \alpha_4^* & -\alpha_1^* & -\alpha_2^* \\ \alpha_4 & -\alpha_3 & \alpha_2 & -\alpha_1 & \alpha_4^* & -\alpha_3^* & \alpha_2^* & -\alpha_1^* \end{pmatrix}. \tag{4.95}$$

Using (4.93), one can decode different symbols separately. We provide the decoding formulas for each symbol for the general case of M receive antennas and maximum ratio combining. The maximum-likelihood decoding amounts to minimizing the decision metric

$$
\left| s_1 - \sum_{m=1}^{M} \left\{ r_{1,m}\alpha_{1,m}^* + r_{2,m}\alpha_{2,m}^* + r_{3,m}\alpha_{3,m}^* + r_{4,m}\alpha_{4,m}^* \right. \right.
$$
$$
\left. \left. + r_{5,m}^*\alpha_{1,m} + r_{6,m}^*\alpha_{2,m} + r_{7,m}^*\alpha_{3,m} + r_{8,m}^*\alpha_{4,m} \right\} \right|^2 \quad (4.96)
$$
$$
+ \left(-1 + 2\sum_{m=1}^{M}\sum_{n=1}^{4} |\alpha_{n,m}|^2 \right) |s_1|^2,
$$

over all possible values of s_1, minimizing the decision metric

$$
\left| s_2 - \sum_{m=1}^{M} \left\{ r_{1,m}\alpha_{2,m}^* - r_{2,m}\alpha_{1,m}^* - r_{3,m}\alpha_{4,m}^* + r_{4,m}\alpha_{3,m}^* \right. \right.
$$
$$
\left. \left. + r_{5,m}^*\alpha_{2,m} - r_{6,m}^*\alpha_{1,m} - r_{7,m}^*\alpha_{4,m} + r_{8,m}^*\alpha_{3,m} \right\} \right|^2 \quad (4.97)
$$
$$
+ \left(-1 + 2\sum_{m=1}^{M}\sum_{n=1}^{4} |\alpha_{n,m}|^2 \right) |s_2|^2,
$$

over all possible values of s_2, minimizing the decision metric

$$
\left| s_3 - \sum_{m=1}^{M} \left\{ r_{1,m}\alpha_{3,m}^* + r_{2,m}\alpha_{4,m}^* - r_{3,m}\alpha_{1,m}^* - r_{4,m}\alpha_{2,m}^* \right. \right.
$$
$$
\left. \left. + r_{5,m}^*\alpha_{3,m} + r_{6,m}^*\alpha_{4,m} - r_{7,m}^*\alpha_{1,m} - r_{8,m}^*\alpha_{2,m} \right\} \right|^2 \quad (4.98)
$$
$$
+ \left(-1 + 2\sum_{m=1}^{M}\sum_{n=1}^{4} |\alpha_{n,m}|^2 \right) |s_3|^2,
$$

over all possible values of s_3, and minimizing the decision metric

$$
\left| s_4 - \sum_{m=1}^{M} \left\{ r_{1,m}\alpha_{4,m}^* - r_{2,m}\alpha_{3,m}^* + r_{3,m}\alpha_{2,m}^* - r_{4,m}\alpha_{1,m}^* \right. \right.
$$
$$
\left. \left. + r_{5,m}^*\alpha_{4,m} - r_{6,m}^*\alpha_{3,m} + r_{7,m}^*\alpha_{2,m} - r_{8,m}^*\alpha_{1,m} \right\} \right|^2 \quad (4.99)
$$
$$
+ \left(-1 + 2\sum_{m=1}^{M}\sum_{n=1}^{4} |\alpha_{n,m}|^2 \right) |s_4|^2,
$$

over all possible values of s_4. Note that the above four metrics have terms $(-1 + 2\sum_{m=1}^{M}\sum_{n=1}^{4}|\alpha_{n,m}|^2)|s_k|^2$ that are the same for all symbols in an equal energy constellation like PSK. Therefore, these terms can be removed from the metrics if

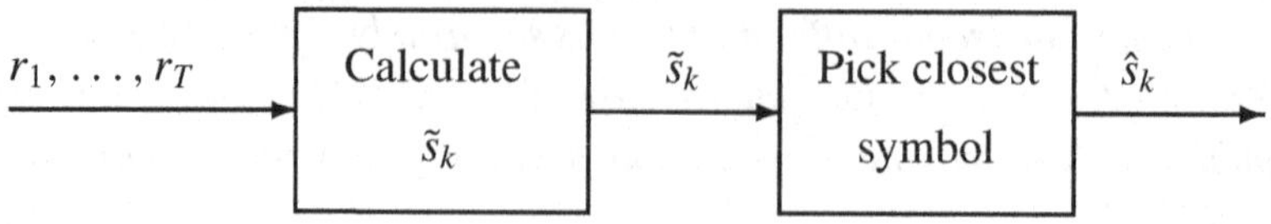

$$\tilde{s}_1 = r_1\alpha_1^* + r_2\alpha_2^* + r_3\alpha_3^* + r_4\alpha_4^* + r_5^*\alpha_1 + r_6^*\alpha_2 + r_7^*\alpha_3 + r_8^*\alpha_4.$$

$$\tilde{s}_2 = r_1\alpha_2^* - r_2\alpha_1^* - r_3\alpha_4^* + r_4\alpha_3^* + r_5^*\alpha_2 - r_6^*\alpha_1 - r_7^*\alpha_4 + r_8^*\alpha_3.$$

$$\tilde{s}_3 = r_1\alpha_3^* + r_2\alpha_4^* - r_3\alpha_1^* - r_4\alpha_2^* + r_5^*\alpha_3 + r_6^*\alpha_4 - r_7^*\alpha_1 - r_8^*\alpha_2.$$

$$\tilde{s}_4 = r_1\alpha_4^* - r_2\alpha_3^* + r_3\alpha_2^* - r_4\alpha_1^* + r_5^*\alpha_4 - r_6^*\alpha_3 + r_7^*\alpha_2 - r_8^*\alpha_1.$$

Fig. 4.5.   Decoder block diagram; Example 4.7.1; $M = 1$ receive antenna.

all symbols in the constellation have the same energy. The resulting minimization problem, for s_1, is to calculate the following linear combination of received signals and find the closest symbol in the constellation.

$$\tilde{s}_1 = \sum_{m=1}^{M} \{ r_{1,m}\alpha_{1,m}^* + r_{2,m}\alpha_{2,m}^* + r_{3,m}\alpha_{3,m}^* + r_{4,m}\alpha_{4,m}^*$$
$$+ r_{5,m}^*\alpha_{1,m} + r_{6,m}^*\alpha_{2,m} + r_{7,m}^*\alpha_{3,m} + r_{8,m}^*\alpha_{4,m} \}. \tag{4.100}$$

Similar simplifications can be done for decoding s_2, s_3, s_4 as depicted in Figure 4.5.

By removing one of the columns of the generator matrix in (4.94), one can design a similar code for $N = 3$ transmit antennas.

Example 4.7.2 *A similar example that utilizes eight transmit antennas is defined by:*

$$\mathcal{G} = \begin{pmatrix}
x_1 & x_2 & x_3 & x_4 & x_5 & x_6 & x_7 & x_8 \\
-x_2 & x_1 & x_4 & -x_3 & x_6 & -x_5 & -x_8 & x_7 \\
-x_3 & -x_4 & x_1 & x_2 & x_7 & x_8 & -x_5 & -x_6 \\
-x_4 & x_3 & -x_2 & x_1 & x_8 & -x_7 & x_6 & -x_5 \\
-x_5 & -x_6 & -x_7 & -x_8 & x_1 & x_2 & x_3 & x_4 \\
-x_6 & x_5 & -x_8 & x_7 & -x_2 & x_1 & -x_4 & x_3 \\
-x_7 & x_8 & x_5 & -x_6 & -x_3 & x_4 & x_1 & -x_2 \\
-x_8 & -x_7 & x_6 & x_5 & -x_4 & -x_3 & x_2 & x_1 \\
x_1^* & x_2^* & x_3^* & x_4^* & x_5^* & x_6^* & x_7^* & x_8^* \\
-x_2^* & x_1^* & x_4^* & -x_3^* & x_6^* & -x_5^* & -x_8^* & x_7^* \\
-x_3^* & -x_4^* & x_1^* & x_2^* & x_7^* & x_8^* & -x_5^* & -x_6^* \\
-x_4^* & x_3^* & -x_2^* & x_1^* & x_8^* & -x_7^* & x_6^* & -x_5^* \\
-x_5^* & -x_6^* & -x_7^* & -x_8^* & x_1^* & x_2^* & x_3^* & x_4^* \\
-x_6^* & x_5^* & -x_8^* & x_7^* & -x_2^* & x_1^* & -x_4^* & x_3^* \\
-x_7^* & x_8^* & x_5^* & -x_6^* & -x_3^* & x_4^* & x_1^* & -x_2^* \\
-x_8^* & -x_7^* & x_6^* & x_5^* & -x_4^* & -x_3^* & x_2^* & x_1^*
\end{pmatrix}. \tag{4.101}$$

This code transmits $K = 8$ symbols at $T = 16$ time slots and the rate of the code is $R = 0.5$. To design a rate half code for $N = 5, 6, 7$ transmit antennas, one can remove the columns of the generator matrix in (4.101).

The above method is not the only way to design a rate half OSTBC. The following example shows that other examples of such codes exist and provide similar properties.

Example 4.7.3 *In this example, we provide a rate $1/2$ OSTBC for $N = 8$ transmit antennas.*

$$G_{848} = \begin{pmatrix} x_1 & x_2 & x_3 & 0 & x_4 & 0 & 0 & 0 \\ -x_2^* & x_1^* & 0 & x_3 & 0 & x_4 & 0 & 0 \\ x_3^* & 0 & -x_1^* & x_2 & 0 & 0 & x_4 & 0 \\ 0 & x_3^* & -x_2^* & -x_1 & 0 & 0 & 0 & x_4 \\ x_4^* & 0 & 0 & 0 & -x_1^* & x_2 & -x_3 & 0 \\ 0 & x_4^* & 0 & 0 & -x_2^* & -x_1 & 0 & -x_3 \\ 0 & 0 & x_4^* & 0 & -x_3^* & 0 & x_1 & x_2 \\ 0 & 0 & 0 & x_4^* & 0 & -x_3^* & -x_2^* & x_1^* \end{pmatrix}. \quad (4.102)$$

It is easy to check that this is an orthogonal STBC that satisfies (4.76). The rate of this code is the same as that of Example 4.7.2; however, its delay $T = 8$ is less than that of Example 4.7.2.

So far we have shown that rate one complex orthogonal designs exists only for two transmit antennas. A similar result is proved in [83] that shows rate one generalized complex orthogonal designs do not exist for more than two transmit antennas. Also, it is shown in [148] that the rate of a generalized complex orthogonal design cannot exceed $R = 3/4$ for more than two antennas. Rate half, $R = 0.5$, generalized complex orthogonal designs exist for any number of transmit antennas. Examples of generalized complex orthogonal designs and the corresponding OSTBCs at rate $R = 3/4$ for three and four transmit antennas are provided in [136]. Therefore, the highest rate OSTBCs for up to four transmit antennas are known. The code for three transmit antennas is derived from the code for four transmit antennas by removing one of the columns. We provide an example of the rate $R = 3/4$ code for four transmit antennas in the following example.

Example 4.7.4 *In this example, we consider the following OSTBC that utilizes four transmit antennas*

$$G_{434} = \begin{pmatrix} x_1 & x_2 & x_3 & 0 \\ -x_2^* & x_1^* & 0 & x_3 \\ x_3^* & 0 & -x_1^* & x_2 \\ 0 & x_3^* & -x_2^* & -x_1 \end{pmatrix}. \quad (4.103)$$

Three symbols are transmitted through four time slots. Therefore, the rate of the code is $R = 3/4$. For three transmit antennas, by removing one of the columns of (4.103), we have

$$
\mathcal{G}_{334} = \begin{pmatrix} x_1 & x_2 & x_3 \\ -x_2^* & x_1^* & 0 \\ x_3^* & 0 & -x_1^* \\ 0 & x_3^* & -x_2^* \end{pmatrix}. \tag{4.104}
$$

Designing STBCs for rates between $R = 0.5$ and $R = 3/4$ and for more than four transmit antennas or showing that they do not exist is an interesting problem. Some sporadic examples are provided in the literature; however, it is an open problem in many cases. In what follows we provide some of the existing examples.

Example 4.7.5 *For $N = 5$ transmit antennas, an OSTBC is provided in [82].*

$$
\mathcal{G} = \begin{pmatrix}
x_1 & x_2^* & x_3^* & x_4^* & 0 \\
x_2 & -x_1^* & 0 & 0 & x_5^* \\
x_3 & 0 & -x_1^* & 0 & -x_6^* \\
0 & x_3 & -x_2 & 0 & x_7 \\
x_4 & 0 & 0 & -x_1^* & x_8^* \\
0 & -x_4 & 0 & x_2 & -x_9 \\
0 & 0 & -x_4 & x_3 & x_{10} \\
x_5 & 0 & -x_7^* & -x_9^* & -x_2^* \\
0 & x_5 & -x_6 & x_8 & x_1 \\
x_6 & -x_7^* & 0 & -x_{10}^* & x_3^* \\
x_7 & x_6^* & x_5^* & 0 & 0 \\
x_8 & x_9^* & -x_{10}^* & 0 & -x_4^* \\
x_9 & -x_8^* & 0 & x_5^* & 0 \\
x_{10} & 0 & x_8^* & x_6^* & 0 \\
0 & -x_{10} & -x_9 & x_7 & 0
\end{pmatrix}. \tag{4.105}
$$

This is an $R = 2/3$ example to transmit $K = 10$ symbols in $T = 15$ time slots.

Example 4.7.6 *An example of an OSTBC for $N = 6$ transmit antennas is provided in [127].*

$$
\mathcal{G} = \begin{pmatrix}
x_1 & x_2 & x_3 & 0 & x_7 & 0 \\
-x_2^* & x_1^* & 0 & x_4^* & 0 & x_{11}^* \\
-x_3^* & 0 & x_1^* & x_5^* & 0 & x_{12}^* \\
0 & -x_3^* & x_2^* & x_6^* & 0 & x_{13}^* \\
0 & -x_4 & -x_5 & x_1 & x_8 & 0 \\
x_4 & 0 & -x_6 & x_2 & x_9 & 0 \\
x_5 & x_6 & 0 & x_3 & x_{10} & 0 \\
-x_6^* & x_5^* & -x_4^* & 0 & 0 & x_{14}^* \\
-x_7^* & 0 & 0 & -x_8^* & x_1^* & x_{15}^* \\
0 & -x_7^* & 0 & -x_9^* & x_2^* & x_{16}^* \\
0 & 0 & -x_7^* & -x_{10}^* & x_3^* & x_{17}^* \\
-x_9^* & x_8^* & 0 & 0 & x_4^* & x_{18}^* \\
-x_{10}^* & 0 & x_8^* & 0 & x_5^* & x_{19}^* \\
0 & -x_{10}^* & x_9^* & 0 & x_6^* & x_{20}^* \\
x_8 & x_9 & x_{10} & -x_7 & 0 & 0 \\
0 & -x_{11} & -x_{12} & 0 & -x_{15} & x_1 \\
x_{11} & 0 & -x_{13} & 0 & -x_{16} & x_2 \\
x_{12} & x_{13} & 0 & 0 & -x_{17} & x_3 \\
0 & 0 & x_{14} & -x_{11} & -x_{18} & x_4 \\
0 & -x_{14} & 0 & -x_{12} & -x_{19} & x_5 \\
x_{14} & 0 & 0 & -x_{13} & -x_{20} & x_6 \\
x_{15} & x_{16} & x_{17} & 0 & 0 & x_7 \\
0 & -x_{18} & -x_{19} & x_{15} & 0 & x_8 \\
x_{18} & 0 & -x_{20} & x_{16} & 0 & x_9 \\
x_{19} & x_{20} & 0 & x_{17} & 0 & x_{10} \\
-x_{13}^* & x_{12}^* & -x_{11}^* & -x_{14}^* & 0 & 0 \\
-x_{16}^* & x_{15}^* & 0 & x_{18}^* & -x_{11}^* & 0 \\
-x_{17}^* & 0 & x_{15}^* & x_{19}^* & -x_{12}^* & 0 \\
0 & -x_{17}^* & x_{16}^* & x_{20}^* & -x_{13}^* & 0 \\
x_{20}^* & -x_{19}^* & x_{18}^* & 0 & x_{14}^* & 0
\end{pmatrix}. \tag{4.106}
$$

For this example, $K = 20$ and $T = 30$ which results in a STBC with rate $R = 2/3$.

Example 4.7.7 *The following rate $R = 5/8$ OSTBC for $N = 7$ transmit antennas is provided in [128].*

$$\mathcal{G} = \begin{pmatrix}
x_1 & x_2 & x_3 & 0 & x_7 & 0 & x_{21} \\
-x_2^* & x_1^* & 0 & x_4^* & 0 & x_{11}^* & 0 \\
-x_3^* & 0 & x_1^* & x_5^* & 0 & x_{12}^* & 0 \\
0 & -x_3^* & x_2^* & x_6^* & 0 & x_{13}^* & 0 \\
0 & -x_4 & -x_5 & x_1 & x_8 & 0 & x_{22} \\
x_4 & 0 & -x_6 & x_2 & x_9 & 0 & x_{23} \\
x_5 & x_6 & 0 & x_3 & x_{10} & 0 & x_{24} \\
-x_6^* & x_5^* & -x_4^* & 0 & 0 & x_{14}^* & 0 \\
-x_7^* & 0 & 0 & -x_8^* & x_1^* & x_{15}^* & 0 \\
0 & -x_7^* & 0 & -x_9^* & x_2^* & x_{16}^* & 0 \\
0 & 0 & -x_7^* & -x_{10}^* & x_3^* & x_{17}^* & 0 \\
-x_9^* & x_8^* & 0 & 0 & x_4^* & x_{18}^* & 0 \\
-x_{10}^* & 0 & x_8^* & 0 & x_5^* & x_{19}^* & 0 \\
0 & -x_{10}^* & x_9^* & 0 & x_6^* & x_{20}^* & 0 \\
x_8 & x_9 & x_{10} & -x_7 & 0 & 0 & x_{25} \\
0 & -x_{11} & -x_{12} & 0 & -x_{15} & x_1 & x_{26} \\
x_{11} & 0 & -x_{13} & 0 & -x_{16} & x_2 & x_{27} \\
x_{12} & x_{13} & 0 & 0 & -x_{17} & x_3 & x_{28} \\
0 & 0 & x_{14} & -x_{11} & -x_{18} & x_4 & x_{29} \\
0 & -x_{14} & 0 & -x_{12} & -x_{19} & x_5 & x_{30} \\
x_{14} & 0 & 0 & -x_{13} & -x_{20} & x_6 & x_{31} \\
x_{15} & x_{16} & x_{17} & 0 & 0 & x_7 & x_{32} \\
0 & -x_{18} & -x_{19} & x_{15} & 0 & x_8 & x_{33} \\
x_{18} & 0 & -x_{20} & x_{16} & 0 & x_9 & x_{34} \\
x_{19} & x_{20} & 0 & x_{17} & 0 & x_{10} & x_{35} \\
-x_{13}^* & x_{12}^* & -x_{11}^* & -x_{14}^* & 0 & 0 & 0 \\
-x_{16}^* & x_{15}^* & 0 & x_{18}^* & -x_{11}^* & 0 & 0 \\
-x_{17}^* & 0 & x_{15}^* & x_{19}^* & -x_{12}^* & 0 & 0 \\
0 & -x_{17}^* & x_{16}^* & x_{20}^* & -x_{13}^* & 0 & 0 \\
x_{20}^* & -x_{19}^* & x_{18}^* & 0 & x_{14}^* & 0 & 0 \\
-x_{21}^* & 0 & 0 & -x_{22}^* & 0 & -x_{26}^* & x_1^* \\
0 & -x_{21}^* & 0 & -x_{23}^* & 0 & -x_{27}^* & x_2^* \\
0 & 0 & -x_{21}^* & -x_{24}^* & 0 & -x_{28}^* & x_3^* \\
-x_{23}^* & x_{22}^* & 0 & 0 & 0 & -x_{29}^* & x_4^* \\
-x_{24}^* & 0 & x_{22}^* & 0 & 0 & -x_{30}^* & x_5^* \\
0 & -x_{24}^* & x_{23}^* & 0 & 0 & -x_{31}^* & x_6^* \\
0 & 0 & 0 & x_{25}^* & -x_{21}^* & -x_{32}^* & x_7^* \\
-x_{25}^* & 0 & 0 & 0 & -x_{22}^* & -x_{33}^* & x_8^* \\
0 & -x_{25}^* & 0 & 0 & -x_{23}^* & -x_{34}^* & x_9^* \\
0 & 0 & -x_{25}^* & 0 & -x_{24}^* & -x_{35}^* & x_{10}^* \\
-x_{27}^* & x_{26}^* & 0 & x_{29}^* & 0 & 0 & x_{11}^* \\
-x_{28}^* & 0 & x_{26}^* & x_{30}^* & 0 & 0 & x_{12}^* \\
0 & -x_{28}^* & x_{27}^* & x_{31}^* & 0 & 0 & x_{13}^* \\
-x_{31}^* & x_{30}^* & -x_{29}^* & 0 & 0 & 0 & x_{14}^* \\
-x_{32}^* & 0 & 0 & -x_{33}^* & x_{26}^* & 0 & x_{15}^* \\
0 & -x_{32}^* & 0 & -x_{34}^* & x_{27}^* & 0 & x_{16}^* \\
0 & 0 & -x_{32}^* & -x_{35}^* & x_{28}^* & 0 & x_{17}^* \\
-x_{34}^* & x_{33}^* & 0 & 0 & x_{29}^* & 0 & x_{18}^* \\
-x_{35}^* & 0 & x_{33}^* & 0 & x_{30}^* & 0 & x_{19}^* \\
0 & -x_{35}^* & x_{34}^* & 0 & x_{31}^* & 0 & x_{20}^* \\
x_{22} & x_{23} & x_{24} & -x_{21} & -x_{25} & 0 & 0 \\
x_{26} & x_{27} & x_{28} & 0 & x_{32} & -x_{21} & 0 \\
0 & -x_{29} & -x_{30} & x_{26} & x_{33} & -x_{22} & 0 \\
x_{29} & 0 & -x_{31} & x_{27} & x_{34} & -x_{23} & 0 \\
x_{30} & x_{31} & 0 & x_{28} & x_{35} & -x_{24} & 0 \\
-x_{33} & -x_{34} & -x_{35} & x_{32} & 0 & x_{25} & 0
\end{pmatrix}. \qquad (4.107)$$

For this example, $K = 35$ and $T = 56$.

A rate $R = (N_0 + 1)/(2N_0)$ OSTBC for $N \leq 18$ transmit antennas and $N = 2N_0$ or $N = 2N_0 - 1$ is provided in [128].

4.7.1 Diagonally orthogonal STBCs

Our definition of orthogonal space-time block codes is based on (4.76) and Definition 4.7.1. If the right-hand side of (4.76) is a diagonal matrix with the nth diagonal element equal to linear combinations of $|x_k|^2$, we still get the nice properties of an orthogonal STBC. Such a code provides full diversity and a simple ML decoding algorithm. In the original STBC paper [136], such a code was called "linear processing orthogonal design." Here, we call it a diagonally orthogonal STBC and define it as follows:

Definition 4.7.2 *A diagonally orthogonal space-time block code is a code with a $T \times N$ generator matrix $\mathcal{G}$ such that the entries of $\mathcal{G}$ are linear combinations of the indeterminate variables $x_1, x_2, \ldots, x_K$ and their conjugates, with*

$$\mathcal{G}^H \cdot \mathcal{G} = D, \tag{4.108}$$

where D is an $N \times N$ diagonal matrix and $D_{n,n} = l_{n,1}|x_1|^2 + l_{n,2}|x_2|^2 + \cdots + l_{n,K}|x_K|^2$.

The encoding and decoding algorithms for diagonally orthogonal STBCs are similar to those of orthogonal STBCs. We provide an example from [127].

Example 4.7.8 *For $N = 5$ transmit antennas, the following diagonally orthogonal STBC is provided in [127].*

$$\mathcal{G} = \begin{pmatrix} x_1 & x_2 & x_3 & 0 & x_4 \\ -x_2^* & x_1^* & 0 & x_3 & x_5 \\ x_3^* & 0 & -x_1^* & x_2 & x_6 \\ 0 & x_3^* & -x_2^* & -x_1 & x_7 \\ x_4^* & 0 & 0 & -x_7^* & -x_1^* \\ 0 & x_4^* & 0 & x_6^* & -x_2^* \\ 0 & 0 & x_4^* & x_5^* & -x_3^* \\ 0 & -x_5^* & x_6^* & 0 & x_1 \\ x_5^* & 0 & x_7^* & 0 & x_2 \\ -x_6^* & -x_7^* & 0 & 0 & x_3 \\ x_7 & -x_6 & -x_5 & x_4 & 0 \end{pmatrix}. \tag{4.109}$$

For this example, $K = 7$ symbols are transmitted over $T = 11$ time slots. Therefore, the rate of this STBC is $R = 7/11$. The 5×5 diagonal matrix $D = \mathcal{G}^H \cdot \mathcal{G}$ contains *the following diagonal elements*

$$D_{1,1} = D_{2,2} = D_{3,3} = D_{4,4} = \sum_{k=1}^{K=7} |x_k|^2,$$

$$D_{5,5} = 2\sum_{k=1}^{K=3} |x_k|^2 + \sum_{k=4}^{K=7} |x_k|^2. \qquad (4.110)$$

Note that the rate of the code in Example 4.7.8 is lower than the rate of the corresponding code in Example 4.7.5.

4.8 Pseudo-orthogonal space-time block codes

An orthogonal design is defined for any arbitrary indeterminate variables $x_1, x_2, \ldots, x_N$. The theory of orthogonal designs and its generalization provide designs that are orthogonal for any real or complex values of $x_1, x_2, \ldots, x_N$. However, in practice, in a communication system, variables $x_1, x_2, \ldots, x_N$ are members of a constellation that has a finite number of signals instead of the real or complex numbers with infinite number of possibilities. Therefore, it is useful to study a class of STBCs that provide all nice properties of orthogonal STBCs only for a finite set of constellation symbols. In fact, while a complex orthogonal design does not exist for more than two transmit antennas, if we restrict the symbols to real numbers $I\!R \subset \mathbb{C}$, real orthogonal designs exist for four and eight transmit antennas. These real orthogonal designs are provided in (4.38) and (4.40). We study the possibility of such designs by defining pseudo-orthogonal designs over a finite subset of numbers. The definition of a pseudo-orthogonal design is identical to that of an orthogonal design with the addition of the condition that the indeterminate variables $x_1, x_2, \ldots, x_N$ are from finite sets of numbers. These finite sets can be any practical constellation, for example PSK or QAM. Note that while the definition of pseudo-orthogonal designs results in separate decoding, unlike the case of orthogonal designs, it does not guarantee maximum diversity [72]. In fact, using Hadamard matrices, it can be shown that pseudo-orthogonal designs exist for binary sets, for example $\{-1, 1\}$, for $N = 2^c$ transmit antennas [72]. However, such designs do not provide full diversity, as we show in the sequel. We define a pseudo-orthogonal STBC as a pseudo-orthogonal design that provides full diversity. This is different from the approach in [72] where the diversity is considered separately.

Definition 4.8.1 *Let $S_n, n = 1, 2, \ldots, N$, be an arbitrary subset of complex numbers $\mathbb{C}$. A pseudo-orthogonal STBC of order N is an N-square matrix $\mathcal{G}_N$ that provides full diversity with polynomial entries of degree one in N variables $x_1 \in S_1, x_2 \in S_2, \ldots, x_N \in S_N$, satisfying the equation:*

$$\mathcal{G}_N^H \cdot \mathcal{G}_N = (|x_1|^2 + |x_2|^2 + \cdots + |x_N|^2)I_N. \qquad (4.111)$$

Note that similar to the case of orthogonal designs, a generalized pseudo-orthogonal STBC can be defined as a $T \times N$ matrix $\mathcal{G}$ with entries that are linear combinations of the indeterminate variables $x_1 \in S_1, x_2 \in S_2, \ldots, x_K \in S_K$ and their conjugates such that

$$\mathcal{G}^H \cdot \mathcal{G} = \kappa (|x_1|^2 + |x_2|^2 + \cdots + |x_K|^2) I_N, \qquad (4.112)$$

where κ is a constant and S_k, $k = 1, 2, \ldots, K$, is an arbitrary subset of $\mathcal{C}$.

The main motivation behind a pseudo-orthogonal STBC is the following. The restriction of x_n assuming specific values from a set $S \in \mathcal{C}$ may relax some of the conditions under which an orthogonal design can exist. For example, a QPSK constellation consists of four symbols, $1, j, -1, -j$. The fact that there is no matrix $\mathcal{G}$ satisfying (4.111) for particular values of N and arbitrary indeterminate variables $x_1, x_2, \ldots, x_N$ might not mean that such a matrix does not exist when $x_1, x_2, \ldots, x_N \in \{1, j, -1, -j\}$. If such a matrix exists, we can use it to design a full-rate pseudo-orthogonal STBC to transmit 2 bits/(s Hz) and provide separate decoding and full diversity.

In [72, 92], it has been shown that if S_n is a fixed subset of $\mathbb{R}$ or $\mathcal{C}$, for any $n = 1, 2, \ldots, N$, one cannot design new interesting pseudo-orthogonal STBCs. In other words, if we limit the definition of pseudo-orthogonal designs to sets $S_1 = S_2 = \ldots = S_N$, as in [72, 92], new pseudo-orthogonal designs do not exist. As a result, the new class of codes in [72, 92] do not add new practical codes that did not exist before. For real constellations, we have the following theorem [72].

Theorem 4.8.1 *If all symbols are taken from an arbitrary real subset S having at least three non-zero real elements, the three standard square orthogonal designs for $N = 2, 4, 8$ are the only real pseudo-orthogonal designs.*

For subsets with two real elements, the above theorem does not hold. In fact, Hadamard matrices can be utilized to achieve generator matrices that satisfy (4.111).

Example 4.8.1 *An $N \times N$ Hadamard matrix is a binary matrix A_N with elements $\{-1, 1\}$ such that*

$$A_N \cdot A_N^T = A_N^T \cdot A_N = N I_N. \qquad (4.113)$$

Hadamard matrices have been constructed for $N = 2^c$, where c is any positive integer [49]. An example of a 2×2 Hadamard matrix is

$$A_2 = \begin{pmatrix} 1 & 1 \\ 1 & -1 \end{pmatrix}. \qquad (4.114)$$

An $N \times N$ Hadamard matrix can be constructed from an $N/2 \times N/2$ Hadamard matrix by

$$A_N = \begin{pmatrix} A_{N/2} & A_{N/2} \\ A_{N/2} & -A_{N/2} \end{pmatrix}. \tag{4.115}$$

For $N = 2^c$ transmit antennas, let us define the generator matrix

$$\mathcal{G} = A_N \cdot \begin{pmatrix} x_1 & 0 & \cdots & 0 \\ 0 & x_2 & \cdots & 0 \\ \vdots & \vdots & \ddots & \vdots \\ 0 & 0 & \cdots & x_N \end{pmatrix}. \tag{4.116}$$

Using (4.113), we can show that

$$\mathcal{G}^H \cdot \mathcal{G} = \begin{pmatrix} Nx_1^2 & 0 & \cdots & 0 \\ 0 & Nx_2^2 & \cdots & 0 \\ \vdots & \vdots & \ddots & \vdots \\ 0 & 0 & \cdots & Nx_N^2 \end{pmatrix}. \tag{4.117}$$

If $x_1, x_2, \ldots, x_N \in \{-1, 1\}$, we have $|x_1|^2 + |x_2|^2 + \cdots + |x_N|^2 = Nx_n^2$ for any $n = 1, 2, \ldots, N$. Therefore, $\mathcal{G}$ is a pseudo-orthogonal design for $x_1, x_2, \ldots, x_N \in \{-1, 1\}$. In other words, square pseudo-orthogonal designs defined over $\{-1, 1\}$ exist for 16, 32, . . . transmit antennas despite the fact that their orthogonal design counterparts do not exist. Unfortunately, the resulting pseudo-orthogonal STBCs do not provide full diversity [72, 92].

A similar theorem for complex constellations is presented below [92].

Theorem 4.8.2 *If all symbols are taken from an arbitrary subset S having at least three non-zero complex elements that do not lie on a straight line and at least one pair of points lies on either a vertical or horizontal line, a pseudo-orthogonal STBC exists only for $N = 2$ transmit antennas.*

For the proof of the above two theorems, we refer the interested reader to [72, 92].

In fact, the constraint of having all symbols from the same subset S may be restrictive. In what follows, we provide an example of pseudo-orthogonal STBCs for $N = 4$ transmit antennas that provides all nice properties of orthogonal STBCs. In this example, different symbols are from different constellations and therefore Theorems 4.8.1 and 4.8.2 do not apply.

Example 4.8.2 *Let us consider the following STBC for four transmit antennas:*

$$\mathcal{G} = \begin{pmatrix} x_1 & x_2 & x_3 & x_4 \\ -x_2^* & x_1^* & -x_4^* & x_3^* \\ -x_3^* & -x_4^* & x_1^* & x_2^* \\ x_4 & -x_3 & -x_2 & x_1 \end{pmatrix}. \tag{4.118}$$

This is not an orthogonal STBC since

$$\mathcal{G}^H \cdot \mathcal{G} = \begin{pmatrix} a & 0 & 0 & b \\ 0 & a & -b & 0 \\ 0 & -b & a & 0 \\ b & 0 & 0 & a \end{pmatrix}, \tag{4.119}$$

where $a = \sum_{k=1}^{4} |x_k|^2$ and $b = 2\Re(x_1 x_4^* - x_2 x_3^*)$. We assume that $x_k \in S_k$, $k = 1, 2, 3, 4$ and S_k is a subset of complex numbers $\mathbb{C}$. The off-diagonal terms in (4.119) contain the real part of the product of a symbol and the conjugate of another symbol. Let us consider one such term as $\Re(c^* d)$. If c is a real number and d is an imaginary number, or if d is a real number and c is an imaginary number, $\Re(c^* d) = 0$. We use this observation to make the off-diagonal terms in (4.119) equal to zero. For the code in (4.118), choosing a real constellation for S_1 and S_2 while selecting an imaginary constellation for S_3 and S_4, or vice versa, results in

$$\mathcal{G}^H \cdot \mathcal{G} = \sum_{k=1}^{4} |x_k|^2 I_4. \tag{4.120}$$

For example, one can pick a PAM constellation for S_1 and S_2 and a $\pi/2$ rotated version of the same PAM constellation for S_3 and S_4. To transmit 2 bits/(s Hz), this example results in $S_1 = S_2 = \{-3, -1, 1, 3\}$ and $S_3 = S_4 = \{-3j, -j, j, 3j\}$. Note that such a pseudo-orthogonal STBC provides full diversity, rate one and separate decoding. Unfortunately, while the form of this code is new, its structure is very similar to that of the real orthogonal STBC for four transmit antennas and do not provide new performances.

The structure in the above example and similar structures are studied in depth in Chapter 5.

4.9 Performance analysis

In this section, we study the performance of the orthogonal STBCs. First, we provide an analysis of the received SNR to compare the case of multiple transmit antennas with the case of multiple receive antennas. Then, we provide exact formulas for pairwise error probability of orthogonal STBCs.

Let us consider two systems with different arrangements of multiple antennas. The first case consists of N transmit antennas and one receive antenna using OSTBCs for transmission. The channel in Case 1 can be considered as a multiple-input single-output (MISO) channel with path gains $\alpha_1, \alpha_2, \ldots, \alpha_N$. The second case contains one transmit antenna and $M = N$ receive antennas with the same set of path gains $\alpha_1, \alpha_2, \ldots, \alpha_N$. For this single-input multiple-output (SIMO) channel, we use MRC for ML decoding as discussed in Section 4.3. We have chosen the same set of path gains to have a fair comparison between the two cases. Also, let us assume an equal energy constellation like PSK for the sake of simplicity. Considering the same noise variance $N_0/2$ per dimension and the same average power of the transmitted symbols, E_s, the average receive SNR is the same for both cases. Note that assuming the same power for transmitted symbols in these two cases, the total transmitted power of the first case is N times that of the second case. Therefore, to have the same performance, the receive SNR of Case 1 should be N times that of Case 2. Let us consider $N = M = 2$ to validate the above claim. In Case 1, the decoder minimizes $|\tilde{s}_k - s_k|^2$ for decoding s_k, where $k = 1, 2$. Note that $\tilde{s}_1$ is

$$\tilde{s}_1 = r_1\alpha_1^* + r_2^*\alpha_2 = \sum_{n=1}^{2} |\alpha_n|^2 s_1 + N_1, \qquad (4.121)$$

where N_1 is a zero-mean complex Gaussian random variable with variance $(N_0/2) \sum_{n=1}^{2} |\alpha_n|^2$ per real dimension. The power of the signal at the receiver is $E_s[\sum_{n=1}^{2} |\alpha_n|^2]^2$ if the power of the transmitted signal is E_s. Therefore, the receive SNR for the first symbol is $\gamma = (E_s/N_0) \sum_{n=1}^{2} |\alpha_n|^2$. The same signal and noise powers are achieved for the second symbol. Note that since we transmit from two antennas, the total average transmit power is $2E_s$.

For the second case using maximum ratio combining, the decoder minimizes $|\tilde{s} - s|^2$ for decoding the transmitted symbol s, where

$$\tilde{s} = \sum_{m=1}^{2} |\alpha_m|^2 s + N'. \qquad (4.122)$$

In this case, N' is a zero-mean complex Gaussian random variable with variance $(N_0/2) \sum_{m=1}^{2} |\alpha_m|^2$ per real dimension. Also, the power of signal at the receiver is $E_s[\sum_{m=1}^{2} |\alpha_m|^2]^2$. Note that in this case, the average transmit power is E_s, which is half of that of the first case. Comparing Case 1 and Case 2, both the power of the signal and noise are the same. Therefore, orthogonal STBCs provide exactly the same performance as receive maximum ratio combining if the transmission power is adjusted correctly. This is another indication of the fact that orthogonal STBCs are optimal in terms of providing the maximum possible diversity.

As we mentioned before, the transmission powers should be the same for a fair comparison. Therefore, the total transmission power of the orthogonal STBCs in Case 1 should be N times more than those of the second case to achieve the same performance. This is consistent with similar analysis for the capacity of MISO and SIMO channels in Chapter 2.

The above analysis shows that it suffices to calculate the performance of a MISO system and derive the performance of the corresponding SIMO system from it or vice versa. In what follows we derive the exact formulas for the pairwise error probability (PEP) of orthogonal STBCs. As in Section 3.2, we define the PEP as the probability of transmitting $\mathbf{C}^1$ and detecting it as $\mathbf{C}^2$, that is $P(\mathbf{C}^1 \rightarrow \mathbf{C}^2)$, when there is no other codeword. In general when there is more than two codewords, the probability of error is upper bounded by the union bound over all possible erroneous codewords. There are different ways to calculate the PEP as explained in [115, 131, 145] and the references therein. These different schemes result in similar results. Therefore, we use one specific technique to calculate PEP from [113]. We rewrite the conditional PEP in (3.9) as

$$P(\mathbf{C}^1 \rightarrow \mathbf{C}^2 | \mathbf{H}) = Q\left(\sqrt{\frac{\gamma}{2} \text{Tr}[\mathbf{H}^H \cdot (\mathbf{C}^2 - \mathbf{C}^1)^H \cdot (\mathbf{C}^2 - \mathbf{C}^1) \cdot \mathbf{H}]} \right). \quad (4.123)$$

Using the definition of an orthogonal design from (4.76), the conditional PEP is

$$P(\mathbf{C}^1 \rightarrow \mathbf{C}^2 | \mathbf{H}) = Q\left(\sqrt{\frac{\gamma}{2} \kappa \sum_{k=1}^{K} |s_k^2 - s_k^1|^2 \, \text{Tr}[\mathbf{H}^H \cdot \mathbf{H}]} \right)$$

$$= Q\left(\sqrt{\kappa \frac{\gamma}{2} \sum_{k=1}^{K} |s_k^2 - s_k^1|^2 \sum_{n=1}^{N} \sum_{m=1}^{M} |\alpha_{n,m}|^2} \right). \quad (4.124)$$

Let us denote the Euclidean distance between the transmitted and detected symbols by

$$d_E = \sqrt{\sum_{k=1}^{K} |s_k^2 - s_k^1|^2}. \quad (4.125)$$

Note that the Euclidean distance in (4.125) is independent of the path gains. The conditional PEP in terms of the Euclidean distance becomes

$$P(\mathbf{C}^1 \rightarrow \mathbf{C}^2 | \mathbf{H}) = Q\left(\sqrt{\kappa \frac{\gamma}{2} d_E^2 \sum_{n=1}^{N} \sum_{m=1}^{M} |\alpha_{n,m}|^2} \right). \quad (4.126)$$

To calculate PEP, one needs to integrate (4.126) weighted by the density of path gains. To make it easier to calculate such an integral, one method is to employ the following formula for the Q-function due to Craig [29]:

$$Q(x) = \frac{1}{\pi} \int_0^{\pi/2} \exp\left(\frac{-x^2}{2\sin^2\theta}\right) d\theta. \qquad (4.127)$$

Replacing (4.127) in (4.126) results in

$$P(\mathbf{C}^1 \to \mathbf{C}^2|\mathbf{H}) = \frac{1}{\pi} \int_0^{\pi/2} \exp\left(\frac{-\kappa\gamma d_E^2 \sum\limits_{n=1}^{N}\sum\limits_{m=1}^{M} |\alpha_{n,m}|^2}{4\sin^2\theta}\right) d\theta$$

$$= \frac{1}{\pi} \int_0^{\pi/2} \prod_{m=1}^{M}\prod_{n=1}^{N} \exp\left(\frac{-\kappa\gamma d_E^2|\alpha_{n,m}|^2}{4\sin^2\theta}\right) d\theta. \qquad (4.128)$$

Note that assuming a Rayleigh fading channel, as we did in Chapter 2, results in a chi-square distribution with two degrees of freedom or equivalently an exponential distribution with unit mean for $|\alpha_{n,m}|^2$. Let us assume that the path gains are independent from each other. Then, integrating (4.128) over the distribution of the path gains is the same as the product of MN equal integrals as follows:

$$P(\mathbf{C}^1 \to \mathbf{C}^2) = \frac{1}{\pi} \int_0^{\pi/2} \left[\int_0^{\infty} \exp\left(\frac{-\kappa\gamma d_E^2 x}{4\sin^2\theta}\right) f_\chi(x)\,dx\right]^{MN} d\theta, \qquad (4.129)$$

where $f_\chi(x) = e^{-x}$, $x > 0$ is the pdf of $|\alpha_{n,m}|^2$. We recall that the moment generating function (MGF) of a random variable X is defined as

$$M(u) = E[e^{uX}]. \qquad (4.130)$$

Therefore, the MGF of an exponential distribution for $u < 1$ is

$$M_\chi(u) = E[e^{uX}] = \int_0^{\infty} e^{ux} f_\chi(x)\,dx = \int_0^{\infty} e^{ux} e^{-x}\,dx = \frac{1}{1-u}. \qquad (4.131)$$

One can rewrite the PEP in (4.129) using the above MGF as follows:

$$P(\mathbf{C}^1 \to \mathbf{C}^2) = \frac{1}{\pi} \int_0^{\pi/2} \left[\frac{\sin^2\theta}{\sin^2\theta + \kappa\frac{\gamma}{4}d_E^2}\right]^{MN} d\theta. \qquad (4.132)$$

Therefore, the PEP depends on the Euclidean distance between the transmitted and detected symbols, the product of the number of transmit and receive antennas and the received SNR. In fact, the integral in (4.132) can be evaluated in closed form

[114]. As a result, the general form of the PEP for orthogonal STBCs is

$$P(\mathbf{C}^1 \rightarrow \mathbf{C}^2) = \frac{1}{2}\left\{1 - \sqrt{\frac{a}{1+a}} \sum_{i=0}^{MN-1} \binom{2i}{i} \left[\frac{1}{4(1+a)}\right]^i\right\}, \quad (4.133)$$

where $a = \kappa \frac{\gamma}{4} d_E^2$. For a given orthogonal STBC, κ is provided from the structure of the code in Definition 4.7.1. For a given pair of codewords, the Euclidean distance d_E is calculated by (4.125) from the corresponding constellation points. A higher Euclidean distance d_E results in a lower PEP. Therefore, the worst case scenario for a given constellation is the pair of symbol sets that are the closest pair among all possible pair of symbols. We consider the following example to elaborate the calculations.

Example 4.9.1 *For the Alamouti code, we transmit $K = 2$ symbols in $T = 2$ time slots from $N = 2$ transmit antennas and have $\kappa = 1$. Let us assume a BPSK constellation. Following the normalization in Section 3.2, the average symbol transmission power from each antenna is $E_S = \frac{1}{N} = \frac{1}{2}$. In other words, the BPSK constellation points are $\frac{-1}{\sqrt{2}}$ and $\frac{1}{\sqrt{2}}$. When corresponding symbols in the transmitted and received codewords are different from each other, the Euclidean distance is $d_E = \sqrt{2+2} = 2$. Therefore, the PEP is*

$$P(\mathbf{C}^1 \rightarrow \mathbf{C}^2) = \frac{1}{2}\left\{1 - \sqrt{\frac{\gamma}{1+\gamma}} \sum_{i=0}^{2M-1} \binom{2i}{i} \left[\frac{1}{4(1+\gamma)}\right]^i\right\}. \quad (4.134)$$

For $M = 1$ receive antenna, the PEP is

$$P(\mathbf{C}^1 \rightarrow \mathbf{C}^2) = \frac{1}{2}\left\{1 - \sqrt{\frac{\gamma}{1+\gamma}} \left(1 + \frac{1}{2(1+\gamma)}\right)\right\}. \quad (4.135)$$

When only one of the corresponding symbols in $\mathbf{C}^1$ and $\mathbf{C}^2$ are different, for example $s_1^1 = s_1^2$ and $s_2^1 = -s_2^2 = 1/\sqrt{2}$, the Euclidean distance is $d_E = \sqrt{0+2} = \sqrt{2}$. Therefore, the PEP is

$$P(\mathbf{C}^1 \rightarrow \mathbf{C}^2) = \frac{1}{2}\left\{1 - \sqrt{\frac{\gamma/2}{1+\gamma/2}} \sum_{i=0}^{2M-1} \binom{2i}{i} \left[\frac{1}{4(1+\gamma/2)}\right]^i\right\}. \quad (4.136)$$

For $M = 1$ receive antenna, the PEP is

$$P(\mathbf{C}^1 \rightarrow \mathbf{C}^2) = \frac{1}{2}\left\{1 - \sqrt{\frac{\gamma}{2+\gamma}} \left(1 + \frac{1}{(2+\gamma)}\right)\right\}. \quad (4.137)$$

The above analysis provides the PEP for a given pair of codewords. Note that we have defined the PEP as the probability of transmitting $\mathbf{C}^1$ and detecting it as $\mathbf{C}^2$, when there is no other codeword. As we discussed before, when the code includes more than two codewords, the probability of error is upper bounded by a union bound. Also, to calculate the average symbol error rate or bit error rate, one needs to consider all possible symbol or bit errors for different PEPs and weight them by the probability of such errors. Due to the use of the union bound, such a formula provides an approximation of the error rate. Inherently, such an approximation is a result of calculating PEP for the case that only codewords $\mathbf{C}^1$ and $\mathbf{C}^2$ exist. In a STBC with more than two codewords, the exact probability of transmitting $\mathbf{C}^1$ and receiving $\mathbf{C}^2$ is slightly different from the above PEP formulas because of the existence of other codewords.

In what follows, we consider a different method of analyzing the error rates for orthogonal STBCs. Consider a $T \times N$ generalized complex orthogonal design $\mathcal{G}$ as defined in Definition 4.7.1. Let us consider a normalized version of the generalized complex orthogonal design, $\mathcal{G}' = \sqrt{\zeta}\,\mathcal{G}$, such that for the codeword $\mathbf{C} = \mathcal{G}'(s_1, s_2, \ldots, s_K)$ we have $||\mathbf{C}||_F^2 = T$. Such a normalization only adjusts the constant κ and the constellation power to make the average sum of the transmitted power from all antennas at each time slot equal to one. This way, our previous normalization assumption of $N_0 = \frac{1}{\gamma}$ is still valid, where the variance of noise is $N_0/2$ and γ is the SNR at each receive antenna.

The general ML decoding cost function in (4.81) shows how different symbols can be decoded separately. For the sake of simplicity, let us assume that all the constellation symbols have equal energies, for example using PSK. Then, the function $f_k(s_k)$, $k = 1, 2, \ldots, K$ that needs to be minimized for the ML decoding of s_k can be calculated as the linear combination of received signals. The formulas for $f_k(s_k)$, $k = 1, 2, \ldots, K$ depend on the structure of the corresponding STBC as we have shown before. These formulas can be utilized to calculate the average symbol error probability. We concentrate on the separate decoding formulas similar to those derived from (4.93). For example, this is possible when x_k or x_k^* appears only once in every row of $\mathcal{G}$, like all examples that we have considered in this chapter so far.

Let us define the $T \times 1$ vector $\mathbf{r}_m, m = 1, 2, \ldots, M$, as the signal vector received at the mth receive antenna. In fact, $\mathbf{r}_m$ is the mth column of $\mathbf{r}$ in (2.5). If only the mth receive antenna exists, for each index k, the function $f_k(s_k)$ can be written as the vector product of $[(\mathbf{r}_m')^k]^T$ and $(\Omega_{mk})^H$, where $(\mathbf{r}_m')^k$ is a $T \times 1$ vector derived from $\mathbf{r}_m$ by replacing some of its elements with their conjugates and Ω_{mk} is a $1 \times T$ vector constructed from the path gains for the kth symbol. In general, $(\mathbf{r}_m')^k$ and Ω_{mk} are defined as follows. For the kth symbol, the tth element of the $T \times 1$ vector $(\mathbf{r}_m')^k$ is defined by

$$(\mathbf{r}'_m)^k(t) = \begin{cases} \mathbf{r}^*_m(t) & \text{if } x^*_k \text{ or } -x^*_k \text{ exists in the } t\text{th row of } \mathcal{G} \\ \mathbf{r}_m(t) & \text{otherwise} \end{cases}. \quad (4.138)$$

For the kth symbol, the tth element of the $1 \times T$ vector Ω_{mk} is defined by

$$\Omega_{mk}(t) = \begin{cases} \alpha_{n,m} & \text{if } \mathcal{G}_{t,n} = x_k \\ \alpha^*_{n,m} & \text{if } \mathcal{G}_{t,n} = x^*_k \\ -\alpha_{n,m} & \text{if } \mathcal{G}_{t,n} = -x_k \\ -\alpha^*_{n,m} & \text{if } \mathcal{G}_{t,n} = -x^*_k \\ 0 & \text{otherwise} \end{cases}. \quad (4.139)$$

A $K \times T$ matrix Ω_m can be constructed by putting Ω_{mk} as its kth row. Note that for the Alamouti code, all rate one real OSTBCs, and all rate $R = 0.5$ OSTBCs presented in Section 4.7, the vector $(\mathbf{r}'_m)^k$ is independent of k and can be denoted by $\mathbf{r}'_m$. This is clear from (4.93) and similar formulas. Also note that the time dimension in (4.86), for the $R = 0.5$ OSTBCs, is $2T$ instead of T as in that construction variable T was used for the time dimension of the corresponding rate one real OSTBC. As an example, $(\mathbf{r}'_m)^k$ for the Alamouti code, as derived in (4.19), is the same for all values of k and we have

$$\mathbf{r}'_m = \begin{pmatrix} r_{1,m} \\ r^*_{2,m} \end{pmatrix}, \quad \Omega_m = \begin{pmatrix} \alpha_{1,m} & \alpha^*_{2,m} \\ \alpha_{2,m} & -\alpha^*_{1,m} \end{pmatrix}. \quad (4.140)$$

Using maximum ratio combining, we have

$$\sum_{m=1}^{M} [(\mathbf{r}'_m)^k]^T \cdot (\Omega_{mk})^H = \sqrt{\zeta}\,\kappa \sum_{m=1}^{M} \sum_{n=1}^{N} |\alpha_{n,m}|^2 s_k + \sum_{m=1}^{M} \mathbf{N}_{mk}, \quad (4.141)$$

where $\mathbf{N}_{mk}$ is an iid zero-mean complex Gaussian random variable with variance equal to $\frac{\kappa}{\gamma} \sum_{n=1}^{N} |\alpha_{n,m}|^2$. Mathematically, the above equation represents an orthogonal STBC as K independent transmission of K separate symbols over equivalent SISO fading channels. The instantaneous SNR of each equivalent SISO channel is equal to $\zeta\kappa \sum_{m=1}^{M} \sum_{n=1}^{N} |\alpha_{n,m}|^2 \gamma$. Therefore, the performance over each of these equivalent channels is the same as the performance of a system transmitting one symbol over $M \times N$ channels using maximum ratio combining for decoding. Also, the probability of error is the same for all equivalent channels and is the same as their average. For different constellations, the probability of symbol error can be calculated for a given SNR, that is,

$$P(\text{symbol error}|\mathbf{H}) = P\left(\text{symbol error}\middle| \text{SNR} = \zeta\kappa \sum_{m=1}^{M} \sum_{n=1}^{N} |\alpha_{n,m}|^2 \gamma\right). \quad (4.142)$$

Then, the average probability of symbol error or the average symbol error rate (SER) can be calculated by computing the expectation of the above formula over the distribution of channel path gains:

$$
\text{SER} = E\left[P\left(\text{symbol error} \,\middle|\, \text{SNR} = \zeta\kappa \sum_{m=1}^{M}\sum_{n=1}^{N} |\alpha_{n,m}|^2\, \gamma \right) \right]. \quad (4.143)
$$

To find the SER of an orthogonal STBC using a specific constellation, we need to calculate the SER of the corresponding modulation over a fading channel. The SER of an L-PSK constellation over a Rayleigh fading channel is calculated as [114]

$$
\begin{aligned}
\text{SER} &= \frac{L-1}{L} - \left(\frac{1}{\pi}\sqrt{\frac{\gamma \sin^2 \frac{\pi}{L}}{1+\gamma \sin^2 \frac{\pi}{L}}} \right) \left\{ \left(\frac{\pi}{2} + \tan^{-1}\beta \right) \sum_{i=0}^{d-1} \frac{\binom{2i}{i}}{\left[4\left(1+\gamma \sin^2 \frac{\pi}{L}\right)\right]^i} \right. \\
&\left. + \sin\left(\tan^{-1}\beta\right) \sum_{i=1}^{d-1}\sum_{j=1}^{i} \frac{T_{ji}}{\left(1+\gamma \sin^2 \frac{\pi}{L}\right)^i} \left[\cos\left(\tan^{-1}\beta\right)\right]^{2(i-j)+1} \right\}, \quad (4.144)
\end{aligned}
$$

where

$$
\beta = \sqrt{\frac{\gamma \sin^2 \frac{\pi}{L}}{1+\gamma \sin^2 \frac{\pi}{L}}} \cot\left(\frac{\pi}{L}\right), \quad (4.145)
$$

$$
T_{ji} = \frac{\binom{2i}{i}}{\binom{2(i-j)}{i-j} 4^j [2(i-j)+1]}, \quad (4.146)
$$

and d is the number of independent paths from transmitter to receiver, or equivalently, the number of maximum ratio combining branches. Therefore, the symbol error rate for an $M \times N$ orthogonal STBC using L-PSK is equal to

$$
\begin{aligned}
\text{SER} &= \frac{L-1}{L} - \left(\frac{1}{\pi}\sqrt{\frac{\zeta\kappa\gamma \sin^2 \frac{\pi}{L}}{1+\zeta\kappa\gamma \sin^2 \frac{\pi}{L}}} \right) \left\{ \left(\frac{\pi}{2} + \tan^{-1}\beta \right) \sum_{i=0}^{MN-1} \frac{\binom{2i}{i}}{\left[4\left(1+\zeta\kappa\gamma \sin^2 \frac{\pi}{L}\right)\right]^i} \right. \\
&\left. + \sin\left(\tan^{-1}\beta\right) \sum_{i=1}^{MN-1}\sum_{j=1}^{i} \frac{T_{ji}}{\left(1+\zeta\kappa\gamma \sin^2 \frac{\pi}{L}\right)^i} \left[\cos\left(\tan^{-1}\beta\right)\right]^{2(i-j)+1} \right\}, \quad (4.147)
\end{aligned}
$$

where

$$
\beta = \sqrt{\frac{\zeta\kappa\gamma \sin^2 \frac{\pi}{L}}{1+\zeta\kappa\gamma \sin^2 \frac{\pi}{L}}} \cot\left(\frac{\pi}{L}\right). \quad (4.148)
$$

Example 4.9.2 *Consider an orthogonal STBC using BPSK constellation. For BPSK, we have*

$$P(symbol\ error|\mathbf{H}) = Q\left(\sqrt{2\zeta\kappa \sum_{m=1}^{M}\sum_{n=1}^{N}|\alpha_{n,m}|^2\,\gamma}\right). \qquad (4.149)$$

Therefore, SER when using BPSK is calculated by

$$SER = E\left[Q\left(\sqrt{2\zeta\kappa \sum_{m=1}^{M}\sum_{n=1}^{N}|\alpha_{n,m}|^2\,\gamma}\right)\right], \qquad (4.150)$$

where expectation is over channel path gains. The above expected value is calculated for a Rayleigh fading channel model in [114] as

$$SER = \frac{1}{2}\left[1 - \sqrt{\frac{\zeta\kappa\gamma}{1+\zeta\kappa\gamma}} \sum_{i=0}^{MN-1}\frac{\binom{2i}{i}}{[4(1+\zeta\kappa\gamma)]^i}\right], \qquad (4.151)$$

where can be calculated from (4.147) when L = 2. For an Alamouti code with one receive antenna, we have N = 2, M = 1, κ = 1, and ζ = 1/2 which results in

$$SER = \frac{1}{2}\left\{1 - \sqrt{\frac{\gamma}{2+\gamma}}\left(1 + \frac{1}{(2+\gamma)}\right)\right\}. \qquad (4.152)$$

This is identical to the SER derived from Monte Carlo simulations (see Problem 8).

4.10 Simulation results

In this section, we provide simulation results for the performance of the OSTBCs given in the previous sections. Figure 4.4 illustrates a block diagram of the STBC encoder. The information source is divided into blocks of Kb bits. Using these Kb bits, the encoder picks K symbols from the constellation. The constellation symbols are replaced in the generator matrix of the OSTBC to generate the codeword. Then, the elements of the tth row of the codeword are transmitted from different antennas at time slot t. We assume a quasi-static flat Rayleigh fading model for the channel. Therefore, the path gains are independent complex Gaussian random variables and fixed during the transmission of one block. The receiver utilizes ML decoding to estimate the transmitted symbols and bits. For a fixed constellation and OSTBC, the transmission bit rate is fixed. Then, we use Monte Carlo simulations to derive the bit error rate (BER) versus the received SNR [135].

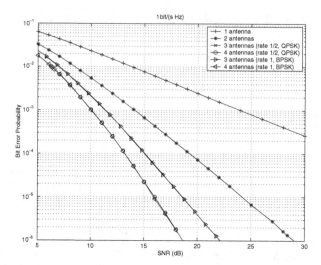

Fig. 4.6.   Bit error probability plotted against SNR for orthogonal STBCs at
1 bit/(s Hz); one receive antenna.

Simulation results are usually reported for a given transmission bit rate and
different number of transmit antennas. To this end, we need to pick the right combination of orthogonal STBC and constellation. If a full rate, $R = 1$, OSTBC exists
for the given number of transmit antennas, we can use the same constellation
that we use for the corresponding one transmit antenna. However, if such a full
rate OSTBC does not exist, we should pick the combination of the orthogonal
STBC with the highest available rate and a constellation that results in the required
transmission bit rate. For example, full rate complex OSTBCs do not exist for
more that two transmit antennas. Therefore, one should use rate 1/2 or 3/4 for
the case of three and four transmit antennas. If a constellation contains 2^b points,
the transmission bit rate of the uncoded system for one transmit antenna or a full
rate OSTBC is b bits/(s Hz). For a STBC with rate R, the transmission bit rate is
Rb bits/(s Hz).

Figure 4.6 provides simulation results for a transmission bit rate of 1 bit/(s Hz)
using one, two, three, and four transmit antennas. The figure shows bit error probability versus the received SNR. For one transmit antenna, a BPSK modulation
scheme is used with no additional coding. The transmission using two transmit
antennas employs the BPSK constellation and the Alamouti code in (4.21). The
QPSK constellation and the code in Example 4.7.1 are utilized for four transmit
antennas. The index assignment for QPSK is depicted in Figure 4.7. For three
transmit antennas, we remove one column of the code for four transmit antennas
as follows:

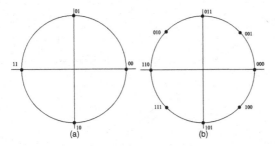

Fig. 4.7. Gray mapping: (a) QPSK; (b) 8-PSK.

$$
\mathcal{G}_{348} = \begin{pmatrix}
x_1 & x_2 & x_3 \\
-x_2 & x_1 & -x_4 \\
-x_3 & x_4 & x_1 \\
-x_4 & -x_3 & x_2 \\
x_1^* & x_2^* & x_3^* \\
-x_2^* & x_1^* & -x_4^* \\
-x_3^* & x_4^* & x_1^* \\
-x_4^* & -x_3^* & x_2^*
\end{pmatrix}.
\tag{4.153}
$$

Since $\mathcal{G}_{448}$ in (4.94) and $\mathcal{G}_{348}$ in (4.153) are rate $1/2$ complex codes, the total transmission bit rate in each case is 1 bit/(s Hz). On the other hand, since BPSK is a real constellation, full rate, $R = 1$, OSTBCs also exist. We include the simulation results of the real OSTBCs for three and four transmit antennas, respectively (4.61) and (4.38), in Figure 4.6 as well. The performance of a rate $1/2$ code using QPSK is identical to that of the corresponding rate one BPSK code. It is seen that at the bit error probability of 10^{-5} the rate $1/2$ QPSK code $\mathcal{G}_{448}$ provides about 7.5 dB gain over the use of a BPSK Alamouti code. If the number of the receive antennas is increased to two, this gain reduces to 3.5 dB as seen in Figure 4.8. The reason is that much of the diversity gain is already achieved using two transmit and two receive antennas.

Bit error probability versus received SNR for transmission of 2 bits/(s Hz) is provided in Figure 4.9. Similar to the previous case, we provide results using two, three, and four transmit antennas in addition to an uncoded system using one transmit antenna. The transmission using two transmit antennas employs the QPSK constellation and the Alamouti code. For three and four transmit antennas, the 16-QAM constellation and the codes $\mathcal{G}_{348}$ and $\mathcal{G}_{448}$ are utilized, respectively. Since $\mathcal{G}_{348}$ and $\mathcal{G}_{448}$ are rate $1/2$ codes, the total transmission rate in each case is 2 bits/(s Hz). We use a Gray mapping for the index assignment of the 16-QAM constellation as depicted in Figure 4.10. A QPSK constellation is more efficient

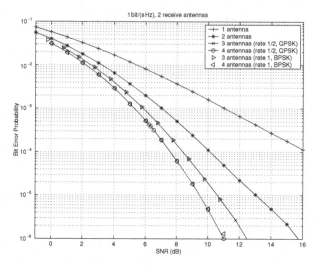

Fig. 4.8. Bit error probability plotted against SNR for orthogonal STBCs at 1 bit/(s Hz); two receive antennas.

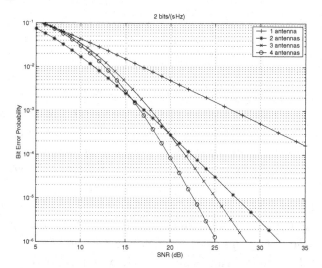

Fig. 4.9. Bit error probability plotted against SNR for orthogonal STBCs at 2 bits/(s Hz); one receive antenna.

than a 16-QAM constellation for the same average power [100]. Therefore, the rate one code for two transmit antennas outperforms the rate half codes for three and four transmit antennas at low SNRs. However, since the diversity gain of the codes for three and four transmit antennas is higher, the slope of their curves is deeper. As a result, the codes for three and four transmit antennas outperform the code for two transmit antennas at high SNRs. It is seen that at the bit error probability of 10^{-5} the rate $1/2$ code $\mathcal{G}_{448}$, utilizing 16-QAM, provides about 5 dB gain over the use of a QPSK Alamouti code. Similar results are provided for two receive antennas in

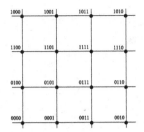

Fig. 4.10.   16-QAM constellation; Gray mapping.

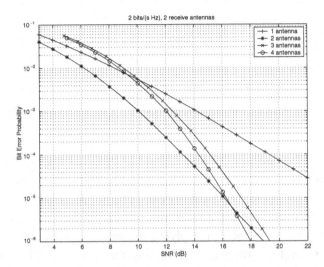

Fig. 4.11.   Bit error probability plotted against SNR for orthogonal STBCs at 2 bits/(s Hz); two receive antennas.

Figure 4.11. Note that the index assignment used for the constellation will affect the results. Using Gray mapping results in the lowest bit error probability.

Figure 4.12 shows bit error probability for transmission of 3 bits/(s Hz). The results are reported for an uncoded 8-PSK and orthogonal space-time block codes using two, three, and four transmit antennas. Simulation results in Figure 4.12 are given for one receive antenna. The transmission using two transmit antennas employs the 8-PSK constellation and the Alamouti code. The index assignment for 8-PSK is depicted in Figure 4.7. The 16-QAM constellation and the code $\mathcal{G}_{434}$ in Example 4.7.4 are utilized for four transmit antennas. To arrive at a rate $3/4$ OSTBC for three transmit antennas, one can remove a column of the corresponding code for four transmit antennas to arrive at

$$\mathcal{G}_{334} = \begin{pmatrix} x_1 & x_2 & x_3 \\ -x_2^* & x_1^* & 0 \\ x_3^* & 0 & -x_1^* \\ 0 & x_3^* & -x_2^* \end{pmatrix}. \tag{4.154}$$

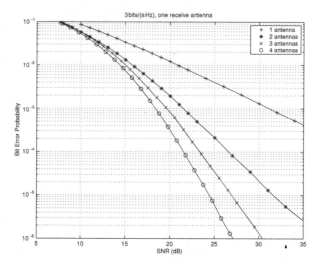

Fig. 4.12.   Bit error probability plotted against SNR for orthogonal STBCs at 3 bits/(s Hz); one receive antenna.

Since $\mathcal{G}_{334}$ and $\mathcal{G}_{434}$ are rate 3/4 codes, using 16-QAM, the total transmission rate in each case is 3 bits/(s Hz). It is seen that at the bit error probability of 10^{-5} the rate 3/4 OSTBC using 16-QAM provides about 7 dB gain over the use of an 8-PSK Alamouti code. Note that as mentioned before the choice of orthogonal STBCs is not unique. In fact, multiplying an orthogonal STBC with a unitary matrix results in another orthogonal STBC with the same properties. For example, one may use the following rate 3/4 OSTBC from [135] instead of the code in Example 4.7.4.

$$
\mathcal{G} = \begin{pmatrix}
x_1 & x_2 & \frac{x_3}{\sqrt{2}} & \frac{x_3}{\sqrt{2}} \\
-x_2^* & x_1^* & \frac{x_3}{\sqrt{2}} & -\frac{x_3}{\sqrt{2}} \\
\frac{x_3^*}{\sqrt{2}} & \frac{x_3^*}{\sqrt{2}} & \frac{(-x_1-x_1^*+x_2-x_2^*)}{2} & \frac{(-x_2-x_2^*+x_1-x_1^*)}{2} \\
\frac{x_3^*}{\sqrt{2}} & -\frac{x_3^*}{\sqrt{2}} & \frac{(x_2+x_2^*+x_1-x_1^*)}{2} & -\frac{(x_1+x_1^*+x_2-x_2^*)}{2}
\end{pmatrix}. \tag{4.155}
$$

Note that the performance of the code remains the same using such a transform. Therefore, the simulation results would be the same for OSTBCs with the same rate and number of antennas using the same constellation.

The above simulation results demonstrate that significant gains can be achieved by increasing the number of transmit antennas with very little decoding complexity. Similar results for two receive antennas are shown in Figure 4.13. It is possible to concatenate an outer trellis code [5] with space-time block codes to achieve even better performance. The additional coding gain provided by the outer code is the same as the gain provided by the code on a Gaussian channel. Also, one can combine

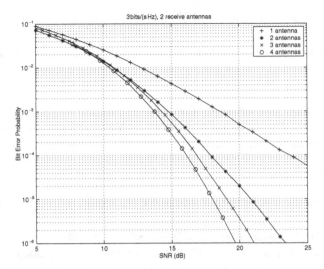

Fig. 4.13. Bit error probability plotted against SNR for orthogonal STBCs at 3 bits/(s Hz); two receive antennas.

space-time block coding and trellis coding in an approach similar to trellis-coded modulation [71]. Adding an outer code provides better performance at the expense of a higher complexity. We investigate such systems in Chapter 7.

4.11 Summary of important results

- Space-time block codes from orthogonal designs provide:
  - simple decoding: maximum-likelihood decoding results in each symbol being decoded separately using only linear processing;
  - maximum diversity: the code satisfies the rank criterion.
- A real orthogonal design of size N is an $N \times N$ orthogonal matrix $\mathcal{G}_N$ with real entries $x_1, -x_1, x_2, -x_2, \ldots, x_N, -x_N$ such that

$$\mathcal{G}_N{}^T \mathcal{G}_N = (x_1^2 + x_2^2 + \cdots + x_N^2) I_N.$$

- A real orthogonal design exists if and only if $N = 2, 4, 8$.
- A generalized real orthogonal design is a $T \times N$ matrix $\mathcal{G}$ with real entries $x_1, -x_1, x_2, -x_2, \ldots, x_K, -x_K$ such that

$$\mathcal{G}^T \mathcal{G} = \kappa \, (x_1^2 + x_2^2 + \cdots + x_K^2) I_N,$$

where I_N is the $N \times N$ identity matrix and κ is a constant. The rate of the code is $R = K/T$.

- For any number of transmit antennas, N, there exists a full rate, $R = 1$, real space-time block code with a block size $T = \min[2^{4c+d}]$, where the minimization is over all possible integer values of c and d in the set $\{c \geq 0, d \geq 0 | 8c + 2^d \geq N\}$.

- A complex orthogonal design of size N is an $N \times N$ orthogonal matrix $\mathcal{G}_N$ with complex entries $x_1, -x_1, x_2, -x_2, \ldots, x_N, -x_N$, their conjugates $x_1^*, -x_1^*, x_2^*, -x_2^*, \ldots, x_N^*, -x_N^*$, and multiples of these indeterminate variables by $j = \sqrt{-1}$ or $-j$ such that

$$\mathcal{G}_N^H \cdot \mathcal{G}_N = (|x_1|^2 + |x_2|^2 + \cdots + |x_N|^2)I_N.$$

- A complex orthogonal design exists if and only if $N = 2$.
- A generalized complex orthogonal design is a $T \times N$ matrix $\mathcal{G}$ with entries that are linear combinations of the indeterminate variables $x_1, x_2, \ldots, x_K$ and their conjugates such that

$$\mathcal{G}^H \cdot \mathcal{G} = \kappa \, (|x_1|^2 + |x_2|^2 + \cdots + |x_K|^2)I_N,$$

where I_N is the $N \times N$ identity matrix and κ is a constant. The rate of the code is $R = K/T$.
- Orthogonal STBCs, built from generalized real or generalized complex orthogonal designs, provide separate maximum-likelihood decoding of the symbols and full diversity, equal to the product of the number of transmit and receive antennas.
- Full rate, $R = 1$, complex OSTBCs do not exist for more than two transmit antennas. For $N = 2$ transmit antennas, Alamouti code is constructed from

$$\begin{pmatrix} x_1 & x_2 \\ -x_2^* & x_1^* \end{pmatrix}.$$

- Rate half, $R = 1/2$, complex OSTBCs exist for any number of transmit antennas.
- Multiplying an orthogonal STBC by a unitary matrix results in another orthogonal STBC with similar properties.
- The pairwise error probability of an orthogonal STBC for a received SNR of γ is given by

$$P(\mathbf{C}^1 \to \mathbf{C}^2) = \frac{1}{2} \left\{ 1 - \sqrt{\frac{a}{1+a}} \sum_{i=0}^{MN-1} \binom{2i}{i} \left[\frac{1}{4(1+a)} \right]^i \right\},$$

where $a = \kappa \frac{\gamma}{4} d_E^2$ and $d_E^2 = \sum_{k=1}^{K} |s_k^2 - s_k^1|^2$.

4.12 Problems

1 Consider a system that uses orthogonal STBCs with QPSK constellation. If the input bitstream is 101101001001, what are the transmitted symbols from each antenna in the following cases?
  (a) Two transmit antennas.
  (b) Three transmit antennas (pick the most efficient STBC for transmission).
2 Derive the decision metrics for the separate maximum-likelihood decoding of the rate 1/2 OSTBC in Example 4.7.3.
3 Derive the decision metrics for the separate maximum-likelihood decoding of the rate 3/4 OSTBC in (4.104). How can we derive these decision metrics from the decision metrics of the rate 3/4 OSTBC in (4.103)?

4 Consider the following STBC

$$
\mathcal{G} = \begin{pmatrix}
x_1^* & 0 & 0 & -x_4^* \\
0 & x_1^* & 0 & x_3^* \\
0 & 0 & x_1^* & x_2^* \\
0 & -x_2^* & x_3^* & 0 \\
x_2^* & 0 & x_4^* & 0 \\
-x_3^* & -x_4^* & 0 & 0 \\
x_4 & -x_3 & -x_2 & x_1
\end{pmatrix}.
$$

(a) What is the rate of this STBC?

(b) Is it possible to have a separate maximum-likelihood decoding for this code? Is this an orthogonal STBC?

5 Using a 4×4 Hadamard matrix, construct a pseudo-orthogonal design for four transmit antennas as in (4.116). What is the rate and diversity of the corresponding code?

6 In (4.128), set $\theta = \pi/2$ in the integrand to derive a bound on the pairwise error probability. Justify the upper bound and explain its relationship with the upper bound in (3.16) and/or (3.26).

7 For a rate $3/4$ OSTBC with four transmit antennas, use 16-PSK and 16-QAM constellations to transmit 3 bits/(s Hz). Use Monte Carlo simulations to derive the bit error rate (BER) versus the received SNR simulation results. Draw the curves for one receive antenna in the same figure and compare the results. Justify the achieved results. What is the effect of index assignment in the results?

8 Consider an Alamouti code with one receive antenna using BPSK constellation. Use Monte Carlo simulations to derive the bit error rate (BER) versus the received SNR simulation results. Compare the simulation results with the calculated error rates in Example 4.9.2.

9 Compare the performance of the rate $R = 0.5$ OSTBCs in Examples 4.7.2 and 4.7.3.

10 Derive the 4×1 vectors $(\mathbf{r}'_m)^k$, $k = 1, 2, 3$ and the 3×4 matrix Ω_m for the OSTBC in (4.104).

5

Quasi-orthogonal space-time block codes

5.1 Pairwise decoding

Full-rate orthogonal designs with complex elements in its transmission matrix are impossible for more than two transmit antennas as discussed in Chapter 4. The only example of a full-rate full-diversity complex space-time block code using orthogonal designs is the Alamouti scheme in Equation (4.21). Here we rewrite the generator matrix for the Alamouti code to emphasize the indeterminate variables x_1 and x_2 in the design:

$$\mathcal{G}(x_1, x_2) = \begin{pmatrix} x_1 & x_2 \\ -x_2^* & x_1^* \end{pmatrix}. \tag{5.1}$$

The main properties of an orthogonal design are simple separate decoding and full diversity. To design full-rate codes, we relax the simple separate decoding property. In this chapter, we consider codes for which decoding pairs of symbols independently is possible. We call this class of codes quasi-orthogonal space-time block codes (QOSTBCs) for reasons that we discuss later.

First, let us consider the following QOSTBC [67, 68]:

$$\mathcal{G} = \begin{pmatrix} \mathcal{G}(x_1, x_2) & \mathcal{G}(x_3, x_4) \\ -\mathcal{G}^*(x_3, x_4) & \mathcal{G}^*(x_1, x_2) \end{pmatrix} = \begin{pmatrix} x_1 & x_2 & x_3 & x_4 \\ -x_2^* & x_1^* & -x_4^* & x_3^* \\ -x_3^* & -x_4^* & x_1^* & x_2^* \\ x_4 & -x_3 & -x_2 & x_1 \end{pmatrix}, \tag{5.2}$$

where a matrix $\mathcal{G}^*$ is the complex conjugate of $\mathcal{G}$, for example

$$\mathcal{G}^*(x_1, x_2) = \mathcal{G}(x_1^*, x_2^*) = \begin{pmatrix} x_1^* & x_2^* \\ -x_2 & x_1 \end{pmatrix}. \tag{5.3}$$

We denote the ith column of $\mathcal{G}$ by V_i. For any indeterminate variables $x_1, x_2, x_3, x_4,$

110

we have

$$< \mathcal{V}_1, \mathcal{V}_2 > = < \mathcal{V}_1, \mathcal{V}_3 > = < \mathcal{V}_2, \mathcal{V}_4 > = < \mathcal{V}_3, \mathcal{V}_4 > = 0, \qquad (5.4)$$

where $< \mathcal{V}_i, \mathcal{V}_j >$ is the inner product of vectors $\mathcal{V}_i$ and $\mathcal{V}_j$. Therefore, the subspace created by $\mathcal{V}_1$ and $\mathcal{V}_4$ is orthogonal to the subspace created by $\mathcal{V}_2$ and $\mathcal{V}_3$. This is the rationale behind the name "quasi-orthogonal" for the code. The minimum rank of the difference matrix $D(\mathbf{C}^i, \mathbf{C}^j)$ could be two. Therefore, following the definitions in Chapter 3, the diversity of the code is two, which is less than the maximum possible diversity of four. In the sequel, we show how to design full-diversity QOSTBCs.

The encoding for QOSTBCs is very similar to the encoding of orthogonal STBCs. To transmit b bits per time slot, we use constellations containing 2^b points. Using $4b$ bits, constellation symbols s_1, s_2, s_3, s_4 are selected. Setting $x_k = s_k$ for $k = 1, 2, 3, 4$ in the generator matrix $\mathcal{G}$, we arrive at a codeword matrix $\mathbf{C} = \mathcal{G}(s_1, s_2, s_3, s_4)$. Then, at time t, the four elements in the tth row of $\mathbf{C}$ are transmitted from the four transmit antennas. Note that since four symbols s_1, s_2, s_3, s_4 are transmitted in four time slots, (5.2) presents a rate one code.

The orthogonality of the subspaces of the generator matrix results in the possibility of decoding pairs of symbols independently. Transmitting the data symbols s_1, s_2, s_3, s_4, we receive $r_{t,m}$ at time t and receive antenna m based on (2.1). Similar to decoding formulas in (4.79), the maximum-likelihood decoding for the QOSTBC in (5.2) amounts to the following minimization problem:

$$\min_{s_1, s_2, s_3, s_4} \left\{ \mathbf{H}^H \cdot \mathbf{C}^H \cdot \mathbf{C} \cdot \mathbf{H} - \mathbf{H}^H \cdot \mathbf{C}^H \cdot \mathbf{r} - \mathbf{r}^H \cdot \mathbf{C} \cdot \mathbf{H} \right\}, \qquad (5.5)$$

where $\mathbf{C}$ is derived by replacing x_k by s_k in (5.2). Simple algebraic manipulation shows that the ML decoding amounts to minimizing the following sum:

$$f_{14}(s_1, s_4) + f_{23}(s_2, s_3), \qquad (5.6)$$

where

$$
\begin{aligned}
f_{14}(s_1, s_4) = \sum_{m=1}^{M} &\left[\left(|s_1|^2 + |s_4|^2 \right) \left(\sum_{n=1}^{4} |\alpha_{n,m}|^2 \right) \right. \\
&+ 2\Re \left\{ \left(-\alpha_{1,m} r_{1,m}^* - \alpha_{2,m}^* r_{2,m} - \alpha_{3,m}^* r_{3,m} - \alpha_{4,m} r_{4,m}^* \right) s_1 \right. \\
&+ \left. \left(-\alpha_{4,m} r_{1,m}^* + \alpha_{3,m}^* r_{2,m} + \alpha_{2,m}^* r_{3,m} - \alpha_{1,m} r_{4,m}^* \right) s_4 \right\} \\
&+ \left. 4\Re \left\{ \alpha_{1,m} \alpha_{4,m}^* - \alpha_{2,m}^* \alpha_{3,m} \right\} \Re \left\{ s_1 s_4^* \right\} \right],
\end{aligned}
\qquad (5.7)
$$

and

$$
\begin{aligned}
f_{23}(s_2, s_3) = \sum_{m=1}^{M} & \left[\left(|s_2|^2 + |s_3|^2 \right) \left(\sum_{n=1}^{4} |\alpha_{n,m}|^2 \right) \right. \\
& + 2\Re \left\{ \left(-\alpha_{2,m} r_{1,m}^* + \alpha_{1,m}^* r_{2,m} - \alpha_{4,m}^* r_{3,m} + \alpha_{3,m} r_{4,m}^* \right) s_2 \right. \\
& \left. + \left(-\alpha_{3,m} r_{1,m}^* - \alpha_{4,m}^* r_{2,m} + \alpha_{1,m}^* r_{3,m} + \alpha_{2,m} r_{4,m}^* \right) s_3 \right\} \\
& \left. + 4\Re \left\{ \alpha_{2,m} \alpha_{3,m}^* - \alpha_{1,m}^* \alpha_{4,m} \right\} \Re \left\{ s_2 s_3^* \right\} \right].
\end{aligned}
\tag{5.8}
$$

Since $f_{14}(s_1, s_4)$ is independent of (s_2, s_3) and $f_{23}(s_2, s_3)$ is independent of (s_1, s_4), the pairs (s_1, s_4) and (s_2, s_3) can be decoded separately. Therefore, the ML decoding results in minimizing $f_{14}(s_1, s_4)$ over all possible values of s_1 and s_4 and minimizing $f_{23}(s_2, s_3)$ over all possible values of s_2 and s_3. If $\Re\{s_1 s_4^*\} = \Re\{s_2 s_3^*\} = 0$, there is no cross-term and we can write $f_{14}(s_1, s_4) = f_1(s_1) + f_4(s_4)$ and $f_{23}(s_2, s_3) = f_2(s_2) + f_3(s_3)$. Therefore, we can decode symbols s_1, s_2, s_3, s_4 separately. This is possible, for example, when s_1 and s_2 are real numbers while s_3 and s_4 are imaginary numbers. In this case, the QOSTBC produces a pseudo-orthogonal STBC as discussed in Section 4.8.

There are many examples of QOSTBCs with the same properties [67, 142]. A popular example is

$$
\mathcal{G} = \begin{pmatrix} \mathcal{G}(x_1, x_2) & \mathcal{G}(x_3, x_4) \\ \mathcal{G}(x_3, x_4) & \mathcal{G}(x_1, x_2) \end{pmatrix}.
\tag{5.9}
$$

In general, to design a QOSTBC, we replace $\mathcal{G}(x_1, x_2)$ and $\mathcal{G}(x_3, x_4)$ in structures (5.2) and (5.9), or any similar structure, with orthogonal STBCs. The characteristics of the resulting QOSTBC depend on the building STBC blocks. In what follows, we provide examples for different number of transmit antennas and different rates. However, first let us derive the properties of the rate one codes for four transmit antennas in (5.2) and (5.9).

5.2 Rotated QOSTBCs

For regular symmetric constellations like PSK and QAM, it is easy to show that the minimum rank of the difference matrix $D(\mathbf{C}^i, \mathbf{C}^j)$ is two for QOSTBCs in (5.2) and (5.9). Therefore, for M receive antennas, a diversity of $2M$ is achieved while the rate of the code is one. The maximum diversity of $4M$ for a rate one complex orthogonal code is impossible in this case if all symbols are chosen from the same constellation. To provide full diversity, we use different constellations for different transmitted symbols. For example, we may rotate symbols x_3 and x_4 before transmission. Let us denote $\tilde{x}_3$ and $\tilde{x}_4$ as the rotated versions of x_3 and

x_4, respectively. We show that it is possible to provide full-diversity QOSTBCs by replacing (x_3, x_4) with $(\tilde{x}_3, \tilde{x}_4)$. Independent examples of such full-diversity QOSTBCs are provided in [70, 112, 126, 141]. The resulting code is very powerful since it provides full diversity, rate one, and simple pairwise decoding with good performance.

In what follows, we derive the general conditions to achieve full diversity for these codes. First, we calculate the CGD for different cases and show that they behave similarly. We use $\mathcal{G}$ to denote the generator matrix with indeterminate variables x_1, x_2, x_3, x_4 and $\mathcal{G}(s_1, s_2, s_3, s_4)$ to denote the same matrix when x_k is replaced by the kth argument in parentheses.

Lemma 5.2.1 *The CGD between a pair of codewords* $\mathbf{C} = \mathcal{G}(s_1, s_2, \tilde{s}_3, \tilde{s}_4)$ *and* $\mathbf{C}' = \mathcal{G}(s_1', s_2', \tilde{s}_3', \tilde{s}_4')$ *from the QOSTBC in (5.2) is given by*

$$CGD(\mathbf{C}, \mathbf{C}') = \det\left[D(\mathbf{C}, \mathbf{C}')^H \cdot D(\mathbf{C}, \mathbf{C}')\right]$$

$$= \left(\left|(s_1 - s_1') - (\tilde{s}_4 - \tilde{s}_4')\right|^2 + \left|(s_2 - s_2') + (\tilde{s}_3 - \tilde{s}_3')\right|^2\right)^2$$

$$\left(\left|(s_1 - s_1') + (\tilde{s}_4 - \tilde{s}_4')\right|^2 + \left|(s_2 - s_2') - (\tilde{s}_3 - \tilde{s}_3')\right|^2\right)^2. \quad (5.10)$$

Proof For the code in (5.2), we have

$$\mathcal{G}^H \cdot \mathcal{G} = \begin{pmatrix} a & 0 & 0 & b \\ 0 & a & -b & 0 \\ 0 & -b & a & 0 \\ b & 0 & 0 & a \end{pmatrix}, \quad (5.11)$$

where $a = \sum_{k=1}^{4} |x_k|^2$ and $b = 2\Re(x_1 x_4^* - x_2 x_3^*)$. Using the determinant equality [62]

$$\det\begin{pmatrix} A & B \\ C & D \end{pmatrix} = \det(A)\det(D - CA^{-1}B), \quad (5.12)$$

we have $\det(\mathcal{G}^H \cdot \mathcal{G}) = (a^2 - b^2)^2$. Simple algebraic manipulation shows that

$$\det(\mathcal{G}^H \cdot \mathcal{G}) = \left(|x_1 - x_4|^2 + |x_2 + x_3|^2\right)^2 \left(|x_1 + x_4|^2 + |x_2 - x_3|^2\right)^2. \quad (5.13)$$

Then, replacing x_1 with $(s_1 - s_1')$, x_2 with $(s_2 - s_2')$, x_3 with $(\tilde{s}_3 - \tilde{s}_3')$, and x_4 with $(\tilde{s}_4 - \tilde{s}_4')$ results in (5.10). $\qquad\square$

A similar calculation for the QOSTBC in (5.9) results in the following lemma.

Lemma 5.2.2 *The CGD between a pair of codewords* $\mathbf{C} = \mathcal{G}(s_1, s_2, \tilde{s}_3, \tilde{s}_4)$ *and* $\mathbf{C}' = \mathcal{G}(s_1', s_2', \tilde{s}_3', \tilde{s}_4')$ *from the QOSTBC in (5.9) is given by*

$$CGD(\mathbf{C}^1, \mathbf{C}^2)$$
$$= \det[D(\mathbf{C}^1, \mathbf{C}^2)^H \cdot D(\mathbf{C}^1, \mathbf{C}^2)] \tag{5.14}$$
$$= \left(\sum_{k=1}^{2} |(s_k - s_k') + (\tilde{s}_{k+2} - \tilde{s}_{k+2}')|^2\right)^2 \left(\sum_{k=1}^{2} |(s_k - s_k') - (\tilde{s}_{k+2} - \tilde{s}_{k+2}')|^2\right)^2.$$

Proof For the code in (5.9), we have

$$\mathcal{G}^H \cdot \mathcal{G} = \begin{pmatrix} a & 0 & b' & 0 \\ 0 & a & 0 & b' \\ b' & 0 & a & 0 \\ 0 & b' & 0 & a \end{pmatrix}, \tag{5.15}$$

where $a = \sum_{k=1}^{4} |x_k|^2$ and $b' = 2\Re(x_1 x_3^* + x_2 x_4^*)$. Using the determinant equality in (5.12), we have $\det(\mathcal{G}^H \cdot \mathcal{G}) = (a^2 - b'^2)^2$. Simple algebraic manipulation shows

$$\det(\mathcal{G}^H \cdot \mathcal{G}) = \left(|x_1 + x_3|^2 + |x_2 + x_4|^2\right)^2 \left(|x_1 - x_3|^2 + |x_2 - x_4|^2\right)^2. \tag{5.16}$$

Then, (5.14) is achieved by replacing x_1 with $(s_1 - s_1')$, x_2 with $(s_2 - s_2')$, x_3 with $(\tilde{s}_3 - \tilde{s}_3')$, and x_4 with $(\tilde{s}_4 - \tilde{s}_4')$. □

Note that if we arrange the codewords of the QOSTBC in (5.2) as $\mathbf{C} = \mathcal{G}(s_1, s_2, -\tilde{s}_4, \tilde{s}_3)$ and $\mathbf{C}' = \mathcal{G}(s_1', s_2', -\tilde{s}_4', \tilde{s}_3')$, (5.10) results in (5.14). Therefore, since (5.14) unifies the coding gain formula for both of the QOSTBCs in (5.2) and (5.9), the two codes would have similar properties. In what follows, with a slight abuse of the notation, we use $\mathbf{C} = \mathcal{G}(s_1, s_2, \tilde{s}_3, \tilde{s}_4)$ along with (5.14) for both codes. However, one should keep in mind that for the code in (5.2), the codeword is in fact $\mathbf{C} = \mathcal{G}(s_1, s_2, -\tilde{s}_4, \tilde{s}_3)$. This one-to-one replacement is the only difference between the two codes as far as the results in this section are concerned. The next lemma outlines the conditions for a full-diversity QOSTBC.

Lemma 5.2.3 *The quasi-orthogonal code provides full diversity if and only if the rotated constellation is such that* $s_k - s_k' = \tilde{s}_{k+2} - \tilde{s}_{k+2}'$, $k = 1, 2$ *is impossible.*

Proof Full diversity is achieved if and only if the CGD in (5.14) is not zero. If the CGD becomes zero, then at least one of the two factors in (5.14) should be zero. To find the minimum CGD, we consider all possible values of $s_k, s_k', \tilde{s}_{k+2}$, and $\tilde{s}_{k+2}'$, $k = 1, 2$. If there is a set of values such that $\sum_{k=1}^{2} |(s_k - s_k') + (\tilde{s}_{k+2} - \tilde{s}_{k+2}')|^2 = 0$, by switching $\tilde{s}_{k+2}$ with $\tilde{s}_{k+2}'$, we have $\sum_{k=1}^{2} |(s_k - s_k') - (\tilde{s}_{k+2} - \tilde{s}_{k+2}')|^2 = 0$ as well. Therefore, without loss of generality, we only need to consider the case that the second sum becomes zero. This sum contains two non-negative terms. Therefore, the sum is zero if and only if both terms are zero. If none of the two pairs of symbols results in $s_k - s'_k = \tilde{s}_{k+2} - \tilde{s}'_{k+2}$, we conclude that the second, and consequently the first, sum are not equal to zero. Also, if the first and the second sums are not zero, $s_k - s'_k \neq \tilde{s}_{k+2} - \tilde{s}'_{k+2}$ for any two pairs of symbols. □

While Lemma 5.2.3 is valid for any constellation, we consider it specifically for the case of L-PSK constellations. In an L-PSK constellation, the symbols can be represented as $e^{j2\pi l/L}$, $l = 0, 1, \ldots, L - 1$. The following theorem proves that if we pick the third and fourth symbols from a rotated constellation, the diversity of the code is four [70].

Theorem 5.2.1 *Let us assume that we pick symbols s_1 and s_2 from an L-PSK constellation with even L and symbols $\tilde{s}_3$ and $\tilde{s}_4$ from a rotated constellation. The corresponding QOSTBC provides full diversity.*

To demonstrate that the above result is not trivial, the following theorem provides a case for which the code is not full diversity [70]. Although, the utilized constellation has no practical significance.

Theorem 5.2.2 *Let us assume that we pick symbols s_1 and s_2 from an L-PSK constellation with odd L and symbols $\tilde{s}_3$ and $\tilde{s}_4$ from a rotated constellation. The code does not provide a full diversity for a rotation $\phi = \pi/L$. The code provides full diversity for any other rotation.*

Theorems 5.2.1 and 5.2.2 show that the proposed rotated QOSTBC provides full diversity and rate one unless we use an L-PSK constellation with odd L and a rotation $\phi = \pi/L$. This now poses the question that among the rotations that provide full diversity, which one provides a higher coding gain.

5.3 Optimal rotation and performance of QOSTBCs

In Section 5.2, we showed how to design a rotated quasi-orthogonal STBC that provides full diversity and rate one while decoding pairs of symbols separately. The structure in Section 5.2 has one degree of freedom in choosing the rotation parameter. In this section, we study the optimal choice of the rotation for different constellations. We use the determinant criterion as defined in Section 3.2 to optimize the rotation parameter. In other words, we find the rotation angle that maximizes the minimum possible value of the CGD in (5.14) among all possible constellation points.

The encoder uses one block of $4b$ input bits to pick symbols (s_1, s_2, s_3, s_4) from a constellation. Then, $\tilde{s}_3 = e^{j\phi}s_3$ and $\tilde{s}_4 = e^{j\phi}s_4$ are the rotated symbols that are used in $\mathcal{G}$, the generator matrix of the QOSTBC. For a given constellation, we find the optimal value of ϕ to maximize the minimum CGD.

We denote the minimum CGD of a rotated QOSTBC by $CGD_{\min}(\phi)$. Note that as expected this minimum CGD is a function of ϕ and the goal is to maximize $CGD_{\min}(\phi)$ among all possible rotations. In [126], it is shown that $CGD_{\min}(\phi)$ can be calculated from the following minimization:

$$CGD_{\min}(\phi) = \min_{(s_1,\tilde{s}_3)\neq(s_1',\tilde{s}_3')} |(s_1 - s_1')^2 - (\tilde{s}_3 - \tilde{s}_3')^2|^4, \tag{5.17}$$

where s_1 and s_1' are symbols from the original constellation and $\tilde{s}_3$ and $\tilde{s}_3'$ are symbols from the rotated constellation. The term $CGD_{\min}(\phi)$ in (5.17) can be used as a distance between a constellation and its rotated version. Let us denote $d_{\min} = \min|s_1 - s_1'|$ as the minimum Euclidean distance among all constellation points. This distance is the same for the original and rotated constellations. Choosing $\tilde{s}_3 = \tilde{s}_3'$ in (5.17) makes the right-hand side equal to the power of eight of the minimum Euclidean distance of the constellation. Therefore,

$$CGD_{\min}(\phi) \leq d_{\min}^8, \tag{5.18}$$

and $d_{\min}^8$ is an upper bound on $CGD_{\min}(\phi)$. If there exists a rotation ϕ^* for which the minimum possible CGD is equal to $d_{\min}^8$, then $\max_\phi[CGD_{\min}(\phi)] = CGD_{\min}(\phi^*) = d_{\min}^8$ and ϕ^* is the optimal rotation. Note that in general the optimal rotation may not be unique. Also, the upper bound is not always achievable. One example for which the upper bound is achievable is the case of QAM constellations. In this case, the optimal rotation that maximizes $CGD_{\min}(\phi)$ is $\phi^* = \pi/4$ [126].

Theorem 5.3.1 *For a symbol constellation drawn from a square lattice, the minimum CGD for $\phi = \pi/4$ is $CGD_{\min}(\pi/4) = d_{\min}^8$.*

We also discuss the case that the upper bound is not achievable, that is $CGD_{\min}(\phi) < d_{\min}^8$. First, we derive another upper bound on $CGD_{\min}(\phi)$ [126].

Theorem 5.3.2 *For a constellation with minimum Euclidean $d_{\min}$, the minimum possible CGD is upper bounded by*

$$CGD_{\min}(\phi) \leq |2\sin\phi|^4 d_{\min}^8. \tag{5.19}$$

Proof Let us assume that the closest points in the constellation are $\hat{s}$ and $\hat{s}'$, that is $d_{\min} = |\hat{s} - \hat{s}'|$. Using the definition of $CGD_{\min}(\phi)$ in (5.17), the minimum CGD cannot exceed the CGD between the pairs $(\hat{s}, \hat{s}\,e^{j\phi})$ and $(\hat{s}', \hat{s}'\,e^{j\phi})$. Therefore,

$$CGD_{\min}(\phi) \leq |(\hat{s} - \hat{s}')^2 - (\hat{s}\,e^{j\phi} - \hat{s}'\,e^{j\phi})^2|^4$$
$$= |1 - e^{j2\phi}|^4 |\hat{s} - \hat{s}'|^8 \tag{5.20}$$
$$= |2\sin\phi|^4 d_{\min}^8.$$

$\square$

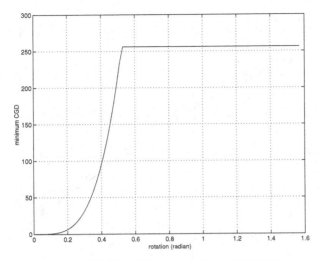

Fig. 5.1.   Minimum CGD for different rotations; QOSTBC using BPSK.

The upper bound of $CGD_{min}(\phi)$ in (5.19) is smaller than the upper bound in (5.18) for $\phi < \pi/6$. In other words, (5.19) is a tighter bound than that of (5.18) for $\phi < \pi/6$. An immediate conclusion for $\phi < \pi/6$ is the fact that $CGD_{min}(\phi) < d_{min}^8$, the upper bound (5.18) is not achievable. Therefore, the upper bound (5.18) is not achievable for an L-PSK constellation for $L > 6$.

Now, we discuss the optimal rotation for PSK constellations. For a BPSK constellation, $\phi = \pi/2$ achieves the upper bound (5.18). In this case, the original constellation consists of $\{-1, 1\}$ and the rotated constellation consists of $\{-j, j\}$. For the optimal $\phi = \pi/2$, we have $CGD_{min}(\pi/2) = 256$, which is the maximum possible $CGD_{min}(\phi)$. Figure 5.1 shows the minimum value of CGD for different rotations, that is $CGD_{min}(\phi)$. As can be seen, the optimal rotation is not unique. In fact, $CGD_{min}(\phi)$ is a non-decreasing function of the rotation $0 \le \phi \le \pi/2$. To transmit one bit/cycle using QOSTBC, we utilize BPSK and different rotation values. Figure 5.2 depicts the bit error probability plotted against the SNR of the resulting codes. Figure 5.2 endorses the fact that as rotation ϕ increases in the range of $0 \le \phi \le \pi/2$, the performance improves. Also, as expected from the saturation of $CGD_{min}(\phi)$, the performance of the QOSTBC for different rotations is similar for values close to $\phi = \pi/2$.

Similar figures for a QPSK constellation transmitting 2 bits/(s Hz) are provided in Figures 5.3 and 5.4. Figure 5.3 shows the minimum value of CGD for different rotations. Figure 5.4 illustrates the performance of the QOSTBC using QPSK and different rotations. Theorem 5.3.1 shows that $\phi = \pi/4$ is optimal for QPSK constellation. Another reason for the optimum nature of $\phi = \pi/4$ is the fact that it achieves the upper bound (5.18). Therefore, the maximum possible $CGD_{min}(\phi)$ is $CGD_{min}(\pi/4) = 16$.

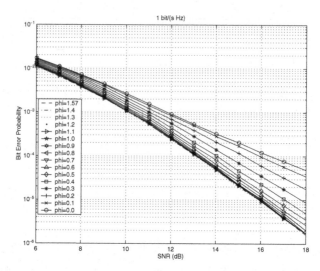

Fig. 5.2.   Bit error probability plotted against SNR for QOSTBCs at 1 bit/(s Hz); different rotations using BPSK; four transmit antennas, one receive antenna.

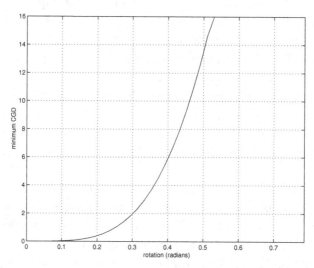

Fig. 5.3.   Minimum CGD for different rotations; QOSTBC using QPSK.

Figure 5.5 illustrates the minimum CGD for different values of rotation, $0 \leq \phi \leq \pi/8$, in the case of 8-PSK. As can be seen, $CGD_{\min}(\phi)$ is an increasing function of the rotation for $0 \leq \phi \leq \pi/8$ and therefore achieves its maximum at $\phi = \pi/8$. Since $\pi/8 < \pi/6$, the upper bound in (5.19) is a tighter upper bound than that of (5.18). As a result, the upper bound in (5.18) is not achievable for 8-PSK.

Figure 5.6 provides simulation results for the transmission of 2 bits/(s Hz) using four transmit antennas and one receive antenna using orthogonal and

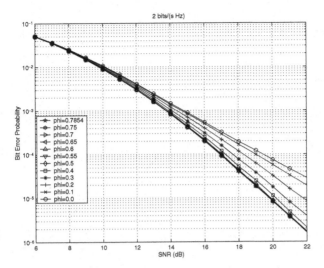

Fig. 5.4. Bit error probability plotted against SNR for QOSTBCs at 2 bits/(s Hz); different rotations using QPSK; four transmit antennas, one receive antenna.

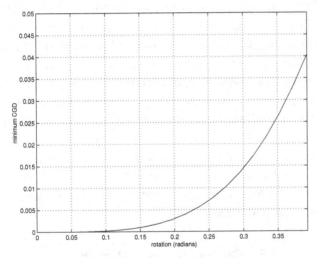

Fig. 5.5. Minimum CGD for different rotations; QOSTBC using 8-PSK.

quasi-orthogonal STBCs. Note that we have used the appropriate modulation schemes to provide the desired transmission rate for different STBCs, that is QPSK for the rate one QOSTBC and the SISO system and 16-QAM for the rate 1/2 orthogonal STBC. We use a rotation of $\phi = \pi/4$ for the QOSTBC in Figure 5.6. The resulting constellation is very similar to a $\pi/4$-QPSK constellation [7].

Simulation results show that full transmission rate is more important for very low SNRs, high BERs, while full diversity is the right choice for high SNRs. A

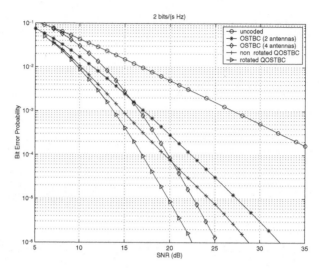

Fig. 5.6.   Bit error probability plotted against SNR for different space-time block codes at 2 bits/(s Hz); four transmit antennas, one receive antenna.

rotated QOSTBC provides both and therefore performs better at all SNRs. This is due to the fact that the degree of diversity dictates the slope of the BER–SNR curve. Although a non full-diversity code with higher rate starts from a better point in the BER–SNR plane, a code with full-diversity benefits more from increasing the SNR. Therefore, the BER–SNR curve of the full-diversity orthogonal STBC passes the curve for the partial diversity QOSTBC at some moderate SNR. The rotated QOSTBC provides the advantages of full diversity and rate one. Also, note that the receiver of the orthogonal STBCs can decode the symbols one by one while the decoding for the rate one quasi-orthogonal STBC is done for pairs of symbols. This means that the decoding complexity of the orthogonal STBCs is lower although both codes have a very low decoding complexity. The encoding complexity of the two systems are small and the same.

5.4 Other examples of QOSTBCs

So far, we have discussed a few structures that provide the properties of rate one, full diversity, and pairwise ML decoding for four transmit antennas. There are other structures with similar properties that have different formats. For example, let us consider the following structure which is similar to the structures in (5.2) and (5.9) [31]:

$$\mathcal{G} = \begin{pmatrix} \mathcal{G}(x_1, x_2) & \mathcal{G}(x_3, x_4) \\ \mathcal{G}(x_1, x_2) & -\mathcal{G}(x_3, x_4) \end{pmatrix} = \begin{pmatrix} x_1 & x_2 & x_3 & x_4 \\ -x_2^* & x_1^* & -x_4^* & x_3^* \\ x_1 & x_2 & -x_3 & -x_4 \\ -x_2^* & x_1^* & x_4^* & -x_3^* \end{pmatrix}. \qquad (5.21)$$

Note that structure (5.21) if directly used as a STBC provides a rate one code with diversity two. In fact, (x_1, x_2) are only transmitted from the first two antennas and (x_3, x_4) are only transmitted from the third and fourth antennas. Therefore, inherently, this structure cannot provide a diversity more than two and is not interesting. Another way of demonstrating the same fact is by calculating $\mathcal{G}^H \cdot \mathcal{G}$ as follows

$$\mathcal{G}^H \cdot \mathcal{G} = \begin{pmatrix} |x_1|^2 + |x_2|^2 & 0 & 0 & 0 \\ 0 & |x_1|^2 + |x_2|^2 & 0 & 0 \\ 0 & 0 & |x_3|^2 + |x_4|^2 & 0 \\ 0 & 0 & 0 & |x_3|^2 + |x_4|^2 \end{pmatrix}. \tag{5.22}$$

The properties of the code in (5.21) can also be captured by dividing the four transmit antennas into two groups of two antennas and using the Alamouti code for each group. The generator matrix of the resulting code is

$$\mathcal{G} = \begin{pmatrix} x_1 & x_2 & 0 & 0 \\ -x_2^* & x_1^* & 0 & 0 \\ 0 & 0 & x_3 & x_4 \\ 0 & 0 & -x_4^* & x_3^* \end{pmatrix}. \tag{5.23}$$

Equation (5.22) is also valid for the rate one code in (5.23). One difference between the code in (5.23) and the other presented QOSTBCs is its peak-to-average power ratio. Because of the use of zeros in (5.23), the code has a high peak-to-average power ratio that may increase its cost of implementation. To achieve full diversity for the codes in (5.21) and (5.23), instead of replacing x_k by s_k in the generator matrix, one can use the following replacements:

$$\begin{aligned} x_1 &= s_1 + s_2, \\ x_2 &= s_3 + s_4, \\ x_3 &= s_1 - s_2, \\ x_4 &= s_3 - s_4. \end{aligned} \tag{5.24}$$

As a result,

$$\mathbf{C}^H \cdot \mathbf{C} = \begin{pmatrix} a+b & 0 & 0 & 0 \\ 0 & a+b & 0 & 0 \\ 0 & 0 & a-b & 0 \\ 0 & 0 & 0 & a-b \end{pmatrix}, \tag{5.25}$$

where $a = \sum_{k=1}^{4} |s_k|^2$ and $b = 2\Re(s_1^* s_2 + s_3^* s_4)$. Calculating the determinant of (5.25), we have $\det[\mathbf{C}^H \cdot \mathbf{C}] = (a^2 - b^2)^2$. The factors a and b in (5.25) are very similar to those of (5.11) and (5.15). All of these equations contain the term $\sum_{k=1}^{4} |s_1|^2$ and the linear combination of some pairs of the symbols. Also, all the determinants are the same after change of variables. Therefore, the behaviors of these codes are the same although the details of transmission and decoding are different. Especially, the conditions for which these codes are full diversity are governed by Theorems

5.2.1 and 5.2.2. Note that in the case of codes in (5.21) and (5.23), instead of s_3, we should pick s_2 from the rotated constellation. Also, pairwise decoding is possible as it is clear from maximum-likelihood decoding in (5.5). Similar to the case of codes in (5.21) and (5.23), when (s_1, s_3) are real and (s_2, s_4) are imaginary, separate decoding of the four symbols is possible. In this case, we have $b = 0$ and $\mathbf{C}^H \cdot \mathbf{C} = \sum_{k=1}^{4} |s_k|^2 I_4$. The resulting code is a special case of pseudo-orthogonal STBCs.

5.5 QOSTBCs for other than four transmit antennas

Like orthogonal STBCs, removing one or more columns of a quasi-orthogonal STBC results in new STBCs for smaller numbers of transmit antennas. The following example discusses the details of such a code for three transmit antennas by removing a column from (5.2).

Example 5.5.1 *For $N = 3$ transmit antennas, the following 4×3 quasi-orthogonal STBC, derived from (5.2), is rate one.*

$$
\mathcal{G} = \begin{pmatrix}
x_1 & x_2 & x_3 \\
-x_2^* & x_1^* & -x_4^* \\
-x_3^* & -x_4^* & x_1^* \\
x_4 & -x_3 & -x_2
\end{pmatrix}. \tag{5.26}
$$

The matrix $\mathcal{G}^H \cdot \mathcal{G}$ can be calculated from the corresponding matrix in (5.11) by removing the fourth column and row. Therefore,

$$
\mathcal{G}^H \cdot \mathcal{G} = \begin{pmatrix}
a & 0 & 0 \\
0 & a & -b \\
0 & -b & a
\end{pmatrix}, \tag{5.27}
$$

where $a = \sum_{k=1}^{4} |x_k|^2$ and $b = 2\Re(x_1 x_4^* - x_2 x_3^*)$. As a result $\det(\mathcal{G}^H \cdot \mathcal{G}) = a(a^2 - b^2)$ and exactly the same conditions for full diversity hold. Therefore, if the rotation is such that the original QOSTBC for $N = 4$ transmit antennas is full diversity, the corresponding $N = 3$ code is also full diversity.

The maximum-likelihood decoding for the code in (5.26) amounts to decoding pairs of symbols separately. In fact, instead of recalculating the terms in (5.5), a simple way of deriving the decoding formulas is to set $\alpha_{4,m} = 0$ in (5.7) and (5.8). Therefore, the ML decoding amounts to minimizing the sum $f_{14}(s_1, s_4) + f_{23}(s_2, s_3)$, where

$$
\begin{aligned}
f_{14}(s_1, s_4) = \sum_{m=1}^{M} &\left[\left(|s_1|^2 + |s_4|^2 \right) \left(\sum_{n=1}^{3} |\alpha_{n,m}|^2 \right) \right. \\
&+ 2\Re\left\{ \left(-\alpha_{1,m} r_{1,m}^* - \alpha_{2,m}^* r_{2,m} - \alpha_{3,m}^* r_{3,m} \right) s_1 \right. \\
&\left. + \left(\alpha_{3,m}^* r_{2,m} + \alpha_{2,m}^* r_{3,m} - \alpha_{1,m} r_{4,m}^* \right) s_4 \right\} \\
&+ \left. 4\Re\left\{ -\alpha_{2,m}^* \alpha_{3,m} \right\} \Re\{s_1 s_4^* \} \right],
\end{aligned} \tag{5.28}
$$

and

$$f_{23}(s_2, s_3) = \sum_{m=1}^{M} \Bigg[\left(|s_2|^2 + |s_3|^2 \right) \left(\sum_{n=1}^{3} |\alpha_{n,m}|^2 \right)$$
$$+ 2\Re \left\{ \left(-\alpha_{2,m} r_{1,m}^* + \alpha_{1,m}^* r_{2,m} + \alpha_{3,m} r_{4,m}^* \right) s_2 \right.$$
$$+ \left(-\alpha_{3,m} r_{1,m}^* + \alpha_{1,m}^* r_{3,m} + \alpha_{2,m} r_{4,m}^* \right) s_3 \}$$
$$+ 4\Re \left\{ \alpha_{2,m} \alpha_{3,m}^* \right\} \Re \{ s_2 s_3^* \} \Bigg]. \tag{5.29}$$

Now, we concentrate on designing codes for the cases that the number of transmit antennas is a power of two. Designing codes for other number of transmit antennas is possible by removing the columns of generator matrices as we discussed before. The main idea behind the structure of the generator matrices for QOSTBCs in (5.2), and other similar structures, is to build a 4×4 matrix from two 2×2 matrices to keep the transmission rate fixed. A similar idea can be used to combine any two $N \times N$ orthogonal STBCs to build a $2N \times 2N$ QOSTBC with the same rate [68]. One can even combine two non-square orthogonal STBCs to build a non-square QOSTBC. As an example, we consider the following QOSTBC for eight transmit antennas:

$$\mathcal{G} = \begin{pmatrix}
x_1 & x_2 & x_3 & 0 & x_4 & x_5 & x_6 & 0 \\
-x_2^* & x_1^* & 0 & -x_3 & x_5^* & -x_4^* & 0 & x_6 \\
x_3^* & 0 & -x_1^* & -x_2 & -x_6^* & 0 & x_4^* & x_5 \\
0 & -x_3^* & x_2^* & -x_1 & 0 & x_6^* & -x_5^* & x_4 \\
-x_4 & -x_5 & -x_6 & 0 & x_1 & x_2 & x_3 & 0 \\
-x_5^* & x_4^* & 0 & x_6 & -x_2^* & x_1^* & 0 & x_3 \\
x_6^* & 0 & -x_4^* & x_5 & x_3^* & 0 & -x_1^* & x_2 \\
0 & x_6^* & -x_5^* & -x_4 & 0 & x_3^* & -x_2^* & -x_1
\end{pmatrix}. \tag{5.30}$$

The QOSTBC in (5.30) provides a rate $R = 3/4$ by transmitting six symbols in eight time slots. A similar code is defined by

$$\mathcal{G} = \begin{pmatrix}
x_1 & x_2 & x_3 & 0 & x_4 & x_5 & x_6 & 0 \\
-x_2^* & x_1^* & 0 & x_3 & -x_5^* & x_4^* & 0 & x_6 \\
x_3^* & 0 & -x_1^* & x_2 & x_6^* & 0 & -x_4^* & x_5 \\
0 & x_3^* & -x_2^* & -x_1 & 0 & x_6^* & -x_5^* & -x_4 \\
x_4 & x_5 & x_6 & 0 & x_1 & x_2 & x_3 & 0 \\
-x_5^* & x_4^* & 0 & x_6 & -x_2^* & x_1^* & 0 & x_3 \\
x_6^* & 0 & -x_4^* & x_5 & x_3^* & 0 & -x_1^* & x_2 \\
0 & x_6^* & -x_5^* & -x_4 & 0 & x_3^* & -x_2^* & -x_1
\end{pmatrix}. \tag{5.31}$$

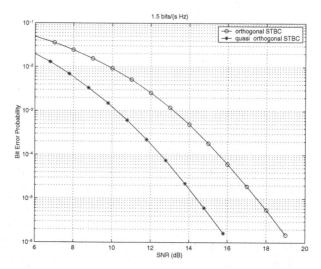

Fig. 5.7.   Bit error probability plotted against SNR for different space-time block codes at 1.5 bits/(s Hz); eight transmit antennas, one receive antenna.

We denote the ith column of $\mathcal{G}$ by $\mathcal{V}_i$. Then, for any indeterminate variables $x_1, x_2, x_3, x_4, x_5, x_6$, we have

$$
\begin{aligned}
< \mathcal{V}_1, \mathcal{V}_i > &= 0, & i &\neq 5, \\
< \mathcal{V}_2, \mathcal{V}_i > &= 0, & i &\neq 6, \\
< \mathcal{V}_3, \mathcal{V}_i > &= 0, & i &\neq 7, \\
< \mathcal{V}_4, \mathcal{V}_i > &= 0, & i &\neq 8, \\
< \mathcal{V}_5, \mathcal{V}_i > &= 0, & i &\neq 1, \\
< \mathcal{V}_6, \mathcal{V}_i > &= 0, & i &\neq 2, \\
< \mathcal{V}_7, \mathcal{V}_i > &= 0, & i &\neq 3, \\
< \mathcal{V}_8, \mathcal{V}_i > &= 0, & i &\neq 4,
\end{aligned}
\tag{5.32}
$$

where $< \mathcal{V}_i, \mathcal{V}_j >$ is the inner product of vectors $\mathcal{V}_i$ and $\mathcal{V}_j$. Therefore, the four subspaces created by $(\mathcal{V}_1, \mathcal{V}_4)$, $(\mathcal{V}_2, \mathcal{V}_5)$, $(\mathcal{V}_3, \mathcal{V}_7)$, and $(\mathcal{V}_4, \mathcal{V}_8)$ are orthogonal to each other. The diversity of these codes are four if we pick all symbols $(x_1, x_2, x_3, x_4, x_5, x_6)$ from the same constellation. Similar to the case of four transmit antennas, a full-diversity QOSTBC is possible by picking (x_4, x_5, x_6) from a rotated constellation. In other words, we replace (x_4, x_5, x_6) with $(\tilde{x}_4, \tilde{x}_5, \tilde{x}_6)$, where $\tilde{x}_k = x_k e^{j\phi}$. Figure 5.7 provides simulation results for the transmission of 1.5 bits/(s Hz) using eight transmit antennas and one receive antenna using orthogonal and quasi-orthogonal STBCs. We use QPSK for the rate 3/4 QOSTBCs and 8-PSK for the rate 1/2 orthogonal STBC. The rotated QOSTBC uses a rotation of $\phi = \pi/4$.

5.6 Summary of important results

- Using the following quasi-orthogonal STBC, one can achieve pairwise decoding, rate one, and full diversity:

$$
\mathcal{G} = \begin{pmatrix} x_1 & x_2 & x_3 & x_4 \\ -x_2^* & x_1^* & -x_4^* & x_3^* \\ -x_3^* & -x_4^* & x_1^* & x_2^* \\ x_4 & -x_3 & -x_2 & x_1 \end{pmatrix}.
$$

- If we pick s_k, $k = 1, 2$ symbols from an L-PSK constellation with even L and $\tilde{s}_{k+2}$ symbols from a rotated constellation, the resulting QOSTBC provides full diversity.
- For an L-PSK constellation, L odd, if the rotation $\phi = \pi/L$, the code is not full diversity. Otherwise, the code provides full diversity.
- The optimal rotations for BPSK, QPSK, 8-PSK, and QAM are $\pi/2$, $\pi/4$, $\pi/8$, and $\pi/4$, respectively.

5.7 Problems

1 Consider the QOSTBC in (5.2) for a system with four transmit antennas and one receive antenna. The encoder uses QPSK and a rotation of $\phi = \pi/4$ for x_3 and x_4.

   (a) What are the transmitted symbols for the following bitstream: 0101100101000111?

   (b) Let us assume that the path gains during the transmission of the first four symbols are $\alpha_1 = 1, \alpha_2 = j, \alpha_3 = -1$, and $\alpha_4 = -j$, where $j = \sqrt{-1}$. Also, let us assume that the noise samples at the receiver are $\eta_1 = 0.1 + 0.1j$, $\eta_2 = 0.4 + 0.3j$, $\eta_3 = -0.4 - 1.4j$, and $\eta_4 = -0.3 - 0.2j$. Find the first eight decoded bits using a maximum-likelihood decoder.

2 Does the following code provide the properties of a quasi-orthogonal STBC? Explain your answer.

$$
\mathcal{G} = \begin{pmatrix} \mathcal{G}_{12} & \mathcal{G}_{34} \\ -\mathcal{G}_{34} & \mathcal{G}_{12} \end{pmatrix} = \begin{pmatrix} x_1 & x_2 & x_3 & x_4 \\ -x_2^* & x_1^* & -x_4^* & x_3^* \\ -x_3 & -x_4 & x_1 & x_2 \\ x_4^* & -x_3^* & -x_2^* & x_1^* \end{pmatrix}.
$$

3 Consider the simulation results in Figures 5.2 and 5.4. Based on Theorem 5.2.1, all codes with $\phi \neq 0$ provide full diversity. On the other hand, the lines in Figures 5.2 and 5.4 are not parallel to each other. Is there a contradiction? Explain your answer.

4 Derive the decision metrics for the pairwise maximum-likelihood decoding of the QOSTBC defined by (5.23) and (5.24).

5 For QPSK constellation, use Monte Carlo simulations to derive the bit error rate (BER) versus the received SNR simulation results for the QOSTBCs in Section 5.4. How does the result compare with the results in Figure 5.6.

6

Space-time trellis codes

6.1 Introduction

Trellis-coded modulation (TCM) [144, 14] combines modulation and coding to achieve higher coding gains. It provides significant coding gain and a better performance for a given bandwidth compared to uncoded modulation schemes. Space-time trellis codes (STTCs) combine modulation and trellis coding to transmit information over multiple transmit antennas and MIMO channels. Therefore, one can think of STTCs as TCM schemes for MIMO channels. We start this chapter with a brief description of TCM design and the corresponding encoding and decoding. The main idea behind coding is to use structured redundancy to reduce the effects of noise. One approach is to restrict the transmitted symbols to a subset of constellation points to have a larger minimum Euclidean distance. This approach, however, results in a rate reduction due to a decrease in the size of available constellation points. To compensate for the rate reduction, one may expand the constellation and use a subset of the expanded constellation at each time slot. A finite-state machine represented by a trellis is utilized to decide which subset should be used at each time slot. The goal is to maximize the coding gain for a given rate. This is the main idea behind TCM and in such a structure the transmitted symbols at different time slots are not independent.

Let us assume that we use a trellis with I states such that 2^l branches leave every state. We enumerate the states from top to bottom starting from zero. Thus, we have states $i = 0, 1, \ldots, I - 1$. To design a TCM for a SISO channel to transmit b bits per channel use, one needs to utilize a modulation scheme with 2^{b+1} points. Let us denote $d_{\min}$ as the minimum Euclidean distance among all points in a set. The first step of TCM design is set partitioning. One needs to partition the set of constellation points into a hierarchy of subsets such that the $d_{\min}$ increases at each partition level. Figure 6.1 shows such a set partitioning for an 8-PSK constellation. To define the Euclidean distance between two sets of constellation points, we consider the set of

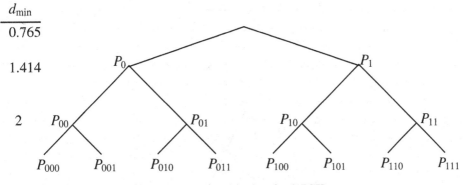

Fig. 6.1.   Set partitioning for 8-PSK.

all possible pairs such that the two symbols of the pair are selected from the two different sets. Then, the Euclidean distance is defined as the minimum Euclidean distance among all such pairs of symbols.

The second step in the design is to assign different subsets to different paths in the trellis. At each state, a collection of l input bits picks one of the 2^l branches originating from the state. The remaining $b - l$ bits, if any, are utilized to pick one of the 2^{b-l} parallel paths. The l bits that select the state transitions are "coded bits" while the $b - l$ remaining bits, if any, are "uncoded bits." Another way of representing the finite-state machine that governs the state transitions is to use the l coded bits as inputs of a convolutional code. When uncoded bits exist, $l \neq b$, one can consider one transition for every constellation point. If separate parallel paths are considered as different paths, then every path in the trellis corresponds to b input bits, one transmitted symbol from the original constellation with 2^{b+1} points, and the states of the TCM before and after the transmission. Note that the actual transmitted symbol depends on the current state and the b input bits. Ungerboeck has suggested the following principles, as a rule of thumb, to maximize the coding gain of the TCM [144]:

• all subsets should be used an equal number of times in the trellis;
• transitions originating from the same state or merging into the same state in the trellis should be assigned subsets that are separated by the largest Euclidean distance;
• parallel paths, if they occur, should be assigned signal points separated by the largest Euclidean distance.

Let us denote the state of the encoder at time t by S_t. The encoder usually starts at state zero, that is $S_1 = 0$. Then at time t, the transmitted symbol depends on the state at time t, S_t, and the b input bits. The b input bits select the next state, S_{t+1}, and the constellation point that is transmitted. This selection is based on the structure of the code and the sets that are assigned to different states. We use Figure 6.2 and

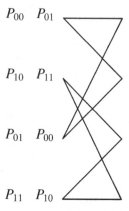

P_{00} P_{01}

P_{10} P_{11}

P_{01} P_{00}

P_{11} P_{10}

Fig. 6.2. A four-state trellis; an example of Ungerboeck's TCM.

the corresponding TCM to explain the details of the selection of the next state and the transmitted symbol. Next to every state, there are two sets corresponding to the two branches originating from that state. The first set corresponds to the top branch and the second set corresponds to the bottom branch. If there were more than two branches originating from each state, the sets from left to right would correspond to branches from top to bottom. The b input bits are divided into two parts of l most significant and $b - l$ least significant bits. For the example in Figure 6.2, $b = 2$ and $l = 1$. Therefore, the most significant bit selects the leaving branch, equivalently the set and next state S_{t+1}, and the least significant bit picks a symbol from the selected set. The encoder's state-machine goes to state S_{t+1} and the same encoding process continues. Enough zero-coded bits should be added to the input bitstream in order to return the state of the encoder to state zero at the end of transmission. If the memory of the convolutional code representing the state-machine is Q, we need to add lQ coded bits at the end. As a result, we have $S_{T+Q+1} = 0$ for all valid codewords. For the example in Figure 6.2, the memory is $Q = 2$. A valid path corresponds to a path that starts from state zero and ends at state zero. Therefore, for a frame length of $T + Q$, a valid path satisfies $S_1 = S_{T+Q+1} = 0$. Note that $2T$ bits, T information symbols, are transmitted over $T + Q$ time slots. At time $t = 1, 2, \ldots, T + Q$, symbol c_t that corresponds to a branch from state S_t to S_{t+1} is transmitted. The last Q symbols do not carry information.[1] Transmitting the symbols over an AWGN channel, the maximum-likelihood decoder finds the most likely valid path that starts from state zero and merges to state zero after $T + Q$ time slots. Let us assume that we receive $r_1, r_2, \ldots, r_{T+Q}$ at time slots $t = 1, 2, \ldots, T + Q$ at the decoder. The distance of a path that transmits symbols $c_1, c_2, \ldots, c_{T+Q}$ from the received signals

[1] If $b > l$, it is possible to transmit $b - l$ bits for each of these Q symbols and still go back to state zero.

is $\sum_{t=1}^{T+Q} |r_t - c_t|^2$. This distance can be considered as a function of the path and is called the "path metric." The most likely path is the one which has the minimum distance from the received signals or equivalently the minimum path metric. The decoder finds the set of constellation symbols that constructs a valid path and solves the following minimization problem:

$$\min_{c_1, c_2, \dots, c_{T+Q}} \sum_{t=1}^{T+Q} |r_t - c_t|^2. \tag{6.1}$$

The number of possible valid paths grows exponentially with the frame length T. This is due to the fact that at each time slot, 2^l branches leave every state. Therefore, there are $2^{l(T+Q)}$ paths per state. Since valid paths start from state zero and go back to state zero, not all of these $I 2^{l(T+Q)}$ paths are valid. However, the number of valid paths grows exponentially with l and T. The Viterbi algorithm can be used to find the most likely path such that the complexity grows linearly with the frame length T. One can associate with every branch of the trellis, transmitting c_t at time t, a branch metric defined as $|r_t - c_t|^2$. Then, for every path, the path metric is the sum of its constructing branch metrics. The Viterbi algorithm, equivalently dynamic programming [13], is a smart search method to find the path with the smallest total metric. The main idea behind the Viterbi algorithm is the optimum nature of the partial parts of the optimal path. We represent a path by the sequence of states that the path goes through. Let us assume that the optimal path consists of $0\hat{S}_2\hat{S}_3 \cdots \hat{S}_{T+Q}0$ and transmits $\hat{c}_1, \hat{c}_2, \dots, \hat{c}_{T+Q}$. Also, let us consider the first T' branches of the optimal path, that is $0\hat{S}_2\hat{S}_3 \cdots \hat{S}_{T'}\hat{S}_{T'+1}$ and its partial path metric $\sum_{t=1}^{T'} |r_t - \hat{c}_t|^2$. Now, we consider another path $0S_2S_3 \cdots S_{T'}\hat{S}_{T'+1}$ and the corresponding partial path metric $\sum_{t=1}^{T'} |r_t - c_t|^2$. Obviously, $\sum_{t=1}^{T'} |r_t - \hat{c}_t|^2 \leq \sum_{t=1}^{T'} |r_t - c_t|^2$ because otherwise the path $0S_2S_3 \cdots S_{T'}\hat{S}_{T'+1}\hat{S}_{T'+2} \cdots \hat{S}_{T+Q}0$ has a path metric smaller than that of the path $0\hat{S}_2\hat{S}_3 \cdots \hat{S}_{T+Q}0$ and this contradicts with the optimality of $0\hat{S}_2\hat{S}_3 \cdots \hat{S}_{T+Q}0$. This is due to the fact that the path metric is an additive metric, that is the path metric of the concatenation of two paths is the sum of their path metrics. Therefore, one can keep track of the optimal path sequentially as time passes. The details of the Viterbi algorithm follow.

At the beginning, the trellis is at state zero. At time $t + 1$, the number of branches merging to every state is 2^l. If we know the optimal partial paths for every state at time t, there are 2^l partial paths at time $t + 1$ merging to every state i. The Viterbi algorithm finds the path with the minimum partial metric among these 2^l paths and removes the other $2^l - 1$ partial paths. Such an optimal partial path is called a survivor path. Repeating the same procedure for every state, we end up with I survivor paths at time $t + 1$. Note that at each time, there is only one survivor path per state. The process is continued until the last time slot. Since the trellis should end at state zero, the solution of the minimization problem in (6.1) corresponds to the survivor path that ends at state zero. The input bitstream that generates this survivor path is the outcome of the maximum-likelihood decoding. Note that the Viterbi algorithm needs to generate all path metrics and find the survival path at time $T + Q + 1$ before it can generate any decoded bit. Therefore, the delay of the Viterbi algorithm is $T + Q$ symbols. One may use traceback or other methods to reduce the delay of decoding [149].

6.2 Space-time trellis coding

Space-time trellis codes (STTCs) combine modulation and trellis coding to transmit information over multiple transmit antennas and MIMO channels. The goal is to construct trellis codes satisfying the space-time coding design criteria. To achieve full diversity, a STTC should satisfy the rank criterion as discussed in Chapter 3. Defining the rate of the code as the number of transmitted symbols per time slot, as we did in Chapter 4, another desirable property is to have a rate one code. Of course among all possible full-rate, full-diversity codes, the one that provides the highest coding gain is preferable. The first example of a rate one full diversity space-time trellis code is the delay diversity scheme in [154]. A similar method is proposed in [110]. Let us assume a constellation that maps b bits of data to a symbol. The delay diversity scheme uses every b input bit to pick one symbol that is transmitted from the first antenna. Then, the second antenna transmits the same symbols with a delay of one symbol. Maximum-likelihood decoding is used at the receiver to recover the transmitted symbols. Since the symbols go through two independent channel path gains, a diversity of two is achieved for one receive antenna (or per receive antenna). Delay diversity is a special case of STTCs presented in [139]. Different examples of STTCs for BPSK, QPSK, 8-PSK, and 16-QAM constellations are designed in [139]. In some cases, the proposed STTCs in [139] are delay diversity codes. In other cases, they provide a higher coding gain and better performance compared to the delay diversity codes.

6.2.1 Encoding

Let us concentrate on two transmit antennas. There are two symbols that are transmitted from these two antennas for every path in the trellis. There is no parallel path in the STTCs of [139]. Therefore, the STTC can be represented by a trellis and a pair of symbols for each trellis path. We use the corresponding indices of the symbols to present the transmitted symbols for each path. For STTCs that send b bits/(s Hz) of information, 2^b branches leave every state. A set of 2^b pairs of indices next to every state represents the 2^b pairs of symbols for the 2^b outgoing branches

00 01 02 03

10 11 12 13

20 21 22 23

30 31 32 33

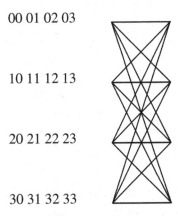

Fig. 6.3.   A four-state STTC; $r = 2$ bits/(s Hz) using QPSK.

from top to bottom. For example, Figure 6.3 illustrates a rate one space-time trellis code to transmit $r = 2$ bits/(s Hz). The code uses a QPSK constellation, $b = 2$, that includes indices 0, 1, 2, 3 to represent $1, j, -1, -j$, respectively. Similar to a TCM, the encoding always starts at state 0. Let us assume that the encoder is at state S_t at time t. Then, $b = 2$ bits arrive at the encoder to pick one of the $2^b = 4$ branches leaving state S_t. The corresponding indices of the selected branch $i_1 i_2$ are used to choose two symbols $c_{t,1} c_{t,2}$ from the constellation. These symbols are respectively sent from the two transmit antennas simultaneously. The encoder moves to state S_{t+1} which is at the right-hand side of the selected branch. At the end, similar to the encoding for a TCM, extra branches are picked to make sure that the encoder stops at state 0. For example, at time t, if the encoder is at state $S_t = 0$ for the STTC in Figure 6.3 and the input bits are 10, the selected indices are $i_1 = 0$ and $i_2 = 2$. The transmitted symbols $c_{t,1} = 1$ and $c_{t,2} = -1$ are sent from the first and second antennas, respectively. The next state of the encoder will be state $S_{t+1} = 2$. Although the code has been designed manually, there is a logic behind it that guarantees full diversity. All branches diverging from a state contain the same symbol for the first antenna while all branches merging to a state contain the same symbol for the second antenna. Using a similar method, one can manually design full-rate full-diversity STTCs for other constellations and trellises [139]. We provide more examples of STTCs in what follows.

In Chapter 3, the CGD between codewords $\mathbf{C}^1$ and $\mathbf{C}^2$ for a full diversity code is defined as $\text{CGD}(\mathbf{C}^1, \mathbf{C}^2) = \det(A(\mathbf{C}^1, \mathbf{C}^2))$. The minimum value of CGD among all possible pairs of codewords is used as an indication of the performance of the code. In the case of STTCs, any valid codeword starts from state zero and ends at state zero. Since the common branches between $\mathbf{C}^1$ and $\mathbf{C}^2$ do not contribute to the CGD, the minimum CGD may correspond to the determinant for any pair of paths

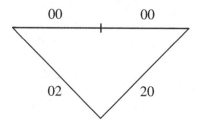

Fig. 6.4.   Two typical paths differing in $P = 2$ transitions.

diverging from a state and merging to the same state after P transitions. Figure 6.4 shows an example for $P = 2$ transitions for the STTC in Figure 6.3. The first path stays in state zero during both transitions, that is 000. The second path goes to state two in the first transition and merges to state zero in the second transition, that is 020. The corresponding codewords are

$$\mathbf{C}^1 = \begin{pmatrix} 1 & 1 \\ 1 & 1 \end{pmatrix}, \quad \mathbf{C}^2 = \begin{pmatrix} 1 & -1 \\ -1 & 1 \end{pmatrix}. \tag{6.2}$$

Therefore, $\mathrm{CGD}(\mathbf{C}^1, \mathbf{C}^2) = 16$. There are many pairs of paths that differ in $P = 2$ transitions. The minimum CGD is calculated by computing the CGD of all these pairs and finding the minimum. In fact, for the example in Figure 6.3, the minimum CGD is 4 that corresponds to paths 000 and 010.

6.2.2 Decoding

The maximum-likelihood decoding finds the most likely valid path that starts from state zero and merges to state zero after $T + Q$ time slots. Let us assume that we receive $r_{1,m}, r_{2,m}, \ldots, r_{T+Q,m}$ at time slots $t = 1, 2, \ldots, T + Q$ and receive antenna m. Similar to the case of TCM, the Viterbi algorithm can be used for the ML decoding of STTCs. If a branch of the trellis transmits symbols s_1 and s_2 from antennas one and two, respectively, the corresponding branch metric is given by

$$\sum_{m=1}^{M} |r_{t,m} - \alpha_{1,m} s_1 - \alpha_{2,m} s_2|^2. \tag{6.3}$$

Then, the path metric of a valid path is the sum of the branch metrics for the branches that form the path. The most likely path is the one which has the minimum path gain. The ML decoder finds the set of constellation symbols that construct a valid path and solves the following minimization problem:

$$\min_{c_{1,1}, c_{1,2}, c_{2,1}, c_{2,2}, \dots, c_{T+Q,1}, c_{T+Q,2}} \sum_{t=1}^{T+Q} \sum_{m=1}^{M} |r_{t,m} - \alpha_{1,m} c_{t,1} - \alpha_{2,m} c_{t,2}|^2. \quad (6.4)$$

The details of the Viterbi algorithm to solve (6.4) and (6.1) are similar to each other. The main difference is using the branch metric in (6.3) for STTCs instead of the Euclidean distance for TCMs.

6.2.3 Design rules and properties of STTCs

To design a good STTC, one needs to consider the design criteria in Chapter 3. The most important criterion is the rank criterion that guarantees full diversity. In [139], the following two design rules have been suggested to achieve full diversity for two transmit antennas:

- transitions diverging from the same state should differ in the second symbol;
- transitions merging to the same state should differ in the first symbol.

The structure of the STTC that we have discussed so far imposes a condition on the minimum number of states in the trellis. The following lemma provides a lower bound on the number of states that is required to achieve a particular diversity order.

Lemma 6.2.1 *For a STTC with rate b bits/(s Hz) and a diversity r, at least $2^{b(r-1)}$ states are needed.*

Proof For a rate of b bits/(s Hz), we need 2^b branches out of each state. At time $r - 1$, there are $2^{b(r-1)}$ paths originating from state 0. To achieve a diversity of r, all these paths should be separated, that is none of them can merge at the same state in any time between 0 and $r - 1$. To prove this fact, we assume that there exists two paths originating at state 0 and merging at time r', where $r' < r$. Let us pick a pair of codewords $(\mathbf{C}^1, \mathbf{C}^2)$ that contain these two paths up to time r' and the same set of branches after time r'. Then, the corresponding difference matrix $D(\mathbf{C}^1, \mathbf{C}^2)$ contains only r' nonzero rows and its rank cannot exceed $r' < r$. This is a contradiction that shows all $2^{b(r-1)}$ paths at time $r - 1$ should be separate and therefore we should have at least $2^{b(r-1)}$ states. $\qquad\square$

As a result of the above lemma, using the discussed STTC structure, we need at least 2^b states to design a full diversity STTC for two transmit antennas. In other words, it is impossible to design STTCs with less than 2^b states and achieve full diversity. In Chapter 7, we provide another structure to design space-time trellis codes that does not suffer from this limitation.

In what follows, we provide some examples of STTCs.

00 01

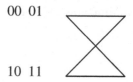

10 11

Fig. 6.5.   A two-state STTC; $r = 1$ bit/(s Hz) using BPSK.

00 01

01 00

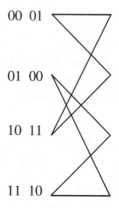

10 11

11 10

Fig. 6.6.   A four-state STTC; $r = 1$ bit/(s Hz) using BPSK.

Example 6.2.1 *An example for two states and $r = 1$ bit/(s Hz) using BPSK is depicted in Figure 6.5. The minimum CGD for this code is 16. One example of two codewords that provide a CGD of 16 corresponds to the following pair of paths. The first path stays in state zero during two transitions, that is 000. The corresponding codeword is*

$$\mathbf{C}^1 = \begin{pmatrix} 1 & 1 \\ 1 & 1 \end{pmatrix}. \tag{6.5}$$

The second path goes to state one in the first transition and merges to state zero in the second transition, that is 010, resulting in

$$\mathbf{C}^2 = \begin{pmatrix} 1 & -1 \\ -1 & 1 \end{pmatrix}. \tag{6.6}$$

Therefore, $CGD(\mathbf{C}^1, \mathbf{C}^2) = 16$.

Example 6.2.2 *An example of a four-state STTC with $r = 1$ bit/(s Hz) using BPSK is depicted in Figure 6.6. The minimum CGD for this code is 32. One pair of*

00 01 02 03

10 11 12 13

20 21 22 23

30 31 32 33

22 23 20 21

32 33 30 31

02 03 00 01

12 13 10 11

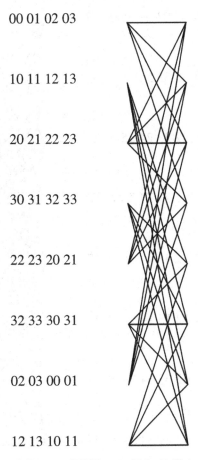

Fig. 6.7.   An eight-state STTC; $r = 2$ bits/(s Hz) using QPSK.

paths that results in a CGD equal to 32 is 0000 and 0120. The corresponding codewords are

$$\mathbf{C}^1 = \begin{pmatrix} 1 & 1 \\ 1 & 1 \\ 1 & 1 \end{pmatrix}, \quad \mathbf{C}^2 = \begin{pmatrix} 1 & -1 \\ 1 & -1 \\ -1 & 1 \end{pmatrix}. \tag{6.7}$$

Example 6.2.3 *An example for eight states and $r = 2$ bits/(s Hz) using QPSK is depicted in Figure 6.7.*

Example 6.2.4 *An example for 8 states and $r = 3$ bits/(s Hz) using 8-PSK is depicted in Figure 6.8.*

00 01 02 03 04 05 06 07

50 51 52 53 54 55 56 57

20 21 22 23 24 25 26 27

70 71 02 73 74 75 76 77

40 41 42 43 44 45 46 47

10 11 12 13 14 15 16 17

60 61 62 63 64 65 66 67

30 31 32 33 34 35 36 37

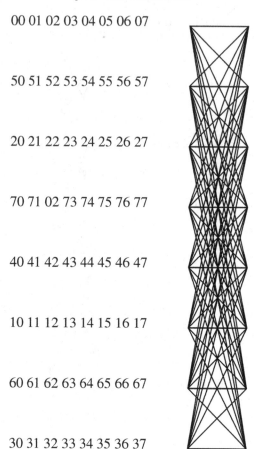

Fig. 6.8.   An eight-state STTC; $r = 3$ bits/(s Hz) using 8-PSK.

6.3 Improved STTCs

The encoder's output of a block code is only a function of the input bits for that block. We call this block of input bits the current block of data. A convolutional or trellis code is different because the code has memory and the encoder's output is a function of the current block of data and the previous blocks of data. Similarly the main difference between STBCs and STTCs is the fact that the encoder of STTCs has memory. For a constellation with $L = 2^b$ points to represent b bits, a STBC encodes a block of Kb bits for every T time slots. The output of the encoder at the next T time slots is only a function of the new Kb input bits. In other words, since there is no memory between different blocks, every block of Kb input bits participates in the encoding process only once. On the other hand, in the case of STTCs, because of the memory, every b input bit participates in the encoding process $Q + 1$ times, once as the input and Q times through the state of the trellis. Therefore, the encoder's output at each time slot can be calculated from the $Q + 1$ blocks of data. Note that the $(Q + 1)b$ input bits that generate the output of the encoder at time t overlap with the $(Q + 1)b$ bits that generate the output symbols at other times. There is no such an overlap for the block codes.

The encoder's output consists of N symbols to be transmitted from N transmit antennas. Let us denote the b input bits at time t as $u_{t,1}, u_{t,2}, \ldots, u_{t,b}$. The indices of the transmitted symbols can be computed from the following formula:

$$(i_{t,1}, i_{t,2}, \ldots, i_{t,N}) = U_t \mathbf{G} \bmod L, \tag{6.8}$$

where

$$U_t = (u_{t,1}, u_{t,2}, \ldots, u_{t,b}, u_{t-1,1}, u_{t-1,2}, \ldots, u_{t-1,b}, \ldots, u_{t-Q,1},$$
$$u_{t-Q,2}, \ldots, u_{t-Q,b}) \tag{6.9}$$

and $\mathbf{G}$ is the generating matrix for the STTC. Note that the generating matrix $\mathbf{G}$ in (6.8) is different from the matrix $\mathcal{G}$ introduced in Chapter 4. In Chapters 4 and 5, $\mathcal{G}$ is utilized to represent the encoder of a STBC. The transmitted symbols for a STBC are derived by replacing the indeterminate variable x_k with the corresponding input symbol s_k for all k. The input symbols affect transmitted symbols, the output of the encoder, only during one block. However, using $\mathbf{G}$ in defining the encoder of STTCs, the input bits define the indices of the transmitted symbols through (6.8). A block of input bits affect the transmitted symbols for $Q + 1$ blocks in this case. In other words, the encoder of a STTC has memory. Also, note that the second subscript index in $i_{t,n}$, n, is an index for the corresponding transmit antenna while the second subscript index in $u_{t,j}$, j, is an index for the corresponding input bit.

Equation (6.8) is just another representation of the STTC. The trellis representation of the previous section and the matrix representation of this section are equivalent. The following example demonstrates these two different approaches for encoding an input bit stream.

Example 6.3.1 *The generating matrix for the code in Figure 6.3 is*

$$\mathbf{G} = \begin{pmatrix} 0 & 2 \\ 0 & 1 \\ 2 & 0 \\ 1 & 0 \end{pmatrix}. \tag{6.10}$$

In this case, $Q = 1$ and four bits generate the two transmitted symbols at each time slot.

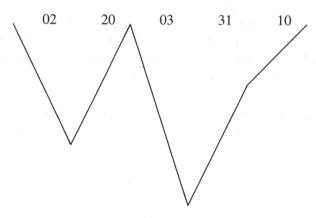

02   20   03   31   10

Fig. 6.9.   Encoder's output for Example 6.3.1.

Let us assume that the input bitstream for the code in Figure 6.3 with generating matrix (6.10) is 01110010. *Figure 6.9 shows the trellis path corresponding to this input bitstream. Note that we have added* 00 *at the end to guarantee that the state-machine returns to state zero. Using (6.8) and the generating matrix (6.10), one can find the same set of transmitted symbols. At time t = 1, we have*

$$(1\ 0\ 0\ 0) \begin{pmatrix} 0 & 2 \\ 0 & 1 \\ 2 & 0 \\ 1 & 0 \end{pmatrix} = (0\ 2). \tag{6.11}$$

Therefore, 1 *and* −1 *are transmitted at time t = 1 from the first and second antennas, respectively. At time t = 2, we have*

$$(0\ 0\ 1\ 0) \begin{pmatrix} 0 & 2 \\ 0 & 1 \\ 2 & 0 \\ 1 & 0 \end{pmatrix} = (2\ 0). \tag{6.12}$$

Therefore, −1 *and* 1 *are transmitted at time t = 2 from the first and second antennas, respectively. At time t = 3, we have*

$$(1\ 1\ 0\ 0) \begin{pmatrix} 0 & 2 \\ 0 & 1 \\ 2 & 0 \\ 1 & 0 \end{pmatrix} = (0\ 3). \tag{6.13}$$

Therefore, 1 *and* −j *are transmitted at time t = 3 from the first and second antennas, respectively. At time t = 4, we have*

$$(0\ 1\ 1\ 1) \begin{pmatrix} 0 & 2 \\ 0 & 1 \\ 2 & 0 \\ 1 & 0 \end{pmatrix} = (3\ 1). \tag{6.14}$$

Therefore, $-j$ *and* j *are transmitted at time* $t = 4$ *from the first and second antennas, respectively. At time* $t = 5$, *we have*

$$(0\ 0\ 0\ 1) \begin{pmatrix} 0 & 2 \\ 0 & 1 \\ 2 & 0 \\ 1 & 0 \end{pmatrix} = (1\ 0). \tag{6.15}$$

Therefore, j *and* 1 *are transmitted at time* $t = 5$ *from the first and second antennas, respectively.*

The algebraic structure of the new presentation of STTCs in (6.8) makes it possible to search for different STTCs using a computer. In fact, many researchers have proposed such a search and have designed new STTCs. The following example is from [9].

Example 6.3.2 *A four-state STTC for transmitting* $b = 2$ *bits/(s Hz) using QPSK is designed from the following generating matrix:*

$$\mathbf{G} = \begin{pmatrix} 2 & 2 \\ 0 & 2 \\ 1 & 0 \\ 3 & 1 \end{pmatrix}. \tag{6.16}$$

The corresponding trellis representation is given in Figure 6.10. The CGD of this code is 8 which is more than that of the STTC in Figure 6.3.

The following example is from [156].

Example 6.3.3 *Another four-state STTC for transmitting* $b = 2$ *bits/(s Hz) using QPSK is designed from the following generating matrix:*

$$\mathbf{G} = \begin{pmatrix} 2 & 2 \\ 0 & 2 \\ 1 & 2 \\ 2 & 1 \end{pmatrix}. \tag{6.17}$$

The corresponding trellis representation is given in Figure 6.11. The CGD of this code is 8 which is more than that of the STTC in Figure 6.3.

00 02 22 20

31 33 13 11

10 12 32 30

01 03 23 21

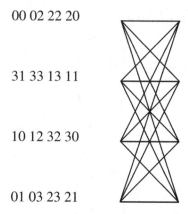

Fig. 6.10.　A four-state STTC; $r = 2$ bits/(s Hz) using QPSK.

00 02 22 20

21 23 03 01

12 10 30 32

33 31 11 13

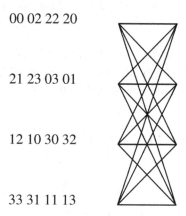

Fig. 6.11.　A four-state STTC; $r = 2$ bits/(s Hz) using QPSK.

Example 6.3.4 *An eight-state STTC for transmitting $b = 2$ bits/(s Hz) using QPSK can be designed from a 5×2 generating matrix. The elements of the generating matrix are QPSK indices. The following generating matrix provides a CGD of 16:*

$$
G = \begin{pmatrix} 0 & 2 \\ 2 & 1 \\ 1 & 0 \\ 0 & 2 \\ 2 & 2 \end{pmatrix}. \tag{6.18}
$$

Example 6.3.5 *A 16-state STTC for transmitting $b = 2$ bits/(s Hz) using QPSK with a CGD of 32 can be designed from the following generating*

00 01

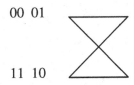

11 10

Fig. 6.12. A two-state STTC; $r = 1$ bit/(s Hz) using BPSK.

matrix:

$$G = \begin{pmatrix} 0 & 2 \\ 2 & 2 \\ 1 & 1 \\ 1 & 2 \\ 2 & 0 \\ 0 & 2 \end{pmatrix}. \tag{6.19}$$

Example 6.3.6 *To design a 32-state STTC for transmitting $b = 2$ bits/(s Hz) using QPSK, a 7×2 matrix of QPSK indices is needed. An example with a CGD of 36 is:*

$$G = \begin{pmatrix} 2 & 2 \\ 0 & 2 \\ 1 & 0 \\ 2 & 1 \\ 1 & 2 \\ 2 & 0 \\ 2 & 2 \end{pmatrix}. \tag{6.20}$$

Example 6.3.7 *A two-state STTC for transmitting $b = 1$ bit/(s Hz) using BPSK can be designed from the following generating matrix:*

$$G = \begin{pmatrix} 0 & 1 \\ 1 & 1 \end{pmatrix}. \tag{6.21}$$

The corresponding trellis representation is given in Figure 6.12. The CGD of this code is 16.

Example 6.3.8 *In this example, we consider a four-state STTC for transmitting $b = 1$ bit/(s Hz) using BPSK. A 3×2 binary matrix can generate such a code. One example is*

$$G = \begin{pmatrix} 0 & 1 \\ 1 & 0 \\ 1 & 1 \end{pmatrix}. \tag{6.22}$$

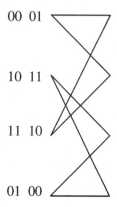

00 01

10 11

11 10

01 00

Fig. 6.13.   A four-state STTC; $r = 1$ bit/(s Hz) using BPSK.

We use this example to derive the corresponding trellis representation. The parameters of the trellis are $Q = 2$ and $b = 1$. Therefore, the input vector is $U_t = (u_{t,1}, u_{t-1,1}, u_{t-2,1}) = (u_t, u_{t-1}, u_{t-2})$ and the branch labels are calculated from (6.8). First, we need a convention to map input bits at time $t - 1$ and $t - 2$ to state numbers. To have a trellis similar to that of Figure 6.6, we use $2u_{t-2} + u_{t-1}$ as the state number. Then, for the branches starting from state zero, we have $u_{t-2} = u_{t-1} = 0$. In this case, an input bit equal to zero and one results in $U_t = (0, 0, 0)$ and $U_t = (1, 0, 0)$, respectively. Similarly, for state one, we have $u_{t-2} = 0$, $u_{t-1} = 1$ while $U_t = (0, 1, 0)$ and $U_t = (1, 1, 0)$ for $u_t = 0$ and $u_t = 1$, respectively. For state two, the input vectors are $U_t = (0, 0, 1)$ and $U_t = (1, 0, 1)$. Finally, the two remaining input vectors $U_t = (0, 1, 1)$ and $U_t = (1, 1, 1)$ correspond to state three. Using (6.8), we calculate the branch labels that are shown in Figure 6.13. The CGD of this code is 48. One pair of paths that results in this CGD is 0000 and 0120. The corresponding codewords are

$$\mathbf{C}^1 = \begin{pmatrix} 1 & 1 \\ 1 & 1 \\ 1 & 1 \end{pmatrix}, \quad \mathbf{C}^2 = \begin{pmatrix} 1 & -1 \\ -1 & 1 \\ -1 & -1 \end{pmatrix}. \tag{6.23}$$

Example 6.3.9 *An eight-state STTC for transmitting $b = 1$ bit/(s Hz) using BPSK can be designed from the following generating matrix:*

$$\mathbf{G} = \begin{pmatrix} 1 & 1 \\ 0 & 1 \\ 1 & 0 \\ 1 & 1 \end{pmatrix}. \tag{6.24}$$

The minimum CGD for this code is 80.

Example 6.3.10 *To design a 16-state STTC for transmitting $b = 1$ bit/(s Hz) using BPSK, a 5×2 binary matrix is needed. One example with a CGD of 112 is:*

$$\mathbf{G} = \begin{pmatrix} 1 & 0 \\ 1 & 1 \\ 0 & 1 \\ 1 & 1 \\ 1 & 1 \end{pmatrix}. \tag{6.25}$$

6.4 Performance of STTCs

We rewrite the conditional PEP in (3.9) as

$$P(\mathbf{C}^1 \rightarrow \mathbf{C}^2|\mathbf{H}) = Q\left(\sqrt{\frac{\gamma}{2}\|(\mathbf{C}^2 - \mathbf{C}^1) \cdot \mathbf{H}\|_F^2}\right). \tag{6.26}$$

Using Craig's formula [29] to calculate the Q-function as we did in (4.127) results in

$$P(\mathbf{C}^1 \rightarrow \mathbf{C}^2|\mathbf{H}) = \frac{1}{\pi} \int_0^{\pi/2} \exp\left(\frac{-\gamma\|(\mathbf{C}^2 - \mathbf{C}^1) \cdot \mathbf{H}\|_F^2}{4 \sin^2 \theta}\right) d\theta. \tag{6.27}$$

Defining

$$\xi = \frac{\gamma\|(\mathbf{C}^2 - \mathbf{C}^1) \cdot \mathbf{H}\|_F^2}{4} = \frac{\gamma}{4} \sum_{m=1}^{M} \mathbf{H}_m^H \cdot (\mathbf{C}^2 - \mathbf{C}^1)^H \cdot (\mathbf{C}^2 - \mathbf{C}^1) \cdot \mathbf{H}_m, \tag{6.28}$$

we have

$$P(\mathbf{C}^1 \rightarrow \mathbf{C}^2|\mathbf{H}) = \frac{1}{\pi} \int_0^{\pi/2} \exp\left(\frac{-\xi}{\sin^2 \theta}\right) d\theta. \tag{6.29}$$

Integrating over the pdf of ξ results in the following PEP [114]

$$P(\mathbf{C}^1 \rightarrow \mathbf{C}^2) = \frac{1}{\pi} \int_0^{\pi/2} M_\xi\left(\frac{-1}{\sin^2 \theta}\right) d\theta, \tag{6.30}$$

where

$$M_\xi(u) = E[e^{u\xi}] \tag{6.31}$$

is the MGF of random variable ξ. To compute the PEP, we need to calculate $M_\xi(u)$ given the statistics of the path gains. Turin has provided formulas that can be used in this calculation when the path gains are independent complex Gaussian random variables [143]. It can be shown that [114]

$$M_\xi(u) = \left\{\det\left[I_N - \frac{u\gamma}{4}(\mathbf{C}^2 - \mathbf{C}^1)^H \cdot (\mathbf{C}^2 - \mathbf{C}^1)\right]\right\}^{-M}. \tag{6.32}$$

Denoting the nth eigenvalue of $\mathbf{C}^2 - \mathbf{C}^1$ by λ_n, we have [115]

$$M_\xi(u) = \left\{ \prod_{n=1}^{N} \left[1 - \frac{u\gamma\lambda_n}{4} \right] \right\}^{-M}. \tag{6.33}$$

Therefore, the PEP can be calculated from

$$P(\mathbf{C}^1 \to \mathbf{C}^2) = \frac{1}{\pi} \int_0^{\pi/2} \left\{ \prod_{n=1}^{N} \left[\frac{4\sin^2\theta}{4\sin^2\theta + \gamma\lambda_n} \right] \right\}^M d\theta. \tag{6.34}$$

As a special case, if for the two given codewords, $\mathbf{C}^1$ and $\mathbf{C}^2$, the columns of $\mathbf{C}^2 - \mathbf{C}^1$ are orthogonal to each other, we only need to calculate the determinant of a diagonal matrix. Since such a determinant is the product of diagonal elements, the PEP for this special case is simplified to

$$P(\mathbf{C}^1 \to \mathbf{C}^2) = \frac{1}{\pi} \int_0^{\pi/2} \left\{ \prod_{n=1}^{N} \left[\frac{4\sin^2\theta}{4\sin^2\theta + \gamma \sum_{t=1}^{T+Q} |c_{t,n}^1 - c_{t,n}^2|^2} \right] \right\}^M d\theta. \tag{6.35}$$

Example 6.4.1 *As an example, let us consider a four-state STTC using QPSK in Figure 6.3. To calculate the PEP, first we need to specify two valid paths. For the first path, we consider transmission of a sequence of zeros that results in staying at state zero and transmission of $c_{t,n}^1 = 1$, $\forall t, n$. Note that Figure 6.3 shows the indices of the symbols and for QPSK a zero index results in the transmission of 1 as the QPSK symbol. The shortest error event, as the second path, includes two transitions. Considering the path that starts at state zero, goes to state one in the first transition and merges to state zero in the second transition, that is 010, as the error event, the two codewords are*

$$\mathbf{C}^1 = \begin{pmatrix} 1 & 1 \\ 1 & 1 \end{pmatrix}, \quad \mathbf{C}^2 = \begin{pmatrix} 1 & j \\ j & 1 \end{pmatrix}. \tag{6.36}$$

The PEP for this case is

$$P(\mathbf{C}^1 \to \mathbf{C}^2) = \frac{1}{\pi} \int_0^{\pi/2} \left[\frac{2\sin^2\theta}{2\sin^2\theta + \gamma} \right]^{2M} d\theta. \tag{6.37}$$

The above integral can be found in closed form as follows

$$P(\mathbf{C}^1 \to \mathbf{C}^2) = \frac{1}{2} \left[1 - \sqrt{\frac{\gamma}{2+\gamma}} \sum_{m=0}^{2M-1} \binom{2m}{m} \left(\frac{1}{4+2\gamma} \right)^m \right]. \tag{6.38}$$

If we consider the path 030 instead of the above second path, the PEP is still calculated from (6.38). On the other hand, if the error event, the second path, is replaced with 020 *as in Figure 6.4, we have the codewords in (6.2). The corresponding PEP is*

$$P(\mathbf{C}^1 \to \mathbf{C}^2) = \frac{1}{\pi} \int_0^{\pi/2} \left[\frac{\sin^2 \theta}{\sin^2 \theta + \gamma} \right]^{2M} d\theta$$

$$= \frac{1}{2} \left[1 - \sqrt{\frac{\gamma}{1+\gamma}} \sum_{m=0}^{2M-1} \binom{2m}{m} \left(\frac{1}{4(1+\gamma)} \right)^m \right]. \quad (6.39)$$

The PEP for any other pair of paths with more than two transitions can be calculated from (6.35) as well.

There are many different methods to calculate the PEP of STTCs, for example references [145] and [131]. For a complete treatment of the subject, the interested reader is referred to [115]. Using the PEP, one can utilize the union bound to calculate an upper bound on the average symbol error probability. Also, the closed form of the PEP and the transfer function of the code provide an approximation to the average bit error probability. Such an approximation is derived by considering the error events of lengths less than or equal to a given number [116]. For the class of uniform error probability codes, the correct codeword $\mathbf{C}^1$ can always be picked as the all zero codeword. Using the symmetry property of this class of codes, averaging over all possible transmitted codewords is not necessary. Therefore, the true union bound is the same as the union bound assuming the transmission of an all zero codeword.

6.5 Simulation results

In this section, we provide simulation results for the performance of the STTCs given in the previous sections. We assume a quasi-static flat Rayleigh fading model for the channel. Therefore, the path gains are independent complex Gaussian random variables and fixed during the transmission of one frame. In all the simulations, similar to the results in [139], a frame consists of 130 transmissions out of each transmit antenna. We use Monte Carlo simulations to derive the frame error rate (FER) versus the received SNR.

Figure 6.14 shows the simulation results for transmitting 2 bits/(s Hz) using one receive antenna for the codes in Figures 6.3, 6.10, and 6.11. All these full-diversity codes use two transmit antennas, four states, and a QPSK constellation. These STTCs provide almost identical results despite the difference between their minimum CGDs. As mentioned before, minimum CGD is not the only parameter that affects the performance. The multiplicity of the different error paths in the trellis and the distance spectrum of the STTC influence the performance as well [3]. The higher minimum CGD of STTCs in Figures 6.10 and 6.11 result in a better

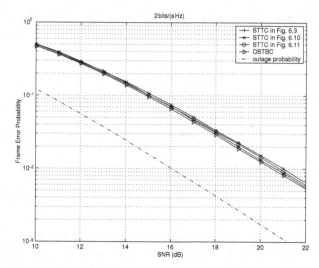

Fig. 6.14.   Frame error probability plotted against SNR for four-state STTCs at 2 bits/(s Hz); one receive antenna.

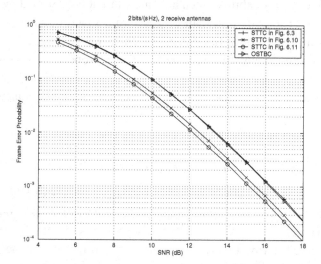

Fig. 6.15.   Frame error probability plotted against SNR for STTCs at 2 bits/(s Hz); two receive antennas.

performance for more than one receive antenna. Figure 6.15 shows the simulation results for transmitting 2 bits/(s Hz) using two receive antennas. As can be seen from Figure 6.15, the STTCs in Figures 6.10 and 6.11 outperform the STTC in Figure 6.3 by about 1 dB. As argued in [3], CGD is a good measure for a large number of receive antennas. For one receive antenna a complete analysis of distance spectrum is needed to claim a coding gain advantage. Simulation results in Figures 6.14 and 6.15 demonstrate an example for this argument. While the CGD of STTCs in

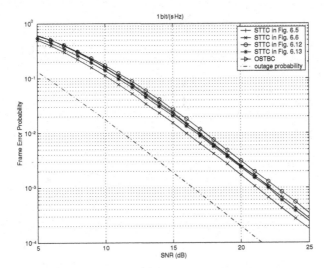

Fig. 6.16.   Frame error probability plotted against SNR for STTCs at 1 bit/(s Hz);
one receive antenna.

Figures 6.10 and 6.11 is better than that of Figure 6.3, their performance for one
receive antenna is almost identical to that of Figure 6.3. On the other hand, for
more than one receive antenna, the CGD difference is evident from the simulation
results.

Figure 6.16 depicts the simulation results for transmitting 1 bit/(s Hz) using one
receive antenna. Simulation results are for STTCs with two and four states from Figures 6.5, 6.12, 6.6, and 6.13. The behaviors of the STTCs at rates 1 and 2 bits/(s Hz)
are very similar to each other. Four-state codes outperform two-state codes with
similar path assignments. In general, there is a trade-off between the number of
states that reflects the complexity of decoding and the performance of the code.
Figure 6.17 shows the simulation results for Figures 6.5, 6.12, 6.6, and 6.13 to
transmit 1 bit/(s Hz) using two receive antennas. Similar to the case of 2 bits/(s Hz),
CGD is a good indication of the performance for a large number of receive antennas.

STTCs are designed based on criteria that try to minimize the pairwise error
probability. In the pairwise error probability bounds of Chapter 3, the codewords
correspond to frames of data in our simulations. Therefore, the STTCs provide a
low frame error probability as it is evident from Figures 6.14–6.17. In these figures,
we only provide simulation results for trellises with up to four states. It is possible to
reduce the frame error probability even further by increasing the number of states.
On the other hand, like most other trellis codes, the errors in STTCs are bursty,
that is when a frame is in error, many of its symbols are in error. Therefore, the
bit error probability of STTCs may not be as good as other competitive structures.
Figure 6.18 compares the bit error probability of STTCs with that of an OSTBC

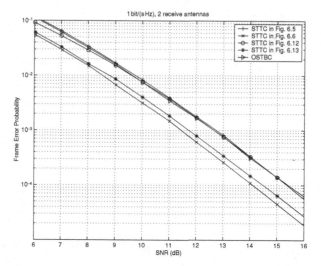

Fig. 6.17.   Frame error probability plotted against SNR for STTCs at 1 bit/(s Hz); two receive antennas.

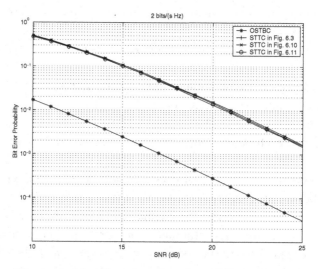

Fig. 6.18.   Bit error probability plotted against SNR for four-state STTCs at 2 bits/(s Hz); one receive antenna.

for two transmit antennas and one receive antenna at 2 bits/(s Hz). Similar results are observed at other transmission rates. As can be seen from the figure, OSTBCs outperform STTCs in terms of bit error probabilities. The small block size of OSTBCs reduces the chance of bursty errors. The choice between FER and BER as performance measures depends heavily on the application. For example, FER is more appropriate for wireless PCS applications. It is possible to improve the

Table 6.1. *Best examples of STTCs*

Rate bits/(s Hz)	Number of states	Minimum CGD
1	2	16
1	4	48
1	8	80
1	16	112
2	4	8
2	8	16
2	16	32
2	32	36

bit error probability of a STTC by modifying its structure. We study one such a structure in the next chapter.

6.6 Summary of important results

- A space-time trellis code for N transmit antennas is designed by assigning N constellation symbols to every transition in a trellis.
- The decoding of STTCs is done using the Viterbi algorithm.
- For a STTC with rate b bits/(s Hz) and a diversity r, at least $2^{b(r-1)}$ states are needed.
- To achieve full diversity for two transmit antennas, the following design rules have been suggested:
  - transitions diverging from the same state should differ in the second symbol;
  - transitions merging to the same state should differ in the first symbol.
- STTCs are easily implementable designed manually or by computer search.
- STTCs can achieve full spatial diversity and full rate.
- It is possible to improve the performance of STTCs by increasing the number of states of the trellis.
- It is possible to represent a STTC using a generating matrix $\mathbf{G}$. Denoting the b input bits at time t as $u_t^1, u_t^2, \ldots, u_t^b$, the indices of the transmitted symbols are calculated by:

$$\left(u_t^1, u_t^2, \ldots, u_t^b, u_{t-1}^1, u_{t-1}^2, \ldots, u_{t-1}^b, \ldots, u_{t-Q}^1, u_{t-Q}^2, \ldots, u_{t-Q}^b\right)\mathbf{G} \bmod L.$$

- The best examples of STTCs for $b = 1, 2$ bits/(s Hz) presented in our examples are shown in Table 6.1.

6.7 Problems

1 Consider the STTC in Figure 6.7. What are the transmitted symbols from each antenna for the following input bitstream: 01110010?

2 Consider the eight-state STTC in Example 6.3.9.

    (a) Find the corresponding trellis representation.

    (b) Find the codeword corresponding to the input bitstream 100.

    (c) Find the coding gain distance (CGD) between the codeword in part (b) and the all-zero codeword.

3 Find the trellis representations of the STTC in Example 6.3.4.

4 Derive the generating matrix $\mathbf{G}$ of the STTC in Example 6.2.3.

5 Consider STTCs in Figures 6.3, 6.10, and 6.11. Compare the minimum CGD and the minimum trace of $\mathbf{A}(\mathbf{C}^i, \mathbf{C}^j) = D(\mathbf{C}^i, \mathbf{C}^j)^H D(\mathbf{C}^i, \mathbf{C}^j)$ among all $i \neq j$. Justify the simulation results in Figures 6.14 and 6.15.

6 Consider the STTC in Figure 6.3 when there is only one receive antenna. Assume a transmitted codeword that stays at state zero at all time.

    (a) For the error event paths with length two, find the formulas for the pairwise error probability (PEP).

    (b) For each error event of length two, calculate the corresponding number of bit errors.

    (c) Calculate an approximation to the average bit error probability by only considering the error events of length two.

7 For a frame consisting of 130 transmissions out of each transmit antenna, use Monte Carlo simulations to derive the frame error rate (FER) versus the received SNR simulation results for the STTC in Figure 6.8.

7

Super-orthogonal space-time trellis codes

7.1 Motivation

STBCs provide full diversity and small decoding complexity. STBCs can be considered as modulation schemes for multiple transmit antennas and as such do not provide coding gains. As we discussed earlier in Chapters 4 and 5, full rate STBCs do not exist for every possible number of transmit antennas. On the other hand, STTCs are designed to achieve full diversity and high coding gains while requiring a higher decoding complexity. The STTCs presented in Chapter 6 are designed either manually or by computer search. In this chapter, we provide a systematic method to design space-time trellis codes.

Another way of achieving high coding gains is to concatenate an outer trellis code that has been designed for the AWGN channel with a STBC. Let us view each of the possible orthogonal matrices generated by a STBC as a point in a high dimensional space. The outer trellis code's task is to select one of these high dimensional signal points to be transmitted based on the current state and the input bits. In [5], it is shown that for the slow fading channel, the trellis code should be based on the set partitioning concepts of "Ungerboeck codes" for the AWGN channel.

The main idea behind super-orthogonal space-time trellis codes (SOSTTCs) is to consider STBCs as modulation schemes for multiple transmit antennas. We assign a STBC with specific constellation symbols to transitions originating from a state. Therefore, in general for a $T \times N$ space-time block code, picking a trellis branch emanating from a state is equivalent to transmitting NT symbols from N transmit antennas in T time intervals. By doing so, it is guaranteed that we get the diversity of the corresponding STBC. Note that in the STTCs of Chapter 6, N constellation symbols are assigned to each trellis branch. So, choosing a trellis branch is equivalent to transmitting N symbols from N transmit antennas in one time slot. What is transmitted at the next time slot depends on the next selected trellis branch and is not determined automatically. These codes are designed such that a maximum diversity and rate are guaranteed. However, it is not clear if the highest possible coding gain is achieved. In this chapter, we present SOSTTCs that provide maximum diversity and rate [66, 71, 119]. In addition, they achieve coding gains higher than those of the codes in Chapter 6. We elaborate on the issues of picking the right STBCs and assigning them to different trellis branches.

7.1.1 Orthogonal matrices and orthogonal designs

An example of a full-rate full-diversity complex STBC is the scheme proposed in Section 4.2 which is defined by the following generator matrix:

$$\mathcal{G}(x_1, x_2) = \begin{pmatrix} x_1 & x_2 \\ -x_2^* & x_1^* \end{pmatrix}. \tag{7.1}$$

The code is designed for $N = 2$ transmit antennas and any number of receive antennas. Using a constellation with $L = 2^b$ points, the code transmits $2b$ bits every two symbol intervals. For each block, $2b$ bits arrive at the encoder and the encoder chooses two modulation symbols s_1 and s_2. Then, replacing x_1 and x_2 by s_1 and s_2, respectively, to arrive at $\mathcal{G}(s_1, s_2)$, the encoder transmits s_1 from antenna one and s_2 from antenna two at time one. Also, the encoder transmits $-s_2^*$ from antenna one and s_1^* from antenna two at time two. This scheme provides diversity gain, but no additional coding gain.

There are other codes which provide behavior similar to those of (7.1) for the same rate and number of transmit antennas. As discussed in Section 4.7, multiplying an orthogonal design by a unitary matrix results in another orthogonal design. The set of all such codes which only use x_1, x_2, and their conjugates with positive or negative signs are listed below:

$$\begin{pmatrix} x_1 & x_2 \\ -x_2^* & x_1^* \end{pmatrix}, \begin{pmatrix} -x_1 & x_2 \\ x_2^* & x_1^* \end{pmatrix}, \begin{pmatrix} x_1 & -x_2 \\ x_2^* & x_1^* \end{pmatrix}, \begin{pmatrix} x_1 & x_2 \\ x_2^* & -x_1^* \end{pmatrix},$$

$$\begin{pmatrix} -x_1 & -x_2 \\ x_2^* & -x_1^* \end{pmatrix}, \begin{pmatrix} -x_1 & x_2 \\ -x_2^* & -x_1^* \end{pmatrix}, \begin{pmatrix} x_1 & -x_2 \\ -x_2^* & -x_1^* \end{pmatrix}, \begin{pmatrix} -x_1 & -x_2 \\ -x_2^* & x_1^* \end{pmatrix}. \tag{7.2}$$

With a small abuse of notation, we call the union of all these codes "super-orthogonal code" set $\mathcal{C}$. Using just one of the constituent codes from $\mathcal{C}$, for example the code in Equation (7.1), one cannot create all possible orthogonal 2×2 matrices for a given constellation. To make this point more evident, let us concentrate on the BPSK constellation for now. It can be shown that one can build all possible 2×2 orthogonal matrices using two of the codes in $\mathcal{C}$. For example, one can generate the following four 2×2 constellation matrices using the code in (7.1):

$$\begin{pmatrix} 1 & 1 \\ -1 & 1 \end{pmatrix}, \begin{pmatrix} -1 & -1 \\ 1 & -1 \end{pmatrix}, \begin{pmatrix} -1 & 1 \\ -1 & -1 \end{pmatrix}, \begin{pmatrix} 1 & -1 \\ 1 & 1 \end{pmatrix}. \tag{7.3}$$

There are four other possible distinct orthogonal 2×2 matrices which are listed below:

$$\begin{pmatrix} -1 & 1 \\ 1 & 1 \end{pmatrix}, \begin{pmatrix} 1 & -1 \\ -1 & -1 \end{pmatrix}, \begin{pmatrix} 1 & 1 \\ 1 & -1 \end{pmatrix}, \begin{pmatrix} -1 & -1 \\ -1 & 1 \end{pmatrix}. \tag{7.4}$$

To create these additional matrices, one can use the following code from the set $\mathcal{C}$:

$$\begin{pmatrix} -x_1 & x_2 \\ x_2^* & x_1^* \end{pmatrix} = \begin{pmatrix} x_1 & x_2 \\ -x_2^* & x_1^* \end{pmatrix} \cdot \begin{pmatrix} -1 & 0 \\ 0 & 1 \end{pmatrix}, \tag{7.5}$$

which represents a phase shift of the signals transmitted from antenna one by π. We denote a set including all 2×2 orthogonal matrices from (7.3) and (7.4) as $\mathcal{O}_2$. It is important to note that the rank of the difference between any two distinct matrices within either (7.3) or (7.4) is 2. However, the rank of a difference matrix between any two elements in (7.3) and (7.4) is 1.

By using more than one code from set $\mathcal{C}$, we can create all possible 2×2 orthogonal matrices from $\mathcal{O}_2$. Therefore, the scheme provides a sufficient number of constellation matrices to design a trellis code with the highest possible rate. Also, it allows a systematic design of space-time trellis codes using the available knowledge about trellis-coded modulation (TCM) [144] and multiple TCM (MTCM) [36] in the literature.

7.2 Super-orthogonal codes

We proved in Section 4.7 that multiplying an orthogonal STBC by a unitary matrix from the left or right results in another orthogonal STBC. In what follows we consider multiplying the generator matrix from the right by the following unitary matrix:

$$U = \begin{pmatrix} e^{j\theta} & 0 \\ 0 & 1 \end{pmatrix}. \tag{7.6}$$

The result is the rotation of the first column of the generator matrix. Note that multiplying the generator matrix from the left results in rotating the first row instead of the first column. However, the properties of the codes that we provide in this chapter remain the same.

7.2.1 A parameterized class of STBCs

Multiplying the Alamouti code in (7.1) from the right by the unitary matrix U in (7.6) results in the following class of orthogonal designs:

$$\mathcal{G}(x_1, x_2, \theta) = \mathcal{G}(x_1, x_2) \cdot U = \begin{pmatrix} x_1 \, \mathrm{e}^{\mathrm{j}\theta} & x_2 \\ -x_2^* \, \mathrm{e}^{\mathrm{j}\theta} & x_1^* \end{pmatrix}. \tag{7.7}$$

Note that $\theta = 0$ provides the code in (7.1). So, with a slight abuse of notation, we have $\mathcal{G}(x_1, x_2, 0) = \mathcal{G}(x_1, x_2)$. The encoder uses $2b$ input bits to select s_1 and s_2. Then, replacing x_1 and x_2 in $\mathcal{G}(x_1, x_2, \theta)$ with s_1 and s_2 respectively, results in $\mathcal{G}(s_1, s_2, \theta)$. First, $s_1 \, \mathrm{e}^{\mathrm{j}\theta}$ and s_2 are transmitted from the first and second transmit antennas, respectively. Second, $-s_2^* \, \mathrm{e}^{\mathrm{j}\theta}$ and s_1^* are transmitted from the two antennas. Let us concentrate on the case for which the set of transmitted signals are the same as the constellation points. In other words, the signal alphabet is not expanded. We need to pick θ such that for any choice of s_1 and s_2 from the original constellation points, the resulting transmitted signals are also from the same constellation. For an L-PSK, the constellation signals can be represented by $\mathrm{e}^{\mathrm{j}2\pi l/L}, l = 0, 1, \ldots, L - 1$. One can pick $\theta = 2\pi l'/L$, where $l' = 0, 1, \ldots, L - 1$ to avoid constellation expansion. In this case, the resulting transmitted signals are also members of the L-PSK constellation. Since the transmitted signals are from a PSK constellation, the peak-to-average power ratio of the transmitted signals is equal to one. So, not only do we not increase the number of signals in the constellation, but also there is no need for an amplifier to provide a higher linear operation region. More specifically, we use $\theta = 0, \pi$ and $\theta = 0, \pi/2, \pi, 3\pi/2$ for BPSK and QPSK, respectively. By using $\mathcal{G}(x_1, x_2, 0)$ and $\mathcal{G}(x_1, x_2, \pi)$ for the BPSK constellation, one can generate all 2×2 orthogonal matrices in $\mathcal{O}_2$ as discussed previously. In fact, $\mathcal{G}(x_1, x_2, 0)$ is the code in (7.1) and $\mathcal{G}(x_1, x_2, \pi)$ is the code in (7.5). By using (7.1) and (7.5), the set of transmitted signals consists of $s_1, s_2, s_1^*, s_2^*, -s_1, -s_2^*$. For any symmetric constellation, the set of transmitted signals is the same as the set of constellation signals. This includes QAM constellations as well as PSK constellations. We call the combination of these two codes a super-orthogonal code. In general, a super-orthogonal code consists of the union of a few orthogonal codes, like the ones in (7.7). A special case is when the super-orthogonal code consists of only one orthogonal code, for example only $\theta = 0$. Therefore, the set of orthogonal codes is a subset of the set of super-orthogonal codes. Obviously, the number of orthogonal matrices that a super-orthogonal code provides is more than, or in the worst case equal to, the number of orthogonal matrices that an orthogonal code provides. Therefore, while the super-orthogonal code does not extend the constellation alphabet of the transmitted signals, it does expand the number of available orthogonal matrices. This is of great benefit and crucial in the design of full-rate,

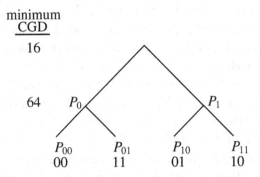

Fig. 7.1.   Set partitioning for BPSK; each pair of binary numbers represents the pair of symbol indices in a STBC.

full-diversity trellis codes. Another advantage of super-orthogonal codes lies in the fact that the code is parameterized. An important example of a super-orthogonal code is the union of $\mathcal{G}(x_1, x_2, \theta)$ for an L-PSK constellation where $\theta = 2\pi l'/L$, and $l' = 0, 1, \ldots, L - 1$. Another example is the union of $\mathcal{G}(x_1, x_2, \theta)$ for a QAM constellation where $\theta = \pi/2, \pi, 3\pi/2$.

Note that one needs to separate "the expansion of the orthogonal matrices," that is a positive result of using super-orthogonal codes, and "the expansion of the signal constellation," that is usually a negative side effect. The super-orthogonal code expands the number of available orthogonal matrices and, as we show in what follows, this is the main reason why we can design full-rate trellis codes that provide full diversity. It has no negative side effect either in terms of expanding the transmitted signal constellation or in terms of increasing the peak-to-average power ratio.

7.2.2 Set partitioning for orthogonal codewords

This section provides a set partitioning for the codewords of an orthogonal code and shows how to maximize the coding gain. Following the definition of CGD from Chapter 3, the minimum of the determinant of the matrix $A(\mathbf{C}^i, \mathbf{C}^j) = D(\mathbf{C}^i, \mathbf{C}^j)^H D(\mathbf{C}^i, \mathbf{C}^j)$ over all possible pairs of distinct codewords $\mathbf{C}^i$ and $\mathbf{C}^j$ corresponds to the coding gain for a full diversity code. Then, we use CGD instead of Euclidean distance to define a set partitioning similar to Ungerboeck's set partitioning [144]. Note that similarly we can use the minimum trace of $A(\mathbf{C}^i, \mathbf{C}^j)$, $i \neq j$, as a distance measure for set partitioning.

First, let us concentrate on the case where we only utilize the code in (7.1). Also, let us assume that we use a BPSK constellation. Consider a four-way partitioning of the codewords of the orthogonal code as shown in Figure 7.1 for BPSK. The numbers at the leaves of the tree in the figure represent BPSK indices. Each pair of indices represents a 2×2 codeword built by the corresponding pair of symbols in the constellation. At the root of the tree, the minimum determinant of $A(\mathbf{C}^i, \mathbf{C}^j)$ among all possible pairs of codewords $(\mathbf{C}^i, \mathbf{C}^j)$ is 16. At the first level of partitioning, the highest determinant that can be obtained is 64. This is obtained by a set partitioning in which subsets P_0 and P_1 use different transmitted signal elements for different transmit antennas. At the next level of partitioning, we have four sets P_{00}, P_{11}, P_{01}, and P_{10} with only one codeword per set.

As shown in Figure 6.2, a four-state trellis code can be constructed based on the set partitioning of Figure 7.1 [5, 120]. The rate of such a code is 0.5 bits/(s Hz) using a BPSK constellation. In general, for a b-bit constellation, 2^{2b} pairs of symbols exist to be partitioned at the leaves of the tree. Half of these pairs can be used for transmission from each state of Figure 6.2 over two time slots. The other half should be reserved for using in other states to avoid a catastrophic code. Therefore, this code can transmit $2^{2b}/2 = 2^{2b-1}$ codewords or $2b - 1$ bits per two time slots. As a result, the rate of the code is $(b - 0.5)$ bits/(s Hz) for a b-bit constellation and the code cannot transmit the maximum possible rate of b bits/(s Hz).

The best set partitioning is the one for which the minimum CGD of the sets at each level of the tree is maximum among all possible cases. To find guidelines for partitioning the sets, we need to derive formulas to calculate CGDs. For an L-PSK constellation, let us represent each signal by $s = e^{jl\omega}, l = 0, 1, \ldots, L - 1$, where $\omega = 2\pi/L$. We consider two distinct pairs of constellation symbols $(s_1^1 = e^{jk_1\omega}, s_2^1 = e^{jl_1\omega})$ and $(s_1^2 = e^{jk_2\omega}, s_2^2 = e^{jl_2\omega})$ and the corresponding 2×2 orthogonal codewords $\mathbf{C}^1$ and $\mathbf{C}^2$. For the sake of brevity, in the calculation of matrices $D(\mathbf{C}^1, \mathbf{C}^2)$ and $A(\mathbf{C}^1, \mathbf{C}^2)$, we omit $(\mathbf{C}^1, \mathbf{C}^2)$ from A and D when there is no ambiguity. For parallel transitions in a trellis, we have

$$D = \begin{pmatrix} e^{jk_1\omega} - e^{jk_2\omega} & e^{jl_1\omega} - e^{jl_2\omega} \\ e^{-jl_2\omega} - e^{-jl_1\omega} & e^{-jk_1\omega} - e^{-jk_2\omega} \end{pmatrix}, \tag{7.8}$$

$$A = \begin{pmatrix} 4 - 2\cos[\omega(k_2 - k_1)] - 2\cos[\omega(l_2 - l_1)] & 0 \\ 0 & 4 - 2\cos[\omega(k_2 - k_1)] - 2\cos[\omega(l_2 - l_1)] \end{pmatrix}. \tag{7.9}$$

Using (7.9), we can calculate CGD by

$$\det(A) = \{4 - 2\cos[\omega(k_2 - k_1)] - 2\cos[\omega(l_2 - l_1)]\}^2. \tag{7.10}$$

So far, we have only considered the codewords corresponding to parallel transitions. These codewords are transmitted over two time slots and are represented by 2×2 matrices. This is good enough for the purpose of set partitioning similar to the case of TCM suggested by Ungerboeck [144]. However, to calculate the minimum CGD of the code, one needs to consider codewords that include more than one trellis transition. If we consider two codewords that diverge from state zero and remerge after P trellis transitions, the size of the corresponding difference matrix D is $2P \times 2$. In fact, such a difference matrix can be represented as the concatenation of P difference matrices corresponding to the P transitions that construct the path. Let us denote the set of constellation symbols for the first codeword by $(s_1^1, s_2^1)^p = (e^{jk_1^p \omega}, e^{jl_1^p \omega})$, $p = 1, 2, \ldots, P$ and for the second codeword by $(s_1^2, s_2^2)^p = (e^{jk_2^p \omega}, e^{jl_2^p \omega})$, $p = 1, 2, \ldots, P$. We also define D^p as the difference matrix of the pth transition and $A^p = D^{pH} \cdot D^p$ for $p = 1, 2, \ldots, P$. For the above two codewords, we have

$$D = \begin{pmatrix} D^1 \\ D^2 \\ \vdots \\ D^P \end{pmatrix}. \tag{7.11}$$

Using (7.11), one can calculate the matrix A as

$$A = D^H \cdot D = \sum_{p=1}^{P} A^p. \tag{7.12}$$

Therefore, matrix A is still a diagonal 2×2 matrix, that is $A_{12} = A_{21} = 0$. The CGD between the above codewords that differ in P transitions can be calculated by

$$\det(A) = \left\{ \sum_{p=1}^{P} 4 - 2 \cos\left[\omega\left(k_2^p - k_1^p\right)\right] - 2 \cos\left[\omega\left(l_2^p - l_1^p\right)\right] \right\}^2. \tag{7.13}$$

Note that (7.13) includes a sum of P terms and each of these terms is non-negative. Therefore, the following inequality holds:

$$\det(A) = \left\{ \sum_{p=1}^{P} 4 - 2 \cos\left[\omega\left(k_2^p - k_1^p\right)\right] - 2 \cos\left[\omega\left(l_2^p - l_1^p\right)\right] \right\}^2$$

$$\geq \sum_{p=1}^{P} \left\{ 4 - 2 \cos\left[\omega\left(k_2^p - k_1^p\right)\right] - 2 \cos\left[\omega\left(l_2^p - l_1^p\right)\right] \right\}^2$$

$$= \sum_{p=1}^{P} \det(A^p). \tag{7.14}$$

Based on the coding distances calculated in (7.10) and (7.13), one can show that the coding gain of such a STTC is dominated by parallel transitions. The optimal set partitioning for BPSK, QPSK, and 8-PSK are demonstrated in Figures 7.1, 7.2,

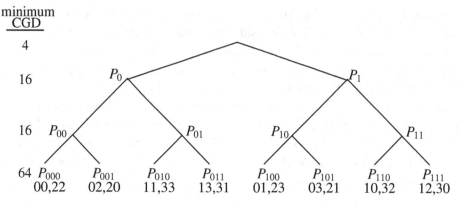

Fig. 7.2.   Set partitioning for QPSK; each pair of binary numbers represents the pair of symbol indices in a STBC.

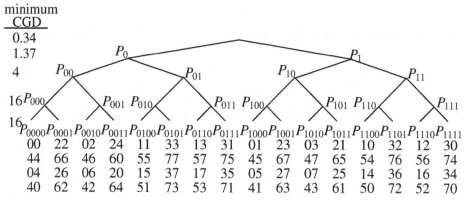

Fig. 7.3.   Set partitioning for 8-PSK; each pair of binary numbers represents the pair of symbol indices in a STBC.

and 7.3, respectively. To calculate the CGD between two codewords, one can use (7.10) in which a codeword is represented by a pair of symbols or the corresponding pair of indices. It is apparent that increasing the Euclidean distance between the first symbols of the first and second symbol pairs will increase the CGD. The CGD also increases as we increase the Euclidean distance between the second symbols. Therefore, a rule of thumb in set partitioning is to choose the codewords that contain signal elements with highest maximum Euclidean distance from each other as the leaves of the set-partitioning tree. For example, in the case of QPSK in Figure 7.2, $s = e^{j\pi l/2}$, $l = 0, 1, 2, 3$ are the QPSK signal constellation elements and $k = 0, 1, 2, 3$ represent $s = 1, j, -1, -j$, respectively. The *maximum minimum* CGD in this case is 64 when $|k_1 - k_2| = 2$ and $|l_1 - l_2| = 2$ in (7.10). This is the justification for the choice of leaf codewords in Figure 7.2. At the second level of the tree from the bottom, it is impossible to have both $|k_1 - k_2|$ and $|l_1 - l_2|$ equal to 2 in all cases. The next highest value for minimum CGD is 16 when $|k_1 - k_2| = 2$, $|l_1 - l_2| = 0$ or $|k_1 - k_2| = 0$, $|l_1 - l_2| = 2$. Therefore, we group the subtrees in the second level such that the worst case is when $|k_1 - k_2| = 2$, $|l_1 - l_2| = 0$ or $|k_1 - k_2| = 0$, $|l_1 - l_2| = 2$. We keep grouping the subtrees to maximize the minimum CGD at each level of set partitioning. Similar strategies are used for other signal constellations.

As can be seen from Figures 7.1, 7.2, and 7.3, the minimum CGD increases (or remains the same) as we go one level down in the tree. The branches at each level can be used to design a trellis code with a specific rate. Higher coding gain necessitates the use of redundancy resulting in reduced rate. In the following sections, we show how to design STTCs without sacrificing the rate.

7.2.3 Set partitioning for super-orthogonal codewords

This section provides set partitioning for super-orthogonal codes and shows how to maximize the coding gain without sacrificing the rate. Code construction based on a super-orthogonal set is as follows. We assign a constituent STBC to all transitions from a state. The adjacent states are typically assigned to one of the other constituent STBCs from the super-orthogonal code. Similarly, we can assign the same STBC to branches that are merging into a state. It is thus assured that every pair of codewords diverging from (or merging to) a state achieves full diversity because the pair is from the same orthogonal code with the same rotation parameter θ. On the other hand, for codewords with different θ, it is possible that they do not achieve full diversity. Since these codewords are assigned to different states, the resulting trellis code would provide full diversity despite the fact that a pair of codewords in a super-orthogonal code may not achieve full diversity. Note that this is just a general method to guarantee full diversity. It is possible to come up with examples that do not follow this rule and still provide full diversity.

Similar to the case of orthogonal designs, it remains to do the set partitioning such that the minimum CGD is maximized at each level of partitioning. This set partitioning should be done for all possible orthogonal 2×2 codewords, for every possible rotation. In other words, we need to partition the set of all possible 2×2 matrices generated by the class of codes in (7.7). Based on the design criteria in Chapter 3, first we need to make sure that we achieve full diversity. Therefore, first, we partition the set of all codewords into subsets with the same rotation. In other words, the first step of the set partitioning is only based on the rotation parameter θ. Then, we should partition the set of all codewords with the same rotation parameter θ as we did for the case of orthogonal designs with $\theta = 0$. In what follows, we show that the set partitioning for the sets with different rotations results in the same set partitioning trees as that of the orthogonal design with $\theta = 0$ in Section 7.2.2.

Similar to the case of $\theta = 0$, we consider two distinct pairs of constellation symbols $(s_1^1 = e^{jk_1\omega}, s_2^1 = e^{jl_1\omega})$ and $(s_1^2 = e^{jk_2\omega}, s_2^2 = e^{jl_2\omega})$. We denote the corresponding codewords by $\mathbf{C}^{\theta 1}$ and $\mathbf{C}^{\theta 2}$ and the corresponding difference matrix by D^θ. For parallel transitions in a trellis, we have

$$D^\theta = \begin{pmatrix} e^{jk_1\omega}e^{j\theta} - e^{jk_2\omega}e^{j\theta} & e^{jl_1\omega} - e^{jl_2\omega} \\ e^{-jl_2\omega}e^{j\theta} - e^{-jl_1\omega}e^{j\theta} & e^{-jk_1\omega} - e^{-jk_2\omega} \end{pmatrix} = D \cdot U, \qquad (7.15)$$

where U is the rotation matrix in (7.6) and D is the difference matrix for $\theta = 0$ in (7.8). To calculate the CGD, we need to compute matrix $A^\theta = D^{\theta H} \cdot D^\theta$ using (7.15).

$$A^\theta = U^H \cdot D^H \cdot D \cdot U = U^H \cdot A \cdot U = A, \qquad (7.16)$$

where $A = D^H \cdot D$ is a diagonal matrix calculated in (7.9) and $U^H \cdot A \cdot U = A$ is easily shown by matrix multiplication. As a result of (7.16), the CGD between two codewords is only a function of the corresponding constellation symbols and the same for any rotation θ. Therefore, we can use the formulas that we have developed in Section 7.2.2 to calculate the CGDs. A good set partitioning for $\theta = 0$ as presented in Section 7.2.2 is also good for any other rotation. Figures 7.1, 7.2, and 7.3 show the set partitioning for BPSK, QPSK, and 8-PSK, respectively.

So far, we have represented the set partitioning for super-orthogonal codes by first partitioning based on the rotation parameter θ and then the constellation pairs. The rationale behind such a two step partitioning is the fact that as we showed in Chapter 3, first one needs to consider the rank criterion to achieve full diversity and then the determinant or trace criterion to obtain the highest possible coding gain. Since the second step of set partitioning is the same for all rotation parameters, we do not show the complete set-partitioning tree. In fact, to specify a subset, we mention the rotation parameter θ or the corresponding orthogonal STBC $\mathcal{G}(x_1, x_2, \theta)$ and a set from the set partitioning for $\theta = 0$. Another way of presenting the codewords is to specify the complete set-partitioning tree. Although the first representation is more compact and easier to use, we provide one example of the latter representation for clarity. Figure 7.4 illustrates the set partitioning of super-orthogonal codewords for a BPSK constellation using rotations $\theta = 0, \pi$. Unlike Figure 7.1, we use the 2×2 matrices instead of the corresponding index pairs. Note that a pair of indices and a rotation identify the 2×2 codeword uniquely. Also, the superscripts in Figure 7.4 represent the rotation. Therefore, the left half of the tree in Figure 7.4 is the same as the tree in Figure 7.1 for $\theta = 0$. Similarly, the right half of the tree in Figure 7.4 is the same as the tree in Figure 7.1

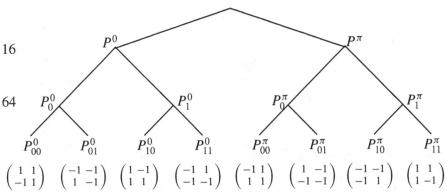

Fig. 7.4. Set partitioning of super-orthogonal codewords for BPSK and $\theta = 0, \pi$; complete tree.

for $\theta = \pi$. Our compact representation in Figures 7.1, 7.2, and 7.3 uses this fact to remove a need for explicitly adding the parameter θ in the set partitioning process.

7.2.4 Super-orthogonal space-time trellis codes

In this section, we show how to use the proposed set-partitioning scheme to design full-diversity full-rate space-time trellis codes. First, we start with a few important examples and then we propose some general rules on how to design a SOSTTC for a given trellis and required rate. For each example of SOSTTC, we assign one subset to each branch of the trellis. Each subset represents a number of 2×2 matrices. Equivalently, as explained before, each subset corresponds to a rotation parameter and a set of possible symbol pairs. The superscript of the subset corresponds to the rotation parameter θ. If the superscript of the subset is θ, we use $\mathcal{G}(x_1, x_2, \theta)$, the STBC in (7.7) with rotation parameter θ, to transmit. The set of possible symbol pairs for a given rotation parameter θ is the same as that of the same subset with superscript $\theta = 0$. These subsets are derived in Section 7.2.2 and depicted in Figures 7.1, 7.2, and 7.3. For example, P_{10}^{π} for QPSK corresponds to $\mathcal{G}(x_1, x_2, \pi)$ and the subset P_{10} in Figure 7.2.

Example 7.2.1 *Figure 7.5 shows a four-state example of SOSTTC. The figure contains two different representation of the same code. In the left representation, the STBC $\mathcal{G}(x_1, x_2, \theta)$ is explicitly mentioned and the subsets are from Figure 7.1 or 7.2. In the right representation, the rotation parameter is the superscript of the corresponding subset.*

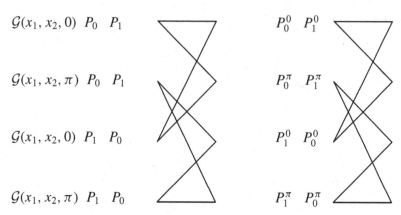

Fig. 7.5.   A four-state SOSTTC; $r = 1$ bit/(s Hz) using BPSK or $r = 2$ bits/(s Hz) using QPSK.

In this example, when we use BPSK and the corresponding set partitioning in Figure 7.1 the rate of the code is one. We use $\mathcal{G}(x_1, x_2, 0)$ when departing from states zero and two and use $\mathcal{G}(x_1, x_2, \pi)$ when departing from states one and three. Note that, as shown in Figures 7.1 and 7.4, the code uses eight possible 2×2 orthogonal matrices instead of four which results in a full-rate code. The minimum CGD of this code is 64 which can be found in Figure 7.1 and Table 7.1. In Section 7.3, we prove that parallel transitions are dominant in calculating the minimum CGD for this code.

If we use a QPSK constellation and the corresponding set partitioning in Figure 7.5, the result is a four-state SOSTTC at rate 2 bits/(s Hz). The minimum CGD for this 2 bits/(s Hz) code is equal to 16 which is greater than 4, the CGD of the corresponding STTC from [139] in Chapter 6. The highest minimum CGD for a 2 bits/(s Hz) code in Chapter 6 is 8 and smaller than that of SOSTTC in Figure 7.5.

Example 7.2.2 *Figure 7.6 shows a 3 bits/(s Hz) SOSTTC using 8-PSK and the corresponding set partitioning of Figure 7.3. The minimum CGD for this code is equal to 2.69. We study the details of the coding gain calculations in Section 7.3. Based on Lemma 6.2.1, there is no four-state STTC for 8-PSK in Chapter 6.*

Example 7.2.3 *Figure 7.7 demonstrates the codes for a two-state trellis providing a minimum CGD of 48 and 16 at rates 1 bit/(s Hz) using BPSK and 2 bits/(s Hz) using QPSK, respectively. There is no equivalent two-state STTC using QPSK due to Lemma 6.2.1.*

Example 7.2.4 *An eight-state, rate 3 bits/(s Hz) (8-PSK) example of SOSTTCs is shown in Figure 7.8. The CGD of this code is 4 while the CGD of a similar STTC in Chapter 6 from [139] is 2.*

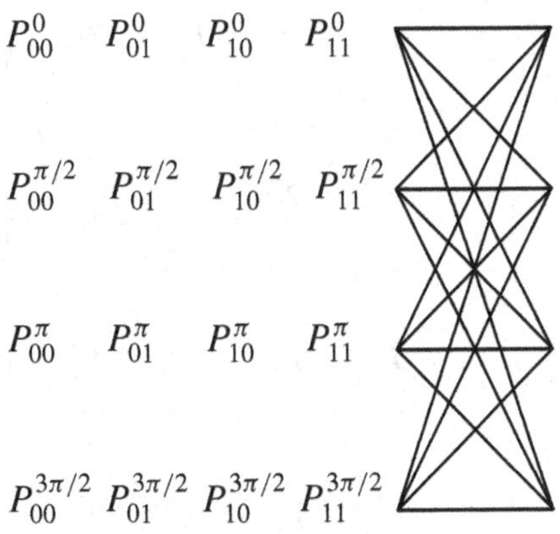

$$P_{00}^0 \quad P_{01}^0 \quad P_{10}^0 \quad P_{11}^0$$

$$P_{00}^{\pi/2} \quad P_{01}^{\pi/2} \quad P_{10}^{\pi/2} \quad P_{11}^{\pi/2}$$

$$P_{00}^{\pi} \quad P_{01}^{\pi} \quad P_{10}^{\pi} \quad P_{11}^{\pi}$$

$$P_{00}^{3\pi/2} \quad P_{01}^{3\pi/2} \quad P_{10}^{3\pi/2} \quad P_{11}^{3\pi/2}$$

Fig. 7.6.   A four-state SOSTTC; $r = 3$ bits/(s Hz) (8-PSK).

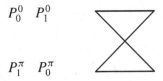

$$P_0^0 \quad P_1^0$$

$$P_1^{\pi} \quad P_0^{\pi}$$

Fig. 7.7.   A two-state SOSTTC; $r = 1$ bit/(s Hz) using BPSK or $r = 2$ bits/(s Hz) using QPSK.

Some degrees of freedom in choosing the rotations and the sets exist in the design of Example 7.2.4. In other words, different codes based on various assignments of the rotations and the sets provide the same coding gain. For example, an eight-state, rate 3 bits/(s Hz) code with similar properties is provided in Figure 7.9. This is due to the fact that the number of available 2×2 orthogonal matrices is more than what is needed in this case. One limitation in picking different options is the possibility of a catastrophic code. To avoid a catastrophic code, a change of a few input bits should not create an infinite number of different symbols. In other words, the same input bits should not create the same codeword when starting from different states. To achieve this goal, either the rotation parameter θ assigned to different states should be different or the assigned subsets should be different. One strategy picks a small number of rotation parameters to create the required number of orthogonal matrices in addition to rearranging the subsets as in Figure 7.8. On the other hand, another strategy selects a different rotation parameter for each state as in Figure 7.9. As the number of states grows, we can pick between these two strategies and the strategies in between to design SOSTTCs. However, there are choices of rotations that make the overall CGD smaller and must be avoided. For example, it is possible to design codes for which the parallel transitions are not dominant in the overall CGD of the code. We study different aspects of this issue in Section 7.3.

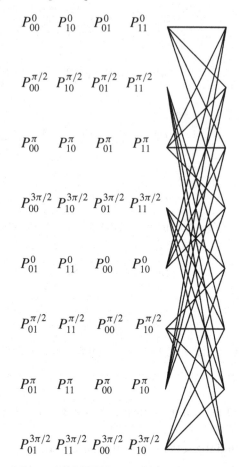

$$P^0_{00} \quad P^0_{10} \quad P^0_{01} \quad P^0_{11}$$

$$P^{\pi/2}_{00} \quad P^{\pi/2}_{10} \quad P^{\pi/2}_{01} \quad P^{\pi/2}_{11}$$

$$P^{\pi}_{00} \quad P^{\pi}_{10} \quad P^{\pi}_{01} \quad P^{\pi}_{11}$$

$$P^{3\pi/2}_{00} \quad P^{3\pi/2}_{10} \quad P^{3\pi/2}_{01} \quad P^{3\pi/2}_{11}$$

$$P^0_{01} \quad P^0_{11} \quad P^0_{00} \quad P^0_{10}$$

$$P^{\pi/2}_{01} \quad P^{\pi/2}_{11} \quad P^{\pi/2}_{00} \quad P^{\pi/2}_{10}$$

$$P^{\pi}_{01} \quad P^{\pi}_{11} \quad P^{\pi}_{00} \quad P^{\pi}_{10}$$

$$P^{3\pi/2}_{01} \quad P^{3\pi/2}_{11} \quad P^{3\pi/2}_{00} \quad P^{3\pi/2}_{10}$$

Fig. 7.8.   An eight-state SOSTTC; $r = 3$ bits/(s Hz) (8-PSK).

Similar to conventional trellis codes, there is always a trade-off between coding gain and the number of states or, equivalently, the complexity. Codes with different number of states and at different rates can be systematically designed using the set partitioning in Figures 7.1, 7.2, and 7.3 or similar set partitioning for other constellations. The rules for assigning different sets to different transitions in the trellis are similar to the general rules of thumb defined in [144] and [36] to design MTCM schemes. First based on the required rate, we select a constellation and use the corresponding set partitioning. The choice of the constellation also defines the valid rotation angles, θ, that would not create an expansion of the transmitted signal constellation. The corresponding set partitioning is used in the design of the trellis code. Since our presentation of the set partitioning was based on a two-step algorithm, we describe the resulting two-step design method in the sequel. Utilizing the valid values of θ in (7.7) defines the valid codewords in the set of super-orthogonal codes. Then, we assign a constituent orthogonal code from the

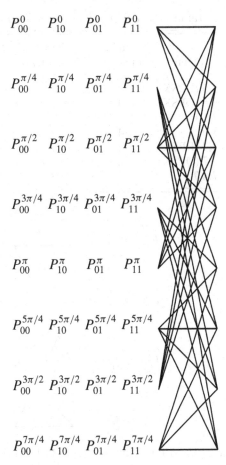

$$P_{00}^0 \quad P_{10}^0 \quad P_{01}^0 \quad P_{11}^0$$

$$P_{00}^{\pi/4} \quad P_{10}^{\pi/4} \quad P_{01}^{\pi/4} \quad P_{11}^{\pi/4}$$

$$P_{00}^{\pi/2} \quad P_{10}^{\pi/2} \quad P_{01}^{\pi/2} \quad P_{11}^{\pi/2}$$

$$P_{00}^{3\pi/4} \quad P_{10}^{3\pi/4} \quad P_{01}^{3\pi/4} \quad P_{11}^{3\pi/4}$$

$$P_{00}^{\pi} \quad P_{10}^{\pi} \quad P_{01}^{\pi} \quad P_{11}^{\pi}$$

$$P_{00}^{5\pi/4} \quad P_{10}^{5\pi/4} \quad P_{01}^{5\pi/4} \quad P_{11}^{5\pi/4}$$

$$P_{00}^{3\pi/2} \quad P_{10}^{3\pi/2} \quad P_{01}^{3\pi/2} \quad P_{11}^{3\pi/2}$$

$$P_{00}^{7\pi/4} \quad P_{10}^{7\pi/4} \quad P_{01}^{7\pi/4} \quad P_{11}^{7\pi/4}$$

Fig. 7.9. An eight-state SOSTTC; $r = 3$ bits/(s Hz) (8-PSK).

set of super-orthogonal codes to all transitions from a state. This is equivalent to assigning a rotation parameter to each state. Typically, we assign another constituent orthogonal code or equivalent rotation parameter to the adjacent states. Similarly, we can assign the same orthogonal code, equivalently the same rotation parameter, to branches that are merging into a state. It is thus assured that any path that diverges from (or merges to) the correct path is full rank. Then, different sets from the set partitioning for $\theta = 0$, or any other rotation, are assigned to different transitions similar to the way that we assign sets in a regular MTCM. To avoid a catastrophic code, the subsets are assigned such that for the same input bits in different states either the rotation angles or the assigned subsets are different. Since the process of set partitioning is done to maximize the minimum CGD at each level, the upper bound on the coding gain of the resulting SOSTTC due to parallel transitions is also maximum.

Table 7.1. *CGD values for different codes*

Figure	Number of states	Rate (bits/(s Hz))	Minimum CGD
7.5 (BPSK)	4	1	64
7.5 (QPSK)	4	2	16
7.10	4	2.5	4
7.6	4	3	2.69
7.7 (BPSK)	2	1	48
7.7 (QPSK)	2	2	16
7.8,7.9	8	3	4

$$P_{00}^0 \quad P_{01}^0$$

$$P_{00}^\pi \quad P_{01}^\pi$$

$$P_{01}^0 \quad P_{00}^0$$

$$P_{01}^\pi \quad P_{00}^\pi$$

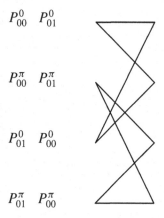

Fig. 7.10.   A four-state code; $r = 2.5$ bits/(s Hz) (8-PSK).

Unlike the manual or computer search design strategies in Chapter 6, one can systematically design a code for an arbitrary trellis and rate using the above set partitioning and code design strategy. While so far examples of full-rate codes have been shown, in general, codes with lower rates can be designed to provide higher coding gains. The design method is exactly the same while using different subsets. Utilizing different levels of the set partitioning to design SOSTTCs provides a trade-off between rate and coding gain as well. One example of a non full-rate SOSTTC with higher coding gain is provided below.

Example 7.2.5 *A four-state rate 2.5 bits/(s Hz) code using 8-PSK with a CGD of 4 is shown in Figure 7.10. The maximum possible rate using 8-PSK is 3 and an example that provides a CGD of 2.69 is shown in Figure 7.6.*

Table 7.1 tabulates the CGD of the different SOSTTC examples. It is evident that using the super-orthogonal space-time trellis coding method, not only can we

systematically design codes that did not exist in Chapter 6, but also the CGD of the
SOSTTCs are consistently better than those of the STTCs. For example, we consider codes at rate $r = 2$ bits/(s Hz). Based on Lemma 6.2.1, there is no two-state
STTC at rate $r = 2$ bits/(s Hz) in Chapter 6. Such a two-state SOSTTC is provided
in Example 7.2.3. Using the methods in Chapter 6, the minimum required number
of states to design a full diversity STTC is four. The highest minimum CGD in
Chapter 6 is 8 from Example 6.3.2. Using SOSTTCs, Example 7.2.1 provides a
minimum CGD of 16.

While minimum CGD is a good indication of the performance of a space-time
code, a one-to-one correspondence between minimum CGD and the performance
is not always the case. Similar to the case of TCM, one should look at the distance
spectrum and not just the minimum distance of a code [3]. In Section 7.7, we
compare the performance of the SOSTTCs with that of the best available STTCs
from Chapter 6.

7.3 CGD analysis

In this section, we derive the coding gain of different SOSTTCs that we introduced
in Section 7.2. We show how to find the dominant path for CGD calculation in the
trellis. The approach is general enough to be easily extended to other trellises in
the literature. We first consider specific examples given in Section 7.2 and calculate
their coding gains. Then, we generalize the methods that we have used in the
calculation of these coding gains to show how to calculate the coding gain of any
SOSTTC.

7.3.1 Error events with path length of three

Let us first consider the trellis of the code in Figure 7.5. Parallel transitions between
two states may be considered as different transitions each containing one possible
2×2 symbol matrix. Two codewords may only differ in $P = 1$ trellis transition if
they both start and end in the same state. In other words, if two codewords belong
to the same parallel transitions, they differ in $P = 1$ trellis transition. However, due
to the structure of the trellis, it is impossible to have two codewords which differ
in $P = 2$ trellis transitions. Because, for example, if two codewords diverge from
state zero, they have to go through at least three transitions to remerge as shown
in Figure 7.11. Therefore, the smallest value of P excluding parallel transitions
is three. For $P = 3$, we consider a typical case where the first codeword stays at
state zero. For the second codeword, the first and third transitions, diverging and
merging to state zero, use $C(x_1, x_2, 0)$ and the second transition uses $C(x_1, x_2, \theta)$

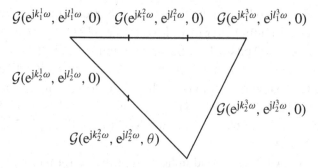

Fig. 7.11.   Two typical paths differing in $P = 3$ transitions.

as in Figure 7.11. Using trigonometric equations, it can be shown that

$$\det(A) = (a + b_1 + c)(a + b_2 + c) - d, \qquad (7.17)$$

where

$$
\begin{aligned}
a &= 4 - 2\cos[\omega(k_2^1 - k_1^1)] - 2\cos[\omega(l_2^1 - l_1^1)], \\
c &= 4 - 2\cos[\omega(k_2^3 - k_1^3)] - 2\cos[\omega(l_2^3 - l_1^3)], \\
b_1 &= 4 - 2\cos[\omega(k_2^2 - k_1^2) + \theta] - 2\cos[\omega(l_2^2 - l_1^2)], \qquad (7.18) \\
b_2 &= 4 - 2\cos[\omega(k_2^2 - k_1^2)] - 2\cos[\omega(l_2^2 - l_1^2) - \theta], \\
d &= (2 - 2\cos\theta)(2 - 2\cos[\omega(k_2^2 - k_1^2 + l_1^2 - l_2^2) + \theta]).
\end{aligned}
$$

For $\theta = \pi$, for example the trellis in Figure 7.5, we have

$$
\begin{aligned}
b_1 &= 4 + 2\cos[\omega(k_2^2 - k_1^2)] - 2\cos[\omega(l_2^2 - l_1^2)], \\
b_2 &= 4 - 2\cos[\omega(k_2^2 - k_1^2)] + 2\cos[\omega(l_2^2 - l_1^2)], \qquad (7.19) \\
d &= 8(1 + \cos[\omega(k_2^2 - k_1^2 + l_1^2 - l_2^2)]).
\end{aligned}
$$

Since $a, b_1, b_2, c, d \geq 0$, we have

$$\min \det(A) \geq (\min a + \min b_1 + \min c)(\min a + \min b_2 + \min c) - \max d. \qquad (7.20)$$

One can use the above results to calculate the CGD of the codes for which it is impossible to diverge from a state and merge to the same state in $P = 2$ transitions. We consider specific codes for the clarity of the presentation while similar approaches apply to other codes as well.

Example 7.3.1 *In this example, we consider the four-state code in Figure 7.5 using BPSK and transmitting $r = 1\,bit/(s\,Hz)$. Without loss of generality we* *assume two codewords diverging from state zero and remerging after P tran-sitions to state zero. For parallel transitions, that is $P = 1$, one can calcu-late $\min \det(A) = 64$ from (7.10). For a BPSK constellation, we have $\omega = \pi$, $\min a = \min b_1 = \min b_2 = \min c = 4$, and $\max d = 16$. Therefore, Inequality (7.20) results in*

$$\min \det(A) \geq 128. \tag{7.21}$$

Also, for $k_2^1 = k_1^1 = l_1^1 = k_2^2 = k_1^2 = l_1^2 = l_2^2 = k_2^3 = k_1^3 = l_1^3 = 0 \, and \, l_2^1 = l_2^3 = 1 \, in (7.17), we have $\det(A) = 128$ which means

$$\min \det(A) \leq 128. \tag{7.22}$$

Combining Inequalities (7.21) and (7.22) provides

$$\min \det(A) = 128 \tag{7.23}$$

which is greater than 64. It is easy to show that the minimum value of the CGD when $P > 3$ is greater than the minimum value of the CGD when $P = 3$. This proves that the minimum CGD for the code is dominated by parallel transitions and is equal to 64.

Example 7.3.2 *In this example, we consider the four-state code in Figure 7.5 using QPSK and transmitting $r = 2$ bits/(s Hz). Again, we assume two codewords diverg-ing from state zero and remerging after P transitions to state zero. Using (7.10), one can show $\min \det (A) = 16$ for parallel transitions. Also, we have $\omega = \pi/2$, $\min a = \min c = 2$, $\min b_1 = \min b_2 = 4$, and $\max d = 16$. Therefore, Inequality (7.20) results in*

$$\min \det(A) \geq 48. \tag{7.24}$$

Also, for $k_2^1 = k_1^1 = l_1^1 = k_2^2 = k_1^2 = l_1^2 = l_2^2 = k_2^3 = k_1^3 = l_1^3 = 0 \, and \, l_2^1 = l_2^3 = 1 \, in (7.17), we have $\det(A) = 48$ which means

$$\min \det(A) \leq 48. \tag{7.25}$$

Combining Inequalities (7.24) and (7.25) shows

$$\min \det(A) = 48 \tag{7.26}$$

which is greater than 16. Again, the minimum value of the CGD when $P > 3$ is greater than the minimum value of the CGD when $P = 3$. Therefore, the minimum CGD of the code is dominated by parallel transitions and is equal to 16.

Note that we have considered specific codes for the clarity of the presentation. One can show similar results for any trellis for which it is impossible to diverge from a

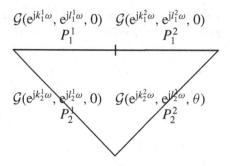

Fig. 7.12.   Two typical paths differing in $P = 2$ transitions.

state and merge to the same state in $P = 2$ transitions. For the trellises that contain a path with $P = 2$ transitions, like Figure 7.12, we provide similar formulas in what follows.

7.3.2 Error events with path length of two

We consider two codewords diverging from state zero and remerging after $P = 2$ transitions to state zero in Figure 7.12. For parallel transitions, that is $P = 1$, one can calculate the CGD from (7.10). We consider a typical case where the first codeword stays at state zero. For the second codeword, the first transition diverging from state zero uses $C(x_1, x_2, 0)$ and the second transition remerging to state zero uses $C(x_1, x_2, \theta)$ as in Figure 7.12. Note that $\theta = \pi/2, \pi, 3\pi/2$ in Figure 7.6, $\theta = \pi$ in Figures 7.7 and 7.8, and $\theta = \pi/2$ in Figure 7.9. It can be shown that

$$\det(A) = (a + b_1)(a + b_2) - d = a^2 + a(b_1 + b_2) + b_1 b_2 - d, \quad (7.27)$$

where

$$
\begin{aligned}
a &= 4 - 2\cos[\omega(k_2^1 - k_1^1)] - 2\cos[\omega(l_2^1 - l_1^1)], \\
b_1 &= 4 - 2\cos[\omega(k_2^2 - k_1^2) + \theta] - 2\cos[\omega(l_2^2 - l_1^2)], \\
b_2 &= 4 - 2\cos[\omega(k_2^2 - k_1^2)] - 2\cos[\omega(l_2^2 - l_1^2) - \theta], \\
d &= (2 - 2\cos\theta)(2 - 2\cos[\omega(k_2^2 - k_1^2 + l_1^2 - l_2^2) + \theta]),
\end{aligned}
\quad (7.28)
$$

and $a, b_1, b_2, d \geq 0$. Since a only depends on $(k_2^1, k_1^1, l_2^1, l_1^1)$, $b_1 + b_2$ is not negative, and b_1, b_2, d are independent of $(k_2^1, k_1^1, l_2^1, l_1^1)$, one can first calculate $(\min a)$ and then calculate $\min \det(A)$. In other words, one can use the following formula to find $\min \det(A)$:

$$\min \det(A) = (\min a)^2 + \min[(\min a)(b_1 + b_2) + b_1 b_2 - d]. \quad (7.29)$$

We use (7.29) to calculate the CGD for trellises with $P = 2$. We tabulate $\min \det(A)$ for different constellations and rotations (θ) in Table 7.2. In Table 7.2, we use the notation of Figure 7.12, where with a slight abuse of the notation $(k_1^1, l_1^1) \in P_1^1$,

Table 7.2. *Minimum* det(A) *from Equation (7.29) for different constellations and rotations*

Constellation	θ	P_1^1, P_1^2	P_2^1	P_2^2	min det(A)	Parallel CGD
BPSK	π	P_0	P_1	P_1	48	64
BPSK	π	P_0	P_1	P_0	48	64
QPSK	π	P_0	P_1	P_1	24	16
QPSK	π	P_0	P_1	P_0	20	16
QPSK	$\pi/2, 3\pi/2$	P_0	P_1	P_1	12	16
QPSK	$\pi/2, 3\pi/2$	P_0	P_1	P_0	12	16
8-PSK	π	P_{00}	P_{10}, P_{11}	P_{00}	5.03	4
8-PSK	π	P_{00}	P_{10}	P_{10}	5.37	4
8-PSK	π	P_{00}	P_{01}	P_{00}	10.75	4
8-PSK	π	P_{00}	P_{01}	P_{01}	10.75	4
8-PSK	$\pi/2, 3\pi/2$	P_{00}	P_{10}, P_{11}	P_{00}	2.69	4
8-PSK	$\pi/2, 3\pi/2$	P_{00}	P_{10}	P_{10}	2.54	4
8-PSK	$\pi/2, 3\pi/2$	P_{00}	P_{01}	P_{01}	5.49	4
8-PSK	$\pi/2, 3\pi/2$	P_{00}	P_{01}	P_{00}	6.06	4
8-PSK	$\pi/4, 7\pi/4$	P_{00}	P_{10}, P_{11}	P_{00}	1.03	4
8-PSK	$\pi/4, 7\pi/4$	P_{00}	P_{10}	P_{10}	1.03	4
8-PSK	$\pi/4, 7\pi/4$	P_{00}	P_{01}	P_{00}	2.75	4
8-PSK	$\pi/4, 7\pi/4$	P_{00}	P_{01}	P_{01}	2.75	4
8-PSK	$3\pi/4, 5\pi/4$	P_{00}	P_{10}, P_{11}	P_{00}	4.34	4
8-PSK	$3\pi/4, 5\pi/4$	P_{00}	P_{10}	P_{10}	4.34	4
8-PSK	$3\pi/4, 5\pi/4$	P_{00}	P_{01}	P_{00}	9.37	4
8-PSK	$3\pi/4, 5\pi/4$	P_{00}	P_{01}	P_{01}	9.37	4

$(k_1^2, l_1^2) \in P_1^2$, $(k_2^1, l_2^1) \in P_2^1$, and $(k_2^2, l_2^2) \in P_2^2$. We use these values to calculate the CGD of SOSTTCs. Again we consider specific examples, for the sake of simplicity.

Example 7.3.3 *In this example, we calculate the coding gain for the BPSK code in Figure 7.7. We assume two codewords diverging from state zero and remerging after P transitions to state zero. For parallel transitions, P = 1, one can calculate the CGD from (7.10) which is 64. For P = 2, min a = 4 and*

$$\min \det(A) = 16 + \min[4(b_1 + b_2) + b_1 b_2 - d] = 48 < 64. \qquad (7.30)$$

Therefore, CGD = 48 and the coding gain is dominated by paths with P = 2 transitions.

Example 7.3.4 *For the QPSK code in Figure 7.7, the CGD for parallel transitions, P = 1, is 16. For P = 2, since min a = 2, we have*

$$\min \det(A) = 4 + \min[2(b_1 + b_2) + b_1 b_2 - d] = 20 > 16. \qquad (7.31)$$

Therefore, the CGD for the QPSK code in Figure 7.7 is 16.

$$P_{00}^0 \quad P_{01}^0 \quad P_{10}^0 \quad P_{11}^0$$

$$P_{00}^{3\pi/4} \quad P_{01}^{3\pi/4} \quad P_{10}^{3\pi/4} \quad P_{11}^{3\pi/4}$$

$$P_{00}^{\pi} \quad P_{01}^{\pi} \quad P_{10}^{\pi} \quad P_{11}^{\pi}$$

$$P_{00}^{5\pi/4} \quad P_{01}^{5\pi/4} \quad P_{10}^{5\pi/4} \quad P_{11}^{5\pi/4}$$

Fig. 7.13.   A four-state SOSTTC; $r = 3$ bits/(s Hz) (8-PSK).

Example 7.3.5 *In the case of the 8-PSK codes in Figures 7.8 and 7.9, the parallel transitions are dominant and the CGD is 4. This can be shown using (7.29) and Table 7.2.*

Example 7.3.6 *The coding gain in Figure 7.6 is dominated by paths with $P = 2$ transitions. Using (7.29) and Table 7.2, one can show that the minimum CGD is 2.69 as is tabulated in Table 7.1.*

We emphasize the importance of picking the right set of rotations and subsets in providing the maximum coding gain. We also note that the minimum CGD is not always enough in comparing two codes with each other. As an example, let us consider the four-state, rate 3 bits/(s Hz) code in Figure 7.13. We have picked $\theta = 3\pi/4, 5\pi/4$ for states one and three in Figure 7.13 instead of $\theta = \pi/2, 3\pi/2$ in Figure 7.6. The code in Figure 7.13 is an interesting example where starting from state zero, parallel transitions are dominant and the CGD is 4. However, one should consider other error events with length two where the parallel transitions are not dominant. For example, starting from state one and staying at state one for two transitions, an error event with $P = 2$ can be a path diverging from state one to state two and remerging to state one in the second transition. Note that this error event can be part of two codewords with length four as illustrated in Figure 7.14. The CGD for such a path is 1.03 which is lower than 2.69 and 4. There are other paths with $P = 2$ providing different CGDs. Therefore, for the code in Figure 7.13, minimum CGD is not a good indicator of the performance. In addition to minimum CGD, one needs to consider the path weights, that is multiplicity, of error events. In this case, study of the distance spectrum is required to find the best code, that is the best assignment of rotations and subsets. Also note that this is not a linear code.

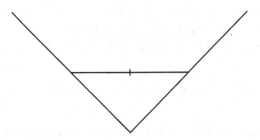

Fig. 7.14. Two paths with length four differing in $P = 2$ transitions.

7.4 Encoding and decoding

So far we have considered SOSTTCs for two transmit antennas. Every path in the trellis corresponds to a block of two symbols for transmission over two time slots. Therefore, if a frame of data includes $2Lb$ bits, L orthogonal blocks corresponding to $2L$ symbols are transmitted. We always start at state zero and return to state zero at the end. Therefore, $2Q$ additional symbols corresponding to Q blocks are transmitted for a trellis with memory Q to guarantee a return to state zero. This is equivalent to adding $2Qb$ zeros to the $2Lb$ input bits. We use $l = 1, 2, \ldots, L + Q$ to index the lth transition or the lth orthogonal block. Therefore, we have $S_1 = S_{L+Q+1} = 0$, where S_l represents the state of the lth block.

To transmit the $2b$ bits of the lth block, first, we pick s_1^l and s_2^l from the constellation using the $2b$ input bits. The rotation parameter of state S_l is denoted by θ^l. The selected symbols and the rotation parameter are used for transmission over two time slots. At the first time slot of the lth block, $s_1^l e^{j\theta^l}$ and s_2^l are transmitted from the first and second transmit antennas, respectively. Then, at the second time slot, $-(s_2^l)^* e^{j\theta^l}$ and $(s_1^l)^*$ are transmitted from the two antennas. The last Q blocks guarantee a return to state zero.

Similar to the case of STTCs in Chapter 6, the ML decoding finds the most likely valid path that starts from state zero and merges to state zero after $L + Q$ blocks. Each state transition corresponds to the transmission of two symbols in two time slots from two transmit antennas. Let us assume that for receive antenna m, we receive $r_{1,m}^l, r_{2,m}^l$ at the two time slots of block l. Assuming fixed path gains $\alpha_{1,m}$ and $\alpha_{2,m}$ throughout the transmission of a codeword, we have

$$\begin{cases} r_{1,m}^l = \alpha_{1,m}s_1^l e^{j\theta^l} + \alpha_{2,m}s_2^l + \eta_{1,m}^l \\ r_{2,m}^l = -\alpha_{1,m}(s_2^l)^* e^{j\theta^l} + \alpha_{2,m}(s_1^l)^* + \eta_{2,m}^l. \end{cases} \tag{7.32}$$

where $\eta_{1,m}^l$ and $\eta_{2,m}^l$ are noise samples for block l. The Viterbi algorithm can be used for the ML decoding of SOSTTCs. Since we allow parallel transitions for SOSTTCs, at each step, we should find the best transition among all parallel transitions. Then, we use the best parallel transition to calculate the path metrics in the Viterbi algorithm. The branch metric of the lth block is given by

$$\sum_{m=1}^{M} \left| r_{1,m}^l - \alpha_{1,m} s_1^l\, \mathrm{e}^{\mathrm{j}\theta^l} - \alpha_{2,m} s_2^l \right|^2 + \left| r_{2,m}^l + \alpha_{1,m} (s_2^l)^* \mathrm{e}^{\mathrm{j}\theta^l} - \alpha_{2,m} (s_1^l)^* \right|^2. \quad (7.33)$$

Expanding the branch metric in (7.33) and removing the constant terms results in the following branch metric:

$$\sum_{m=1}^{M} 2\Re \left\{ r_{2,m}^l \alpha_{1,m}^* (s_2^l) \mathrm{e}^{-\mathrm{j}\theta^l} - r_{2,m}^l \alpha_{2,m}^* (s_1^l) - r_{1,m}^l \alpha_{1,m}^* (s_1^l)^* \mathrm{e}^{-\mathrm{j}\theta^l} - r_{1,m}^l \alpha_{2,m}^* (s_2^l)^* \right\}.$$

$$(7.34)$$

For each block, the rotation, the path gains, and the received signals are known at the receiver. Therefore, the branch metric in (7.34) is only a function of transmitted signals and can be denoted by $2J^l(s_1^l, s_2^l)$. Therefore, the first step is to find the valid pair (s_1^l, s_2^l) among all parallel transitions that minimizes $J^l(s_1^l, s_2^l)$. To use the orthogonality of the STBC, one may utilize the fact that

$$J^l(s_1^l, s_2^l) = J_1^l(s_1^l) + J_2^l(s_2^l), \quad (7.35)$$

where

$$J_1^l(s_1^l) = -\Re \left\{ \left[r_{2,m}^l \alpha_{2,m}^* + (r_{1,m}^l)^* \alpha_{1,m} \, \mathrm{e}^{\mathrm{j}\theta^l} \right] s_1^l \right\} \quad (7.36)$$

is only a function of s_1^l and

$$J_2^l(s_2^l) = \Re \left\{ \left[r_{2,m}^l \alpha_{1,m}^* \, \mathrm{e}^{-\mathrm{j}\theta^l} - (r_{1,m}^l)^* \alpha_{2,m} \right] s_2^l \right\} \quad (7.37)$$

is only a function of s_2^l. Similar to the case of an orthogonal STBC, this can be used to simplify the search for the transition with the minimum branch metric among all parallel transitions. However, while the orthogonality of the blocks results in a simpler ML decoding, the derivation of such a simpler ML decoding is not the same as that of a block code. Note that although $J_1^l(s_1^l)$ is only a function of s_1^l since not all possible (s_1^l, s_2^l) pairs are allowed for each trellis transition, s_1^l and s_2^l are not independent. Similarly, $J_2^l(s_2^l)$ is a function of s_1^l because of the correlation between s_2^l and s_1^l. In other words, the input symbols of the orthogonal STBC are not independent and therefore the separate decoding is not possible. For each branch, among all possible parallel transitions, we should find the valid pair (s_1^l, s_2^l) that provides the minimum $J^l(s_1^l, s_2^l)$. If we find the symbol s_1^l that minimizes $J_1^l(s_1^l)$ and the symbol s_2^l that minimizes $J_2^l(s_2^l)$, the resulting pair of symbols (s_1^l, s_2^l) may not be a valid pair for the considered transition. Therefore, such a pair does not provide the best pair among all possible parallel transitions. On the other hand, it is possible to divide the set of all parallel pairs to subsets for which the symbol pairs (s_1^l, s_2^l) are independent. Then, the simple separate decoding, or equivalently separate minimization of $J_1^l(s_1^l)$ and $J_2^l(s_2^l)$, is done for each subset. Finally, we find the best pair that minimizes $J^l(s_1^l, s_2^l)$ among the best pairs of different subsets. In other words, to utilize the reduced complexity, we need to combine the set partitioning and the separate decoding for each orthogonal STBC building block. To elaborate the simplified decoding method, we provide the subsets for which the symbols s_1^l and s_2^l are independent for different PSK constellations. Later, we provide the details of decoding for different SOSTTC examples.

For a BPSK constellation and the corresponding set partitioning in Figure 7.1, the sets $P_0 = \{00, 11\}$ and $P_1 = \{01, 10\}$ only contain two pairs and there is no need for simplification. The root set $P = \{00, 11, 01, 10\}$ provides independent symbols s_1 and s_2. In other words, since all possible combinations of indices 0 and 1 exist in this set, we can minimize $J_1^l(s_1^l)$ and $J_2^l(s_2^l)$ separately. Any resulting symbol pair (s_1^l, s_2^l) will be a valid pair.

For a QPSK constellation and the corresponding set partitioning in Figure 7.2, the sets P_{00}, P_{01}, P_{10}, and P_{11} provide independent pairs. If P_0 is used, first we need to divide it into two subsets $P_0 = P_{00} \cup P_{01}$. Then, we find the best pairs in P_{00} and P_{01} using the separate minimization of $J_1^l(s_1^l)$ and $J_2^l(s_2^l)$. Finally, we compare the branch metric $J^l(s_1^l, s_2^l)$ for these two pairs and find the minimum among the two. Similarly for P_1, we first divide it into two subsets $P_1 = P_{10} \cup P_{11}$ and find the minimum branch metric of each of these two subsets. Then, the best parallel transition between the two selected transitions is the one that has a lower branch metric.

For an 8-PSK constellation and the corresponding set partitioning in Figure 7.3, any subset P_{abcd} with four subscripts provides independent pairs. For example, P_{0000} includes all combinations of indices 0 and 4. Dealing with any set P_{abc} with three subscripts, first, we divide it in two subsets $P_{abc} = P_{abc0} \cup P_{abc1}$ and using separate decoding, we find the best pair in each subset. Then, the pair with minimum branch metric among the two pairs provides the best parallel transition. Note that if the parallel transitions include the sets P_{00}, P_{01}, P_{10}, and P_{11}, with two subscripts, we do not need to divide the sets. The set P_{00} includes all independent pairs where the indices are from $\{0, 2, 4, 6\}$. Similarly, the set P_{01} includes all independent pairs where the indices are odd. For the set P_{10}, we have all independent pairs with an even first index and an odd second index. Finally, the set P_{11} includes every possible pair of indices with an odd first index and an even second index. If P_0 and P_1 are used as parallel transitions, one can easily divide them as $P_0 = P_{00} \cup P_{01}$ and

$P_1 = P_{10} \cup P_{11}$. Then, the simplified minimization is done using what we explained above for previous cases.

For any other constellation, one can combine the set partitioning and the separate decoding for each orthogonal STBC building block in a similar way. Utilizing such a combination makes it possible to reduce the complexity of the branch metric calculations for parallel transitions. After finding the best parallel branch for each transition, the ML decoder finds the most likely path. First, we need to calculate the path metrics using the best branch metrics $J^l(s_1^l, s_2^l)$. The path metric of a valid path is the sum of the branch metrics for the branches that form the path. The most likely path is the one which has the minimum path gain. The ML decoder finds the set of constellation symbols (s_1^l, s_2^l), $l = 1, 2, \ldots, L + Q$ that constructs a valid path and solves the following minimization problem:

$$\min_{s_1^1, s_2^1, s_1^2, s_2^2, \ldots, s_1^{L+Q}, s_2^{L+Q}} \sum_{l=1}^{L+Q} J^l(s_1^l, s_2^l). \tag{7.38}$$

Note that a valid path starts at state zero and returns to state zero after $L + Q$ transitions. The details of the Viterbi algorithm to solve (7.38) and (6.1) are similar to each other. The main difference is using the branch metric in (7.34), equivalently (7.35), instead of the Euclidean distance. The orthogonality of STBC building blocks makes it possible to reduce the complexity of the search as explained above.

Example 7.4.1 *In this example, we provide the details of a simple decoding for the SOSTTC in Examples 7.2.1 and 7.3.2 that uses the trellis in Figure 7.5 and a QPSK constellation. ML decoding using the Viterbi algorithm is done in two steps: (i) finding the best branch among all parallel transitions in the trellis; and (ii) searching for the best path with the smallest path metric among all valid paths.*

Let us begin with the first step. The sets that are assigned to parallel transitions of the trellis are P_0 and P_1 with rotations $\theta^l = 0, \pi$. They contain the following QPSK indices as it is evident from Figure 7.2:

$$P_0 = \{00, 22, 02, 20, 11, 33, 13, 31\} = \{00, 22, 02, 20\} \cup \{11, 33, 13, 31\},$$
$$P_1 = \{01, 23, 03, 21, 10, 32, 12, 30\} = \{01, 23, 03, 21\} \cup \{10, 32, 12, 30\}.$$
$$\tag{7.39}$$

For $P_{00} = \{00, 22, 02, 20\}$, the two symbols s_1^l and s_2^l can be 1 or -1, corresponding to indices 0 and 2, respectively. Therefore, we need to find

$$\hat{s}_{10}^l = \operatorname*{argmin}_{s_1^l \in \{1, -1\}} J_1^l(s_1^l),$$
$$\hat{s}_{20}^l = \operatorname*{argmin}_{s_2^l \in \{1, -1\}} J_2^l(s_2^l).$$
$$\tag{7.40}$$

Similarly, for $P_{01} = \{11, 33, 13, 31\}$, we find

$$\hat{s}_{11}^l = \underset{s_1^l \in \{1, -1\}}{\operatorname{argmin}} J_1^l(s_1^l),$$

$$\hat{s}_{21}^l = \underset{s_2^l \in \{1, -1\}}{\operatorname{argmin}} J_2^l(s_2^l). \tag{7.41}$$

Then, the best parallel pair is given by

$$(\hat{s}_1^l, \hat{s}_2^l) = \underset{\{(\hat{s}_{10}^l, \hat{s}_{20}^l), (\hat{s}_{11}^l, \hat{s}_{21}^l)\}}{\operatorname{argmin}} \{J_1^l(s_1^l) + J_2^l(s_2^l)\}. \tag{7.42}$$

We should find the optimal pairs for four cases of P_0^0, P_1^0, P_0^π, and P_1^π. We use the resulting best branch metrics in the second step to find the best path.

In the second step, we utilize the Viterbi algorithm to find the best path. This is similar to the Viterbi algorithm for any other code as explained in Chapter 6.

To compare the decoding complexity of the four-state SOSTTC in this example and that of the four-state STTC in Figure 6.3 or 6.10, we count the number of operations. To simplify the calculation, we compute and save the common terms that can be used in different formulas. For example, we need to compute $J_1^l(s_1^l)$ and $J_2^l(s_2^l)$ for different values of s_1^l and s_2^l. To save the number of operations, first, we calculate $r_{2,m}^l \alpha_{2,m}^* + (r_{1,m}^l)^* \alpha_{1,m} e^{j\theta^l}$ and $r_{2,m}^l \alpha_{1,m}^* e^{-j\theta^l} - (r_{1,m}^l)^* \alpha_{2,m}$ that are used in (7.36) and (7.37), respectively. Note that $e^{j\theta^l} = e^{-j\theta^l} = 1$, for $\theta = 0$ and $e^{j\theta^l} = e^{-j\theta^l} = -1$, for $\theta = \pi$. Therefore, $e^{j\theta^l}$ and $e^{-j\theta^l}$ act as different signs for different rotations. Each of these terms needs eight real multiplications and six real additions for given rotation parameters. Calculating them for both rotation parameters only adds four extra real additions. Therefore, we need 16 real multiplications and 16 real additions to calculate all different cases of these two terms. Then, we need to compute and save $J_1^l(s_1^l)$ and $J_2^l(s_2^l)$ for every possible values of s_1^l and s_2^l. Since QPSK constellation includes $\{1, -1, j, -j\}$, this calculation, including the real part operation, requires 16 multiplications for both rotation parameters. Therefore, calculating all possible cases of (7.36) and (7.37) needs 32 real multiplications and 16 real additions. Then, we need to find the minimum $J_1^l(s_1^l)$ and $J_2^l(s_2^l)$ as in (7.40) and (7.41) for different cases. It requires eight real comparisons. Finally, to pick the best pair for each transition following (7.42), eight real additions and four real comparisons are needed. Overall, the branch metric calculation step requires 32 real multiplications, 24 real additions, and 12 real comparisons for a transition that includes the transmission of two symbols. This is much lower than, almost one third of, the number of operations in the branch metric calculation step for the four-state STTC in Figure 6.3 or 6.10. Table 7.3 tabulates different stages of the branch metric calculation step in more details while Table 7.4 counts the corresponding number of operations.

The number of operations for the Viterbi algorithm to find the best path using the calculated branch metrics is the same for codes with similar trellises. It is a function

Table 7.3. *Different stages of the branch-metric*
calculation step in Example 7.4.1

Stage	Calculation	For
1	$r^l_{2,m}\alpha^*_{2,m}$	once
2	$(r^l_{1,m})^*\alpha_{1,m}$	once
3	$r^l_{2,m}\alpha^*_{1,m}$	once
4	$-(r^l_{1,m})^*\alpha_{2,m}$	once
5	$r^l_{2,m}\alpha^*_{2,m} + (r^l_{1,m})^*\alpha_{1,m}\,\mathrm{e}^{j\theta^l}(\theta^l = 0)$	once
6	$r^l_{2,m}\alpha^*_{2,m} + (r^l_{1,m})^*\alpha_{1,m}\,\mathrm{e}^{j\theta^l}(\theta^l = \pi)$	once
7	$r^l_{2,m}\alpha^*_{1,m}\,\mathrm{e}^{-j\theta^l} - (r^l_{1,m})^*\alpha_{2,m}(\theta^l = 0)$	once
8	$r^l_{2,m}\alpha^*_{1,m}\,\mathrm{e}^{-j\theta^l} - (r^l_{1,m})^*\alpha_{2,m}(\theta^l = \pi)$	once
9	$J^l_1(s^l_1)$ in (7.36)	$\forall s^l_1, \theta^l = 0, \pi$
10	$J^l_2(s^l_2)$ in (7.37)	$\forall s^l_2, \theta^l = 0, \pi$
11	$\min\{J^l_1(s^l_1)\}$ in (7.40) & (7.41)	all
12	$\min\{J^l_2(s^l_2)\}$ in (7.40) & (7.41)	all
13	$\min\{J^l_1(s^l_1) + J^l_2(s^l_2)\}$ in (7.42)	each transition

Table 7.4. *Number of operations corresponding to different*
stages in Table 7.3 for Example 7.4.1

Stage	Number of multiplications	Number of additions	Number of comparison
1	4	2	0
2	4	2	0
3	4	2	0
4	4	2	0
5	0	2	0
6	0	2	0
7	0	2	0
8	0	2	0
9	8	0	0
10	8	0	0
11	0	0	4
12	0	0	4
13	0	8	4
Total	32	24	12

of the number of branches diverging from or merging to each state of the trellis. This number, and consequently the complexity, of the second step is smaller for a trellis with a smaller number of branches leaving or merging each state. Therefore, even the complexity of the second step for the trellis used in Figure 7.5 is lower than that of Figure 6.3 or 6.10.

The decoding complexity of a SOSTTC is smaller than that of a STTC with the same number of states at a given rate. This is due to the fact that the orthogonality of STBC building blocks can be utilized to reduce the decoding complexity. We compare the performance of STTCs and SOSTTCs in Section 7.7.

7.5 Performance analysis

Following the derivations in Section 6.4, the PEP of a SOSTTC is also evaluated using (6.30) [117]. The difference is in the calculation of ξ in (6.28). Since the path gains are independent complex Gaussian random variables, to calculate the MGF of the random variable ξ, we use Turin's formula [143] to show that

$$M_\xi(u) = \left\{ \det\left[I_N - \frac{u\gamma}{4}(C^2 - C^1)^H \cdot (C^2 - C^1) \right] \right\}^{-M}. \tag{7.43}$$

Therefore, the PEP is

$$P(C^1 \to C^2) = \frac{1}{\pi} \int_0^{\pi/2} \left\{ \det\left[I_N + \frac{\gamma}{4\sin^2\theta}(C^2 - C^1)^H \cdot (C^2 - C^1) \right] \right\}^{-M} d\theta. \tag{7.44}$$

If C^1 and C^2 differ in $P = L + Q$ transitions, $C^2 - C^1$ is a $2P \times 2$ matrix consisting of P blocks with 2×2 sizes. If we denote the pth block of C^i by C^{ip}, we have

$$C^2 - C^1 = \begin{pmatrix} C^{21} - C^{11} \\ C^{22} - C^{12} \\ \vdots \\ C^{2P} - C^{1P} \end{pmatrix}. \tag{7.45}$$

This is similar to the result in (7.11). Using (7.45), one can calculate the matrix $(C^2 - C^1)^H \cdot (C^2 - C^1)$ as

$$(C^2 - C^1)^H \cdot (C^2 - C^1) = \sum_{p=1}^{P} (C^{2p} - C^{1p})^H \cdot (C^{2p} - C^{1p}). \tag{7.46}$$

Defining the 2×2 matrix $A^p = \frac{\gamma}{4\sin^2\theta}(C^{2p} - C^{1p})^H \cdot (C^{2p} - C^{1p})$, we have

$$P(C^1 \to C^2) = \frac{1}{\pi} \int_0^{\pi/2} \left\{ \det\left[I_N + \sum_{p=1}^{P} A^p \right] \right\}^{-M} d\theta. \tag{7.47}$$

We follow our notation in Section 7.4, to represent C^{1p} and C^{2p} as follow

$$C^{1p} = \begin{pmatrix} s_1^{1p} e^{j\theta^{1p}} & s_2^{1p} \\ -(s_2^{1p})^* e^{j\theta^{1p}} & (s_1^{1p})^* \end{pmatrix}, \quad C^{2p} = \begin{pmatrix} s_1^{2p} e^{j\theta^{2p}} & s_2^{2p} \\ -(s_2^{2p})^* e^{j\theta^{2p}} & (s_1^{2p})^* \end{pmatrix}. \tag{7.48}$$

If we denote the elements of the 2×2 matrix $\mathbf{A}^P$ as

$$\mathbf{A}^P = \begin{pmatrix} A_{11}^P & A_{12}^P \\ A_{21}^P & A_{22}^P \end{pmatrix}, \tag{7.49}$$

they can be easily calculated by

$$A_{11}^P = |s_1^{1P} e^{j\theta^{1P}} - s_1^{2P} e^{j\theta^{2P}}|^2 + |s_2^{1P} e^{-j\theta^{1P}} - s_2^{2P} e^{-j\theta^{2P}}|^2,$$

$$A_{12}^P = (s_1^{1P} e^{j\theta^{1P}} - s_1^{2P} e^{j\theta^{2P}})^*(s_2^{1P} - s_2^{2P}) - (s_2^{1P} e^{-j\theta^{1P}} - s_2^{2P} e^{-j\theta^{2P}})(s_1^{1P} - s_1^{2P})^*,$$

$$A_{21}^P = (s_1^{1P} e^{j\theta^{1P}} - s_1^{2P} e^{j\theta^{2P}})(s_2^{1P} - s_2^{2P})^* - (s_2^{1P} e^{-j\theta^{1P}} - s_2^{2P} e^{-j\theta^{2P}})^*(s_1^{1P} - s_1^{2P}),$$

$$A_{22}^P = |s_1^{1P} - s_1^{2P}|^2 + |s_2^{1P} - s_2^{2P}|^2. \tag{7.50}$$

Note that $A_{12}^P = (A_{21}^P)^*$. Using (7.47), one can calculate the PEP of two SOSTTC codewords as

$$P(\mathbf{C}^1 \to \mathbf{C}^2) = \frac{1}{\pi} \int_0^{\pi/2} \left\{ 1 + \sum_{p=1}^P \mathrm{Tr}[\mathbf{A}^P] + \det\left[\sum_{p=1}^P \mathbf{A}^P\right] \right\}^{-M} d\theta. \tag{7.51}$$

Example 7.5.1 *Let us consider the two-state SOSTTC in Figure 7.7 using BPSK. Unlike STTCs in Chapter 6, SOSTTCs may include parallel paths. For the BPSK SOSTTC in Figure 7.7, the PEP for two parallel paths is given by*

$$P(\mathbf{C}^1 \to \mathbf{C}^2) = \frac{1}{\pi} \int_0^{\pi/2} \left[\frac{\sin^2 \theta}{\sin^2 \theta + 2\gamma} \right]^{2M} d\theta. \tag{7.52}$$

The above PEP can be calculated from the following closed form formula:

$$P(\mathbf{C}^1 \to \mathbf{C}^2) = \frac{1}{2}\left[1 - \sqrt{\frac{2\gamma}{1 + 2\gamma}} \sum_{m=0}^{2M-1} \binom{2m}{m} \left(\frac{1}{4(1 + 2\gamma)}\right)^m \right]. \tag{7.53}$$

For the case of one receive antenna, $M = 1$, the PEP of parallel paths are calculated by

$$P(\mathbf{C}^1 \to \mathbf{C}^2) = \frac{1}{2}\left[1 - \sqrt{\frac{2\gamma}{1 + 2\gamma}} \left(1 + \frac{1}{2 + 4\gamma}\right) \right]. \tag{7.54}$$

For a pair of paths that are not parallel, we consider the following two paths resulting from transitions 000 *and* 010, *respectively:*

$$\mathbf{C}^1 = \begin{pmatrix} 1 & 1 \\ -1 & 1 \\ 1 & 1 \\ -1 & 1 \end{pmatrix}, \quad \mathbf{C}^2 = \begin{pmatrix} 1 & -1 \\ 1 & 1 \\ 1 & 1 \\ 1 & -1 \end{pmatrix}. \tag{7.55}$$

Using the equations in (7.50) results in

$$A_{11}^1 = A_{11}^2 = A_{22}^1 = A_{22}^2 = \frac{\gamma}{\sin^2 \theta},$$

$$A_{12}^1 = A_{21}^1 = 0, \tag{7.56}$$

$$A_{12}^2 = A_{21}^2 = -\frac{\gamma}{\sin^2 \theta}.$$

The corresponding PEP can be written as

$$P(\mathbf{C}^1 \to \mathbf{C}^2) = \frac{1}{\pi} \int_0^{\pi/2} \left[\frac{\sin^2 \theta}{\sin^2 \theta + 3\gamma} \right]^M \left[\frac{\sin^2 \theta}{\sin^2 \theta + \gamma} \right]^M d\theta. \tag{7.57}$$

The above PEP can be calculated in a closed form. For the case of one receive antenna, $M = 1$, the closed form of the PEP is given by

$$P(\mathbf{C}^1 \to \mathbf{C}^2) = \frac{1}{2} \left[1 - \frac{3}{2}\sqrt{\frac{3\gamma}{1 + 3\gamma}} + \frac{1}{2}\sqrt{\frac{\gamma}{1 + \gamma}} \right]. \tag{7.58}$$

7.6 Extension to more than two antennas

In this section, we extend the general approach for designing SOSTTCs to more than two transmit antennas. We follow the same principles to systematically design SOSTTCs using orthogonal designs, constellation rotations, and set partitioning. We provide a few examples to show how the general approach works.

7.6.1 Real constellations

A rate one real $N \times N$ orthogonal design only exists for $N = 2, 4, 8$ as shown in Chapter 4. An example of a 4×4 real orthogonal design is given in (4.38). To expand the orthogonal matrices, similar to the case of two antennas in (7.7), we use phase rotations as follows:

$$\mathcal{G}(x_1, x_2, x_3, x_4, \theta_1, \theta_2, \theta_3, \theta_4) = \begin{pmatrix} x_1 \, e^{j\theta_1} & x_2 \, e^{j\theta_2} & x_3 \, e^{j\theta_3} & x_4 \, e^{j\theta_4} \\ -x_2 \, e^{j\theta_1} & x_1 \, e^{j\theta_2} & -x_4 \, e^{j\theta_3} & x_3 \, e^{j\theta_4} \\ -x_3 \, e^{j\theta_1} & x_4 \, e^{j\theta_2} & x_1 \, e^{j\theta_3} & -x_2 \, e^{j\theta_4} \\ -x_4 \, e^{j\theta_1} & -x_3 \, e^{j\theta_2} & x_2 \, e^{j\theta_3} & x_1 \, e^{j\theta_4} \end{pmatrix}. \tag{7.59}$$

Since the constellation is real, for example BPSK or PAM, and we do not want to expand the constellation symbols, we pick $\theta_i = 0, \pi$, where $i = 1, 2, 3, 4$. This

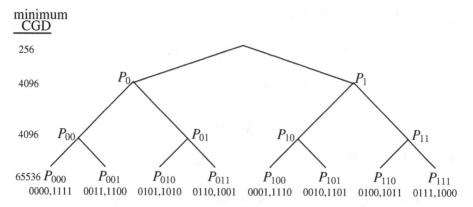

Fig. 7.15.   Set partitioning for BPSK; each group of binary numbers represents the four symbol indices in a STBC with four transmit antennas.

means that we only potentially use a sign change for each column. Equation (4.38) is a member of the super-orthogonal sets in (7.59) for $\theta_1 = \theta_2 = \theta_3 = \theta_4 = 0$.

Similar to the case of two transmit antennas, we use CGD instead of Euclidean distance to define a set partitioning similar to Ungerboeck's set partitioning [144]. An example for a BPSK constellation is shown in Figure 7.15. At the root of the tree, the minimum CGD is 256. At the first level of partitioning, the highest minimum CGD that can be obtained is 4096, which is obtained by creating subsets P_0 and P_1 with transmitted symbol elements differing at least in two positions. At the last level of partitioning, we have eight sets P_{000}, P_{001}, P_{010}, P_{011}, P_{100}, P_{101}, P_{110}, and P_{111} with two elements per set that differ in all positions. The resulting minimum CGD is 65 536.

Code construction based on super-orthogonal sets is as follows. We assign a constituent STBC to all transitions from a state. The adjacent states are typically assigned to one of the other constituent STBC from the super-orthogonal code. Similarly, we can assign the same STBC to branches that are merging into a state. It is thus assured that any path that diverges from (or merges to) the correct path differs by rank 4. In other words, every pair of codewords diverging from (or merging to) a state achieves full diversity because the pair is from the same orthogonal code (same parameters θ_1, θ_2, θ_3, θ_4). On the other hand, for codewords with different θ_1, θ_2, θ_3, θ_4, it is possible that they do not achieve full diversity. Since these codewords are assigned to different states, the resulting trellis code would provide full diversity despite the fact that a pair of codewords in a super-orthogonal code may not achieve full diversity.

Figures 7.16, 7.17, and 7.18 show the examples of two-state, four-state, and eight-state codes, respectively, for transmitting $r = 1$ bit/(s Hz) using BPSK.

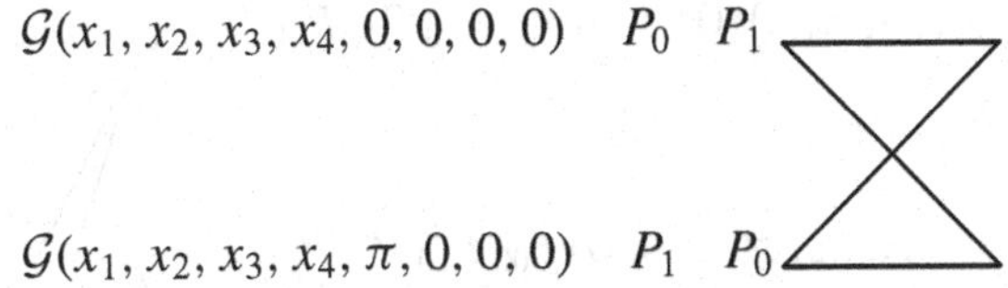

$\mathcal{G}(x_1, x_2, x_3, x_4, 0, 0, 0, 0) \quad P_0 \quad P_1$

$\mathcal{G}(x_1, x_2, x_3, x_4, \pi, 0, 0, 0) \quad P_1 \quad P_0$

Fig. 7.16.   A two-state code; four transmit antennas; $r = 1$ bit/(s Hz) (BPSK).

$\mathcal{G}(x_1, x_2, x_3, x_4, 0, 0, 0, 0) \quad P_0 \quad P_1$

$\mathcal{G}(x_1, x_2, x_3, x_4, \pi, 0, 0, 0) \quad P_0 \quad P_1$

$\mathcal{G}(x_1, x_2, x_3, x_4, 0, 0, 0, 0) \quad P_1 \quad P_0$

$\mathcal{G}(x_1, x_2, x_3, x_4, \pi, 0, 0, 0) \quad P_1 \quad P_0$

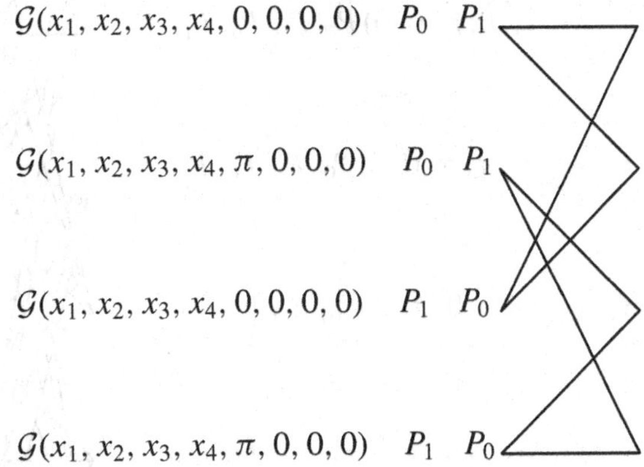

Fig. 7.17.   A four-state code; four transmit antennas; $r = 1$ bit/(s Hz) (BPSK).

7.6.2 Complex constellations

In this section, we discuss how to design complex SOSTTCs for more than two transmit antennas. One can achieve a STBC for three transmit antennas by removing one column from a STBC for four transmit antennas. We concentrate on the case of four transmit antennas since the case of three transmit antennas is very similar and simpler. As we discussed in Chapter 4, the maximum possible rate of an OSTBC for four transmit antennas is 3/4. An example of a 4×4 rate 3/4 complex orthogonal design is given in (4.103). To expand the orthogonal matrices, similar to the case of real constellations, we use the following rotations:

$$\mathcal{G}(x_1, x_2, x_3, \theta_1, \theta_2, \theta_3, \theta_4) = \begin{pmatrix} x_1 e^{j\theta_1} & x_2 e^{j\theta_2} & x_3 e^{j\theta_3} & 0 \\ -x_2^* e^{j\theta_1} & x_1^* e^{j\theta_2} & 0 & x_3 e^{j\theta_4} \\ x_3^* e^{j\theta_1} & 0 & -x_1^* e^{j\theta_3} & x_2 e^{j\theta_4} \\ 0 & x_3^* e^{j\theta_2} & -x_2^* e^{j\theta_3} & -x_1 e^{j\theta_4} \end{pmatrix}. \tag{7.60}$$

Again, we only use rotations that do not expand the constellation. For example, $\theta_1, \theta_2, \theta_3, \theta_4$ can be 0, $\pi/2$, π, $3\pi/2$ for QPSK. Then, we can systematically design rate 3/4 SOSTTCs for any trellis and complex constellation through set partitioning and assigning sets to different branches of the trellis. On the other hand, the

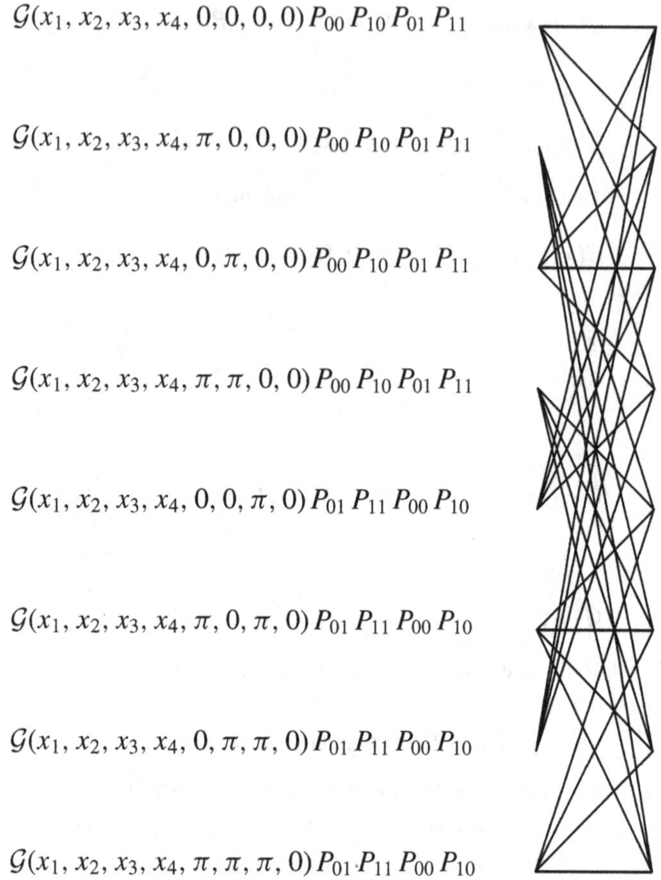

$\mathcal{G}(x_1, x_2, x_3, x_4, 0, 0, 0, 0)\,P_{00}\,P_{10}\,P_{01}\,P_{11}$

$\mathcal{G}(x_1, x_2, x_3, x_4, \pi, 0, 0, 0)\,P_{00}\,P_{10}\,P_{01}\,P_{11}$

$\mathcal{G}(x_1, x_2, x_3, x_4, 0, \pi, 0, 0)\,P_{00}\,P_{10}\,P_{01}\,P_{11}$

$\mathcal{G}(x_1, x_2, x_3, x_4, \pi, \pi, 0, 0)\,P_{00}\,P_{10}\,P_{01}\,P_{11}$

$\mathcal{G}(x_1, x_2, x_3, x_4, 0, 0, \pi, 0)\,P_{01}\,P_{11}\,P_{00}\,P_{10}$

$\mathcal{G}(x_1, x_2, x_3, x_4, \pi, 0, \pi, 0)\,P_{01}\,P_{11}\,P_{00}\,P_{10}$

$\mathcal{G}(x_1, x_2, x_3, x_4, 0, \pi, \pi, 0)\,P_{01}\,P_{11}\,P_{00}\,P_{10}$

$\mathcal{G}(x_1, x_2, x_3, x_4, \pi, \pi, \pi, 0)\,P_{01}\,P_{11}\,P_{00}\,P_{10}$

Fig. 7.18.   An eight-state code; four transmit antennas; $r = 1$ bit/(s Hz) (BPSK).

nice structure of SOSTTCs makes it possible to design rate one codes for specific trellises. We explain why such an increase in rate is possible in the sequel.

To design a rate one 4×4 STBC for a constellation with $L = 2^b$ symbols, we need L^4 codeword matrices to transmit b bits/(s Hz). For example, using QPSK, one needs $4^4 = 256$ codeword matrices to leave every state. A rate 3/4 STBC only provides L^3, for QPSK $4^3 = 64$, codeword matrices. Therefore, it is impossible to have all codewords leaving a state from the same STBC. The number of branches leaving each state of a trellis is two. Therefore, the maximum number of parallel branches in a rate one SOSTTC is $L^4/2$, which is 128 for QPSK. To have full diversity for the SOSTTC, every possible pair among the codewords assigned to parallel branches should provide full diversity. Therefore, it is impossible, for example, to design a two-state SOSTTC like the one in Figure 7.16. On the other hand, by increasing the number of branches leaving each state of a trellis, the number of required parallel branches decreases. While a STBC cannot provide all codewords leaving a state, it can provide the codewords needed for parallel branches by increasing the number of branches leaving a state. For example, in the case of QPSK constellation, if four branches leave every state, we need 64 parallel branches. Then, a rate 3/4 STBC can provide all required codeword matrices for these parallel branches.

Note that so far, in all of our examples, we have assigned the same STBC to all branches leaving from a state. Such an assignment will not provide a rate one SOSTTC for more than two antennas. Instead, by assigning different members of a "set of super-orthogonal codes" to branches leaving from the same state, one can design a rate one SOSTTC [119]. We provide the details of such a design in the following example.

Example 7.6.1 *As we discussed in Section 4.7, OSTBCs are not unique. In fact, multiplying the generator matrix of an OSTBC by a unitary matrix from right or left results in another OSTBC. We have used this property to expand the codeword matrices and design SOSTTCs. We have used the following unitary matrix to rotate the symbols in each column, when multiplied from the right, or rotate the symbols in each row, when multiplied from the left:*

$$U_1(a_1, a_2, a_3, a_4) = \begin{pmatrix} a_1 & 0 & 0 & 0 \\ 0 & a_2 & 0 & 0 \\ 0 & 0 & a_3 & 0 \\ 0 & 0 & 0 & a_4 \end{pmatrix}. \tag{7.61}$$

For example, the rotated generator matrix in (7.60) is the result of multiplying the generator matrix in (4.103) by $U_1(e^{j\theta_1}, e^{j\theta_2}, e^{j\theta_3}, e^{j\theta_4})$ from the right. In this example, in addition to rotating the columns, or rows, of the generator matrix, we swap rows, or columns. The following unitary matrices can rotate and swap at the same time:

$$U_2(a_1, a_2, a_3, a_4) = \begin{pmatrix} 0 & 0 & 0 & a_1 \\ 0 & 0 & a_2 & 0 \\ 0 & a_3 & 0 & 0 \\ a_4 & 0 & 0 & 0 \end{pmatrix},$$

$$U_3(a_1, a_2, a_3, a_4) = \begin{pmatrix} 0 & a_1 & 0 & 0 \\ a_2 & 0 & 0 & 0 \\ 0 & 0 & 0 & a_3 \\ 0 & 0 & a_4 & 0 \end{pmatrix}, \tag{7.62}$$

$$U_4(a_1, a_2, a_3, a_4) = \begin{pmatrix} 0 & 0 & a_1 & 0 \\ 0 & 0 & 0 & a_2 \\ a_3 & 0 & 0 & 0 \\ 0 & a_4 & 0 & 0 \end{pmatrix}.$$

We use these rotation and swapping transforms in designing the example SOSTTC. We denote the generator matrix in (4.103) by $\mathcal{G}$. Note that this generator matrix is a function of (x_1, x_2, x_3). Since the code has too many parameters that cannot be put in a single figure, we define the following abbreviations to represent the generator matrix for each trellis transition:

$$
\begin{aligned}
\mathcal{G}_1 &= U_1(1, 1, 1, j) \cdot \mathcal{G}, \\
\mathcal{G}_2 &= U_2(1, 1, j, 1) \cdot \mathcal{G}, \\
\mathcal{G}_3 &= U_3(1, j, 1, 1) \cdot \mathcal{G}, \\
\mathcal{G}_4 &= U_4(j, 1, 1, 1) \cdot \mathcal{G}.
\end{aligned}
\tag{7.63}
$$

Also, we define $\mathcal{G}_5, \mathcal{G}_6, \ldots, \mathcal{G}_{16}$ as

$$
\begin{aligned}
\mathcal{G}_{4+i} &= U_1(1, 1, -1, -1) \cdot \mathcal{G}_i, && i = 1, 2, 3, 4, \\
\mathcal{G}_{8+i} &= U_1(1, -1, -1, 1) \cdot \mathcal{G}_i, && i = 1, 2, 3, 4, \\
\mathcal{G}_{12+i} &= U_1(1, -1, 1, -1) \cdot \mathcal{G}_i, && i = 1, 2, 3, 4.
\end{aligned}
\tag{7.64}
$$

Note that the transforms used in (7.64) only rotate some of the symbols. In addition to defining the generator matrix for each trellis transition, we need to perform set partitioning on QPSK symbols. For this example, we only need the first layer of the set partitioning tree, that is the sets P_0 and P_1. They include the following triplets of symbol indices from QPSK:

$P_0 = \{000, 002, 020, 022, 200, 202, 220, 222, 110, 112, 130, 132, 310, 312, 330, 332,$
$\quad 011, 013, 031, 033, 211, 213, 231, 233, 101, 103, 121, 123, 301, 303, 321, 323\},$
$$\tag{7.65}$$

$P_1 = \{001, 003, 021, 023, 201, 203, 221, 223, 111, 113, 131, 133, 311, 313, 331, 333,$
$\quad 010, 012, 030, 032, 210, 212, 230, 232, 100, 102, 120, 122, 300, 302, 320, 322\}.$
$$\tag{7.66}$$

Figure 7.19 depicts an example of an eight-state SOSTTC for four transmit antennas. In this example, a generator matrix and set of symbols pair are assigned to each trellis transition using parentheses.

Another method to design full-diversity rate one codes for four transmit antennas is the use of QOSTBCs. Similar ideas based on constellation rotations, set partitioning, and set assignment work for the QOSTBCs from Chapter 5. The method works for any trellis and is called super-quasi-orthogonal space-time trellis code (SQOSTTC) [70].

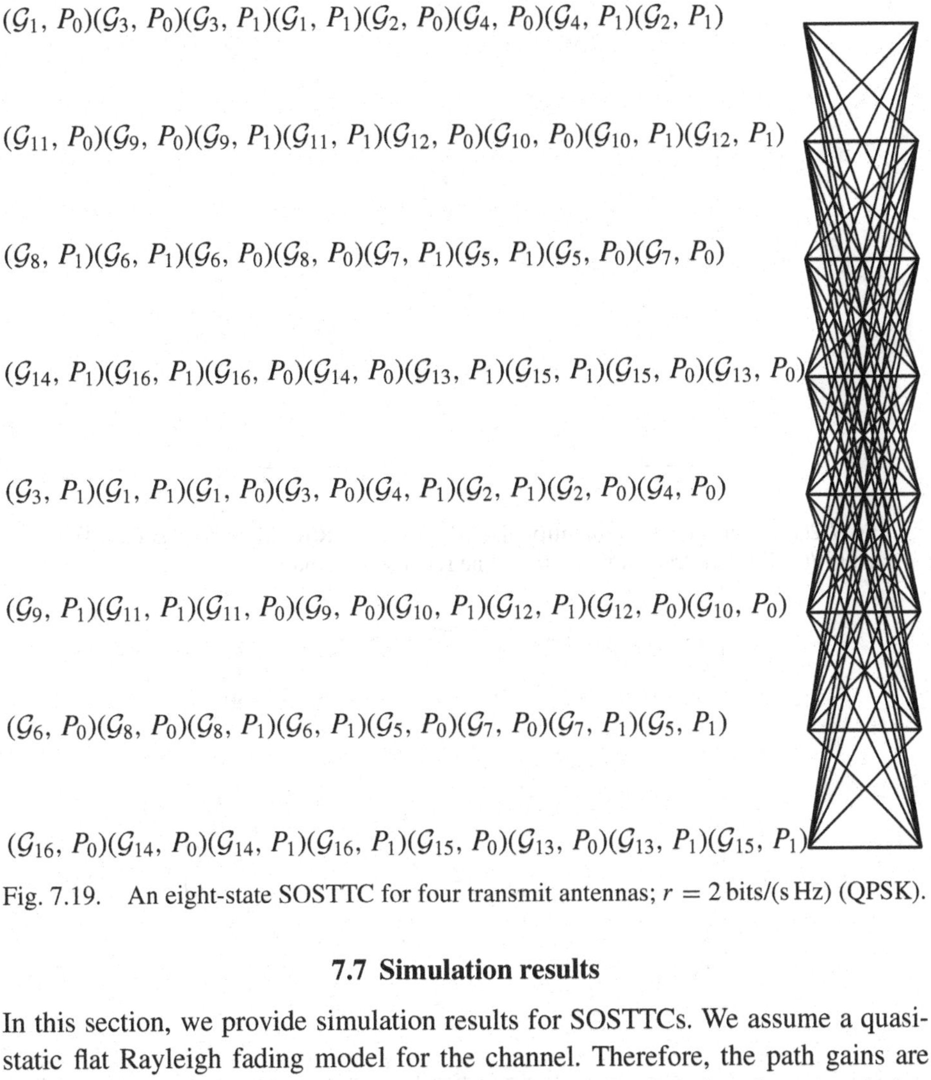

$(\mathcal{G}_1, P_0)(\mathcal{G}_3, P_0)(\mathcal{G}_3, P_1)(\mathcal{G}_1, P_1)(\mathcal{G}_2, P_0)(\mathcal{G}_4, P_0)(\mathcal{G}_4, P_1)(\mathcal{G}_2, P_1)$

$(\mathcal{G}_{11}, P_0)(\mathcal{G}_9, P_0)(\mathcal{G}_9, P_1)(\mathcal{G}_{11}, P_1)(\mathcal{G}_{12}, P_0)(\mathcal{G}_{10}, P_0)(\mathcal{G}_{10}, P_1)(\mathcal{G}_{12}, P_1)$

$(\mathcal{G}_8, P_1)(\mathcal{G}_6, P_1)(\mathcal{G}_6, P_0)(\mathcal{G}_8, P_0)(\mathcal{G}_7, P_1)(\mathcal{G}_5, P_1)(\mathcal{G}_5, P_0)(\mathcal{G}_7, P_0)$

$(\mathcal{G}_{14}, P_1)(\mathcal{G}_{16}, P_1)(\mathcal{G}_{16}, P_0)(\mathcal{G}_{14}, P_0)(\mathcal{G}_{13}, P_1)(\mathcal{G}_{15}, P_1)(\mathcal{G}_{15}, P_0)(\mathcal{G}_{13}, P_0)$

$(\mathcal{G}_3, P_1)(\mathcal{G}_1, P_1)(\mathcal{G}_1, P_0)(\mathcal{G}_3, P_0)(\mathcal{G}_4, P_1)(\mathcal{G}_2, P_1)(\mathcal{G}_2, P_0)(\mathcal{G}_4, P_0)$

$(\mathcal{G}_9, P_1)(\mathcal{G}_{11}, P_1)(\mathcal{G}_{11}, P_0)(\mathcal{G}_9, P_0)(\mathcal{G}_{10}, P_1)(\mathcal{G}_{12}, P_1)(\mathcal{G}_{12}, P_0)(\mathcal{G}_{10}, P_0)$

$(\mathcal{G}_6, P_0)(\mathcal{G}_8, P_0)(\mathcal{G}_8, P_1)(\mathcal{G}_6, P_1)(\mathcal{G}_5, P_0)(\mathcal{G}_7, P_0)(\mathcal{G}_7, P_1)(\mathcal{G}_5, P_1)$

$(\mathcal{G}_{16}, P_0)(\mathcal{G}_{14}, P_0)(\mathcal{G}_{14}, P_1)(\mathcal{G}_{16}, P_1)(\mathcal{G}_{15}, P_0)(\mathcal{G}_{13}, P_0)(\mathcal{G}_{13}, P_1)(\mathcal{G}_{15}, P_1)$

Fig. 7.19.   An eight-state SOSTTC for four transmit antennas; $r = 2$ bits/(s Hz) (QPSK).

7.7 Simulation results

In this section, we provide simulation results for SOSTTCs. We assume a quasi-static flat Rayleigh fading model for the channel. Therefore, the path gains are independent complex Gaussian random variables and fixed during the transmission of one frame. We use Monte Carlo simulations to derive the frame error rate (FER) versus the received SNR. We compare the results with those of the STTCs in Chapter 6 when a comparable code exists. We use frames consisting of 130 transmissions out of each transmit antenna unless mentioned otherwise.

Figure 7.20 shows the frame error probability results plotted against SNR for the codes in Figures 7.5 and 7.7 using BPSK and the corresponding set partitioning in Figure 7.1. The codes in Figures 7.5 and 7.7 are denoted by "four-state SOSTTC" and "two-state SOSTTC," respectively. Both of these codes are rate one and transmit 1 bit/(s Hz). For comparison, we also include the results for STTCs from Figures 6.5 and 6.6. The codes in Figures 6.5 and 6.6 are denoted by "two-state STTC" and

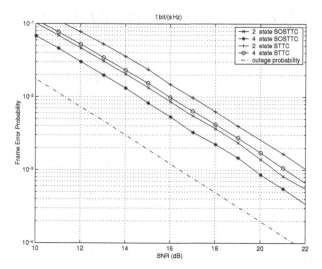

Fig. 7.20.   Frame error probability plotted against SNR for SOSTTCs and STTCs at 1 bit/(s Hz); two transmit antennas, one receive antenna.

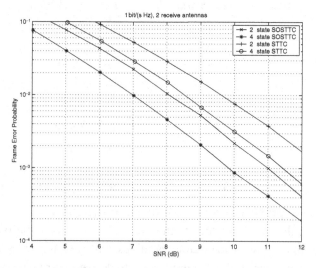

Fig. 7.21.   Frame error probability plotted against SNR for SOSTTCs and STTCs at 1 bit/(s Hz); two transmit antennas, two receive antennas.

"four-state STTC," respectively. The performance of a SOSTTC is about 1.5 dB better than that of the corresponding STTC. The decoding complexity of SOSTTC is lower than the decoding complexity of STTC. We also provide similar results for $M = 2$ receive antennas in Figure 7.21. We only include the results of the best available STTCs in the figure. Similar to the case of one receive antenna, the performance of a SOSTTC is about 1.5 dB better than that of the STTC with the same number of antennas. Figure 7.22 shows the simulation results for

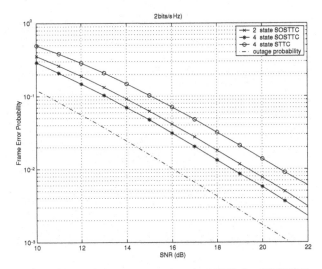

Fig. 7.22. Frame error probability plotted against SNR for SOSTTCs and STTCs at 2 bits/(s Hz); two transmit antennas, one receive antenna.

transmitting 2 bits/(s Hz) using a QPSK constellation. The codes in Figures 7.5 and 7.7 are denoted by "four-state SOSTTC" and "two-state SOSTTC" respectively. The corresponding results for a code with the same rate and four states in Figure 6.3 is also provided for comparison. In Figure 7.22, we denote this code by "four-state STTC." Note that the simulation results for all four-state STTCs in Figure 6.14 are very close. Also, a two-state STTC with rate 2 bits/(s Hz) is impossible as proved in Lemma 6.2.1. As can be seen from the figure, a four-state SOSTTC outperforms the corresponding STTC by more than 2 dB. The performance of the four-state SOSTTC is better than that of a 32-state STTC in [139] and is very close to that of a 64-state STTC in [139]. The decoding complexity of a SOSTTC is lower than that of a STTC with similar number of states. This is due to the orthogonality of the STBC building block as shown in Section 7.4. Similar behavior can be seen for $M = 2$ receive antennas as demonstrated in Figure 7.23. The STTC simulation results in Figure 7.23 is for the four-state STTC in Figure 6.11. This is the best performing STTC for two receive antennas as shown in Figure 6.15 and outperforms the four-state STTC in [139] by about 1 dB. Compared to the already improved STTC in Figure 6.11, a four-state SOSTTC provides an additional 1.5 dB improvement for two receive antennas.

Figure 7.24 shows the simulation results for transmitting 3 bits/(s Hz) using an 8-PSK constellation. An eight-state code outperforms a four-state code by about 0.75 dB. Of course, increasing the number of states improves the performance while increasing the decoding complexity. As can be seen from the figure, SOSTTCs with the same number of states have similar performances. This is despite the difference

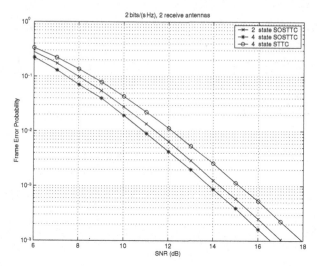

Fig. 7.23.   Frame error probability plotted against SNR for SOSTTCs and STTCs at 2 bits/(s Hz); two transmit antennas, two receive antennas.

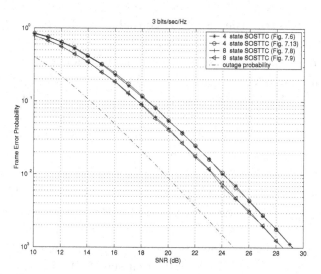

Fig. 7.24.   Frame error probability plotted against SNR for SOSTTCs at 3 bits/(s Hz); two transmit antennas, one receive antenna. (Courtesy of Yun Zhu.)

in the CGD of codes with the same number of states. The CGD difference results in performance differentiation for more than one receive antenna. This is evident from Figure 7.25 that depicts the performance results for $M = 2$ receive antennas. In general, CGD is a good indicator of performance for a large number of receive antennas. In the case of one receive antenna, the distribution of error patterns dictates the performance. A complete analysis of the distance spectrum of different error

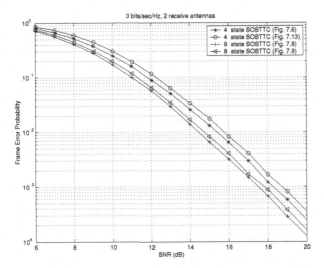

Fig. 7.25. Frame error probability plotted against SNR for SOSTTCs at 3 bits/(s Hz); two transmit antennas, two receive antennas. (Courtesy of Yun Zhu.)

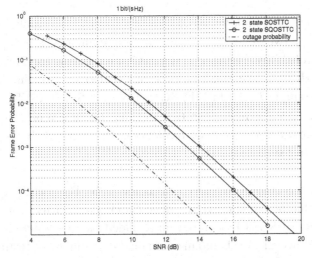

Fig. 7.26. Frame error probability plotted against SNR for SOSTTCs and SQOSTTCs at 1 bit/(s Hz); four transmit antennas, one receive antenna. (Courtesy of Yun Zhu.)

paths is usually very complex. It is easier to perform Monte Carlo simulations, as we have done in this section, to compare different codes.

Figure 7.26 shows the simulation results for 1 bit/(s Hz) using a BPSK constellation and four transmit antennas. When we have four transmit antennas, each frame consists of 132 transmissions out of each transmit antenna. The SOSTTC is from Figure 7.16 and the SQOSTTC is from a similar structure in [70] using QOSTBCs

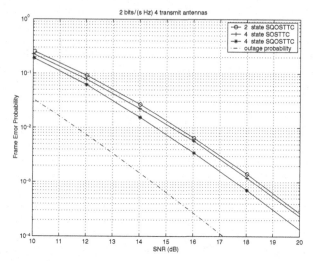

Fig. 7.27. Frame error probability plotted against SNR for SOSTTCs and SQOSTTCs at 2 bits/(s Hz); four transmit antennas, one receive antenna. (Courtesy of Yun Zhu.)

instead of OSTBCs as building blocks (see Problem 7). As can be seen from the figure, with only two states, the results are approximately less than 4 dB away from outage probability. Figure 7.27 depicts similar results for 2 bits/(s Hz) using a QPSK constellation. The SOSTTC is from a four-state code similar to the code discussed in Example 7.6.1 while the SQOSTTCs are from [70]. The figure shows that a rate one code for more than two antennas and a complex constellation is possible and provides very good performances. Although not shown in the figure, going from a four-state SOSTTC to a 16-state SOSTTC improves the results by about 2 dB.

Throughout this chapter, we have considered SOSTTCs that do not expand the original constellation while providing huge coding gains. This is one of the interesting properties of SOSTTCs and we have added the constraint to be fair in our comparisons. However, we should mention that removing this constraint results in better performance without increasing the complexity of decoding.

7.8 Summary of important results

- The following parameterized class of orthogonal STBCs is an example of super-orthogonal codes that expands the number of available orthogonal matrices without extending the constellation alphabet of the transmitted signals:

$$\mathcal{G}(x_1, x_2, \theta) = \begin{pmatrix} x_1\, e^{j\theta} & x_2 \\ -x_2^*\, e^{j\theta} & x_1^* \end{pmatrix}.$$

- Similar to MTCMs, a super-orthogonal space-time trellis code is designed systematically using set partitioning and the assignment of the resulting sets to different branches of a trellis.
- Typically, we assign the same rotation parameter to all transitions from a state.
- SOSTTCs provide full spatial diversity and full rate.
- SOSTTCs can be designed for any trellis, with any number of states, and any rate.
- SOSTTCs outperform STTCs in terms of coding gain and performance.
- The decoding of SOSTTCs is done using the Viterbi algorithm.
- The orthogonality of STBC building blocks is utilized to reduce the decoding complexity of SOSTTCs. The decoding complexity of SOSTTCs is smaller than that of the STTCs.

7.9 Problems

1 Consider the two SOSTTCs in Figure 7.5 using BPSK and QPSK. Assume that the input bitstream is 01110010. What are the transmitted symbols from each antenna for each code?

2 In (7.7), if we change the definition of $\mathcal{G}(x_1, x_2, \theta)$ to $U \cdot \mathcal{G}(x_1, x_2)$, show that we can still generate all 2×2 orthogonal matrices in (7.3) and (7.4). How would it affect the SOSTTCs in Figure 7.5?

3 Repeat the set partitioning for QPSK in Figure 7.2 based on a minimum trace criterion.

4 For the SOSTTCs in Figures 7.5 and 7.7, calculate the minimum trace of $\mathbf{A}(\mathbf{C}^i, \mathbf{C}^j) = D(\mathbf{C}^i, \mathbf{C}^j)^H D(\mathbf{C}^i, \mathbf{C}^j)$ for a pair of codewords $(\mathbf{C}^i, \mathbf{C}^j)$ that generates the minimum CGD.

5 Consider the following class of orthogonal designs

$$\mathcal{G}(x_1, x_2, \phi, \theta) = \begin{pmatrix} x_1 \, e^{j\phi} \, e^{j\theta} & x_2 \, e^{j\phi} \\ -x_2^* \, e^{-j\phi} \, e^{j\theta} & x_1^* \, e^{-j\phi} \end{pmatrix},$$

in which the constellation points are rotated. To design a two-state SOSTTC based on this class, we use the set partitioning of BPSK in Figure 7.1 and the following trellis diagram:

$$\mathcal{G}(x_1, x_2, 0, 0) \; P_0 \; P_1$$

$$\mathcal{G}(x_1, x_2, \tfrac{\pi}{2}, 0) \; P_1 \; P_0$$

Note that the transmitted symbols are not BPSK any more and are expanded.

(a) Show that the resulting SOSTTC provides full diversity.

(b) Assume that the input bitstream is 01110010. What are the transmitted symbols from each antenna?

(c) For a frame consisting of 130 transmissions out of each transmit antenna, use Monte

Carlo simulations to derive the frame error rate (FER) versus the received SNR simulation results and compare them with the results in Figure 7.20.

6 Using QPSK and rate 3/4 OSTBCs, design a two-state SOSTTC for four transmit antennas to transmit $r = 2$ bits/(s Hz).

7 Consider the following class of QOSTBCs, represented by $\mathcal{G}(x_1, x_2, \phi_1, x_3, x_4, \phi_2, \theta_1, \theta_2)$

$$
\begin{pmatrix}
e^{j\theta_1} e^{j\phi_1} x_1 & e^{j\theta_2} e^{j\phi_1} x_2 & e^{j\phi_2} x_3 & e^{j\phi_2} x_4 \\
-e^{j\theta_1} e^{-j\phi_1} x_2^* & e^{j\theta_2} e^{-j\phi_1} x_1^* & -e^{-j\phi_2} x_4^* & e^{-j\phi_2} x_3^* \\
e^{j\theta_1} e^{j\phi_2} x_3 & e^{j\theta_2} e^{j\phi_2} x_4 & e^{j\phi_1} x_1 & e^{j\phi_1} x_2 \\
-e^{j\theta_1} e^{-j\phi_2} x_4^* & e^{j\theta_2} e^{-j\phi_2} x_3^* & -e^{-j\phi_1} x_2^* & e^{-j\phi_1} x_1^*
\end{pmatrix}.
$$

(a) Using a BPSK constellation for symbols x_1, x_2, x_3, x_4, partitioning the set of all possible quadruplets.

(b) Based on the above set partitioning, design a two-state SQOSTTC to transmit $r = 1$ bit/(s Hz) over four transmit antennas.

(c) Double check the simulation results in Figure 7.26.

8

Differential space-time modulation

8.1 Introduction

The primary focus of the codes that we have discussed so far has been on the case when only the receiver knows the channel. This is the case for most practical systems. The transmitter sends pilot signals and the receiver uses them to estimate the channel [138]. Then, the receiver uses the estimated path gains to coherently decode the data symbols during the same frame. In such a coherent system, the underlying assumption is that the channel does not change during one frame of data. In other words, the frame length is chosen such that the path gain change during one frame is negligible. This is basically the quasi-static fading assumption that we have used so far. There is a bandwidth penalty due to the number of transmitted pilot symbols. Of course, choosing a longer frame reduces this bandwidth penalty; however, on the other hand, the quasi-static assumption is less valid for longer frames. Therefore, there is a trade-off between the frame length and the accuracy of the channel estimation. This is an interesting research topic and the interested reader is referred to [56] and the references therein.

For one transmit antenna, differential detection schemes exist that neither require the knowledge of the channel nor employ pilot symbol transmission. These differential decoding schemes are used, for instance, in the IEEE IS-54 standard [37]. This motivates the generalization of differential detection schemes for the case of multiple transmit antennas. A partial solution to this problem is proposed in [133], where it is assumed that the channel is not known. However, this scheme requires that the transmission of symbols is known to the receiver at the beginning and hence is not truly differential. It can be thought of as joint channel and data estimation which can lead to error propagation. The detected sequence at time $t - 1$ is utilized to estimate the channel at the receiver and these estimates are used to detect the transmitted data at time t. Here, we present truly differential detection schemes for multiple transmit antennas. Before presenting the details of different differential

195

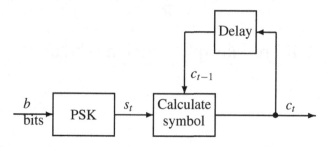

Fig. 8.1.   DPSK encoder block diagram.

space-time coding methods for multiple transmit antennas, we review the concept of differential coding for one transmit antenna. To be precise, we present the details of differential PSK (DPSK) modulation and capture the main ideas behind it (Figure 8.1).

In DPSK modulation, the decoder does not need to perform channel estimation. The transmitted signals depend on each other and the decoder detects the data from successive symbols. Let us assume that we transmit c_t at time t and the path gain between the transmitter and the receiver is α. Note that, for now, we assume one transmit and one receive antenna and if the noise is η_t, the received signal is

$$r_t = \alpha c_t + \eta_t. \tag{8.1}$$

For DPSK modulation, the transmitter sends an arbitrary symbol c_0 at time zero. Then, at time t, the transmitter picks a symbol s_t using the b input data bits and the corresponding L-PSK constellation, where $L = 2^b$. Using the transmitted signal at time $t-1$ and the selected constellation point, $c_t = c_{t-1}s_t$ is transmitted at time t. Note that if the initial transmitted symbol is an L-PSK point, c_t is also an L-PSK point for all time indices. Also, for a unit-energy L-PSK constellation, $|c_t|$ is always equal to one. The transmitted signal, c_t, is affected by the fade and noise as presented in (8.1). To detect the transmitted data at time t, first the receiver computes $r_t r_{t-1}^*$, then it finds the closest symbol of the L-PSK constellation to $r_t r_{t-1}^*$ as the estimate of the transmitted symbol. To investigate the reasoning behind the above detection scheme, we calculate the term $r_t r_{t-1}^*$ as follows:

$$\begin{aligned} r_t r_{t-1}^* &= |\alpha|^2 c_t c_{t-1}^* + \alpha c_t \eta_{t-1}^* + \eta_t \alpha^* c_{t-1}^* + \eta_t \eta_{t-1}^* \\ &\approx |\alpha|^2 c_{t-1} s_t c_{t-1}^* + \alpha c_t \eta_{t-1}^* + \eta_t \alpha^* c_{t-1}^* \\ &= |\alpha|^2 s_t + \mathcal{N}, \end{aligned} \tag{8.2}$$

where $\mathcal{N}$ is a Gaussian noise, the path gain α is assumed to remain the same at times $t-1$ and t, and $\eta_t \eta_{t-1}^*$ is ignored. Therefore, the optimal estimate of s_t using

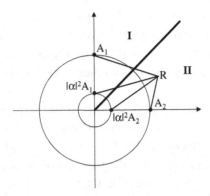

Fig. 8.2.   Two decoding regions of a PSK modulation and its scaled version.

approximation (8.2) is

$$\hat{s}_t = \operatorname*{argmin}_{s_t} \left| r_t r_{t-1}^* - |\alpha|^2 s_t \right|^2. \tag{8.3}$$

Since all L-PSK points are on the same circle, the scaling factor $|\alpha|^2$ will not change the geometry of the detection regions in (8.3) as shown in Figure 8.2. Therefore, an equivalent optimization problem is

$$\hat{s}_t = \operatorname*{argmin}_{s_t} |r_t r_{t-1}^* - s_t|^2. \tag{8.4}$$

In other words, one should find the closet point in the L-PSK constellation to $r_t r_{t-1}^*$.

The non-coherent detection of symbol s_t by ignoring $|\alpha|^2$, as above, results in some performance degradation. To quantify this performance loss, we calculate the receiver's SNR for the same unit transmit power in both cases. For the above non-coherent detection scheme, the power of the signal at the receiver is $|\alpha|^4$. This is equal to $|\alpha|^2$ times the power of signal for the coherent detection scheme. The power of noise at the receiver for the above non-coherent detection scheme is approximately $2|\alpha|^2$ times the power of noise for the coherent detection. Therefore, the received SNR of the differential detection scheme is approximately half of that of the coherent detection scheme for the same transmission power. This translates to a 3 dB difference in the performance of the two systems for a Rayleigh fading channel.

The properties of a DPSK modulation scheme are summarized below:

• the information is embedded between successive symbols;
• since symbols are equi-energy, scaling does not change the geometry of detection regions;
• when a point is closer to a scaled symbol, among all scaled symbols, it is also closer to the corresponding symbol among all original symbols;
• to find the closest symbol, we do not need to know the fade;
• 3 dB is lost due to non-coherent detection for a Rayleigh fading channel.

The goal of this chapter is to design differential modulation schemes for MIMO channels that provide similar properties. While different approaches have been proposed in the literature, we concentrate on low complexity schemes based on orthogonal STBCs. A brief summary of other related schemes is provided for the interested reader.

Unitary space-time codes are proposed in [59]. Subsequently, unitary space-time codes with lower complexity have been proposed [60]. The first differential modulation scheme based on orthogonal STBCs is suggested in [133]. This scheme is the first scheme that provides simple encoding and decoding algorithms. The approach is generalized to more than two transmit antennas [73]. A unified presentation of the results of [73, 133] is provided in [69]. A different construction based on group codes is presented in [63]. Other related constructions are presented in [61] and [25] with the goal of reducing the exponential complexity of the decoding in [59]. A second approach to designing differential detection based on orthogonal STBCs is suggested in [46, 47]. As we show later, this is a special case of the scheme proposed in [73, 133]. The extension of differential space-time modulations to multiple amplitudes is provided in [23, 130].

8.2 Differential encoding

Let us start with a system consisting of two transmit antennas. Also, let us assume that we use a signal constellation with 2^b elements. For each block of $2b$ bits, the encoder calculates two symbols and transmits them using an orthogonal STBC. The 2×2 transmitted codeword of the orthogonal STBC, or the corresponding pair of transmitted symbols also depends on the codeword and symbols transmitted in the previous block. This is similar to the case of DPSK where the transmitted symbol at each time depends on the input bits and the previously transmitted symbol. The main challenge is how to generate the two symbols or corresponding orthogonal codeword such that the receiver can decode them without knowing the path gains. There are two main approaches to address this challenge. One is to replace the PSK modulation in the block diagram of Figure 8.1 with an orthogonal design. We consider this approach later in this chapter. The other approach is to add an intermediate step and generate the pair of transmitted symbols first and then send them by an orthogonal design. The block diagram of such a system is provided in Figure 8.3 and we describe it first.

We group the two symbols for the lth block in a vector S^l as follows

$$S^l = \begin{pmatrix} s_1^l \\ s_2^l \end{pmatrix}. \tag{8.5}$$

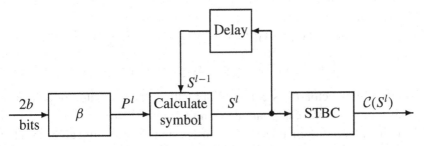

Fig. 8.3.   Differential space-time encoder block diagram.

The symbol vector for block l, S^l, is generated from S^{l-1}, the symbol vector for block $l-1$, and the $2b$ input bits. To describe how we generate S^l, let us consider the following two vectors that construct an orthogonal basis:

$$V_1(S^l) = S^l = \begin{pmatrix} s_1^l \\ s_2^l \end{pmatrix}, \quad V_2(S^l) = \begin{pmatrix} (s_2^l)^* \\ -(s_1^l)^* \end{pmatrix}. \tag{8.6}$$

Note that for a unit-length vector S^l, the lengths of $V_1(S^l)$ and $V_2(S^l)$ will be one too. Otherwise, an orthonormal basis is also possible by a simple normalization:

$$V_1(S^l) = \frac{1}{|S^l|} \begin{pmatrix} s_1^l \\ s_2^l \end{pmatrix}, \quad V_2(S^l) = \frac{1}{|S^l|} \begin{pmatrix} (s_2^l)^* \\ -(s_1^l)^* \end{pmatrix}. \tag{8.7}$$

We fix a set $\mathcal{V}$ which consists of 2^{2b} unit-length distinct vectors $P_1, P_2, \ldots, P_{2^{2b}}$, where each vector P_v is a 2×1 vector, $P_v = (P_{v1}, P_{v2})^T$. We define an arbitrary one-to-one mapping $\beta(\cdot)$ which maps $2b$ bits onto $\mathcal{V}$. Note that the choice of the set $\mathcal{V}$ and the mapping $\beta(\cdot)$ is completely arbitrary as long as vectors $P_1, P_2, \ldots, P_{2^{2b}}$ are unit-length and the mapping $\beta(\cdot)$ is one-to-one. However, there exists structured mappings that do not require an exponential amount of memory to be saved. A few examples will be provided in the sequel.

The encoding starts with the transmission of an arbitrary vector S^0. For block l, we use the $2b$ input bits to pick the corresponding vector P^l in $\mathcal{V}$ using the one-to-one mapping $\beta(\cdot)$. Note that P^l is one of the vectors $P_1, P_2, \ldots, P_{2^{2b}}$ and when we need to refer to its index in set $\mathcal{V}$, we use the notation of P_v. Assuming that S^{l-1} is transmitted for the $(l-1)$th block, we calculate S^l by

$$S^l = P_1^l V_1(S^{l-1}) + P_2^l V_2(S^{l-1}), \tag{8.8}$$

where P_1^l and P_2^l are the first and second elements of the vector P^l, respectively. Note that since $V_1(S^{l-1})$ and $V_2(S^{l-1})$ create an orthogonal basis, P_1^l and P_2^l can be derived from multiplying S^l by $[V_1(S^{l-1})]^H$ and $[V_2(S^{l-1})]^H$, respectively. Therefore, we have

$$P_1^l = [V_1(S^{l-1})]^H \cdot S^l = s_1^l(s_1^{l-1})^* + s_2^l(s_2^{l-1})^*$$
$$P_2^l = [V_2(S^{l-1})]^H \cdot S^l = s_1^l s_2^{l-1} - s_2^l s_1^{l-1}.$$

(8.9)

The length of the vector P^l is constant and with an appropriate normalization of the initial vector S^0, one can guarantee a unit-length vector P^l. Figure 8.3 shows the block diagram of the encoder. Before discussing the decoding algorithm for the suggested differential encoding, we provide a few examples.

Example 8.2.1 *In this example, we consider the encoding of differential space-time modulation using two transmit antennas and BPSK modulation. To have a unit-length vector, we assume that the constellation consists of the normalized points $-1/\sqrt{2}$ and $1/\sqrt{2}$. In other words, the elements of S^l are $-1/\sqrt{2}$ or $1/\sqrt{2}$. We provide an example of one-to-one mapping $\beta(\cdot)$ and the set $\mathcal{V}$. We define the set $\mathcal{V} = \{(1,0)^T, (0,1)^T, (0,-1)^T, (-1,0)^T\}$. Note that all elements of $\mathcal{V}$ have unit lengths. The mapping $\beta(\cdot)$ maps two input bits onto $\mathcal{V}$ and is given by*

$$\beta(00) = \begin{pmatrix} 1 \\ 0 \end{pmatrix}, \qquad \beta(10) = \begin{pmatrix} 0 \\ 1 \end{pmatrix},$$

(8.10)

$$\beta(01) = \begin{pmatrix} 0 \\ -1 \end{pmatrix}, \qquad \beta(11) = \begin{pmatrix} -1 \\ 0 \end{pmatrix}.$$

The encoder begins the transmission by sending arbitrary symbols s_1^0 and s_2^0 at time 1 and symbols $-(s_2^0)^*$ and $(s_1^0)^*$ at time 2 unknown to the receiver. These two transmissions do not convey any information. If the elements of S^0, that is s_1^0 and s_2^0, are chosen from $-1/\sqrt{2}$ and $1/\sqrt{2}$, then using (8.8), the elements of S^l are always either $-1/\sqrt{2}$ or $1/\sqrt{2}$. The differential encoding follows the block diagram in Figure 8.3 and Equation (8.8).

As an example, we assume at block $l-1$, $S^{l-1} = (1/\sqrt{2}, -1/\sqrt{2})^T$ is the symbol vector that has been used in the orthogonal STBC for transmission. If the two-bit input to the encoder at block l is 10, since $\beta(10) = (0,1)^T$, we can calculate S^l as follows:

$$S^l = 0 \cdot \begin{pmatrix} \frac{1}{\sqrt{2}} \\ \frac{-1}{\sqrt{2}} \end{pmatrix} + 1 \cdot \begin{pmatrix} \frac{-1}{\sqrt{2}} \\ \frac{-1}{\sqrt{2}} \end{pmatrix} = \begin{pmatrix} \frac{-1}{\sqrt{2}} \\ \frac{-1}{\sqrt{2}} \end{pmatrix}.$$

(8.11)

Similarly, we have $S^l = (1/\sqrt{2}, -1/\sqrt{2})^T$ for input bits 00, $S^l = (1/\sqrt{2}, 1/\sqrt{2})^T$ for input bits 01, and $S^l = (-1/\sqrt{2}, 1/\sqrt{2})^T$ for input bits 11.

Example 8.2.2 *In this example, we consider the differential space-time modulation encoding for $b = 2$ bits/(s Hz). For $b = 2$ bits/(s Hz), we use QPSK constellation* *consisting of $\{1/\sqrt{2}, j/\sqrt{2}, -1/\sqrt{2}, -j/\sqrt{2}\}$ for S^0 elements. As an example, let us consider the following one-to-one mapping $\beta(\cdot)$:*

$$
\begin{aligned}
\beta(0000) &= (1, 0)^T, \\
\beta(0001) &= (0, 1)^T, \\
\beta(0010) &= (0, -1)^T, \\
\beta(0011) &= (-1, 0)^T, \\
\beta(0100) &= (j, 0)^T, \\
\beta(0101) &= (0, j)^T, \\
\beta(0110) &= (0, -j)^T, \\
\beta(0111) &= (-j, 0)^T, \\
\beta(1000) &= (0.5 + 0.5j, -0.5 + 0.5j)^T, \\
\beta(1001) &= (-0.5 + 0.5j, 0.5 + 0.5j)^T, \\
\beta(1010) &= (-0.5 - 0.5j, 0.5 - 0.5j)^T, \\
\beta(1011) &= (0.5 - 0.5j, -0.5 - 0.5j)^T, \\
\beta(1100) &= (0.5 - 0.5j, 0.5 + 0.5j)^T, \\
\beta(1101) &= (-0.5 - 0.5j, -0.5 + 0.5j)^T, \\
\beta(1110) &= (-0.5 + 0.5j, -0.5 - 0.5j)^T, \\
\beta(1111) &= (0.5 + 0.5j, 0.5 - 0.5j)^T.
\end{aligned}
\tag{8.12}
$$

The corresponding set $\mathcal{V}$ includes the 16 vectors on the right-hand side of (8.12). More precisely, we have

$$
\mathcal{V} = \left\{ \begin{pmatrix} 1 \\ 0 \end{pmatrix}, \begin{pmatrix} 0 \\ 1 \end{pmatrix}, \begin{pmatrix} 0 \\ -1 \end{pmatrix}, \begin{pmatrix} -1 \\ 0 \end{pmatrix}, \begin{pmatrix} j \\ 0 \end{pmatrix}, \begin{pmatrix} 0 \\ j \end{pmatrix}, \begin{pmatrix} 0 \\ -j \end{pmatrix}, \begin{pmatrix} -j \\ 0 \end{pmatrix}, \right.
$$
$$
\begin{pmatrix} 0.5 + 0.5j \\ -0.5 + 0.5j \end{pmatrix}, \begin{pmatrix} -0.5 + 0.5j \\ 0.5 + 0.5j \end{pmatrix}, \begin{pmatrix} -0.5 - 0.5j \\ 0.5 - 0.5j \end{pmatrix}, \begin{pmatrix} 0.5 - 0.5j \\ -0.5 - 0.5j \end{pmatrix}, \tag{8.13}
$$
$$
\left. \begin{pmatrix} 0.5 - 0.5j \\ 0.5 + 0.5j \end{pmatrix}, \begin{pmatrix} -0.5 - 0.5j \\ -0.5 + 0.5j \end{pmatrix}, \begin{pmatrix} -0.5 + 0.5j \\ -0.5 - 0.5j \end{pmatrix}, \begin{pmatrix} 0.5 + 0.5j \\ 0.5 - 0.5j \end{pmatrix} \right\}.
$$

Note that all vectors in $\mathcal{V}$ have a unit length. The mapping β maps four input bits onto $\mathcal{V}$. The differential encoding follows the block diagram in Figure 8.3 and Equation (8.8).

Example 8.2.3 *In this example, we consider another differential space-time modulation encoding for $b = 2\,bits/(s\,Hz)$. This is an example of a special case that is discussed in what follows. In this example, the mapping $\beta(\cdot)$ utilizes the four input bits to pick two QPSK constellation points for any block l. Then, these two constellation points construct the vector P^l. The differential encoding follows the block diagram in Figure 8.3 and Equation (8.8). Since the QPSK constellation points have the same energy, the corresponding vector P^l will have the same length for any four input bits. Using a correct normalization for the QPSK constellation* *results in a unit-length vector P^l. In other words, if the QPSK constellation consists of $\{1/2 + j/2, 1/2 - j/2, -1/2 + j/2, -1/2 - j/2\}$, the vector P^l is a unit-length vector. The corresponding set $\mathcal{V}$ includes the following 16 vectors:*

$$
\begin{aligned}
\mathcal{V} = \Bigg\{ &\begin{pmatrix} 1/2 + j/2 \\ 1/2 + j/2 \end{pmatrix}, \begin{pmatrix} 1/2 + j/2 \\ 1/2 - j/2 \end{pmatrix}, \begin{pmatrix} 1/2 + j/2 \\ -1/2 + j/2 \end{pmatrix}, \begin{pmatrix} 1/2 + j/2 \\ -1/2 - j/2 \end{pmatrix}, \\
&\begin{pmatrix} 1/2 - j/2 \\ 1/2 + j/2 \end{pmatrix}, \begin{pmatrix} 1/2 - j/2 \\ 1/2 - j/2 \end{pmatrix}, \begin{pmatrix} 1/2 - j/2 \\ -1/2 + j/2 \end{pmatrix}, \begin{pmatrix} 1/2 - j/2 \\ -1/2 - j/2 \end{pmatrix}, \\
&\begin{pmatrix} -1/2 + j/2 \\ 1/2 + j/2 \end{pmatrix}, \begin{pmatrix} -1/2 + j/2 \\ 1/2 - j/2 \end{pmatrix}, \begin{pmatrix} -1/2 + j/2 \\ -1/2 + j/2 \end{pmatrix}, \begin{pmatrix} -1/2 + j/2 \\ -1/2 - j/2 \end{pmatrix}, \\
&\begin{pmatrix} -1/2 - j/2 \\ 1/2 + j/2 \end{pmatrix}, \begin{pmatrix} -1/2 - j/2 \\ 1/2 - j/2 \end{pmatrix}, \begin{pmatrix} -1/2 - j/2 \\ -1/2 + j/2 \end{pmatrix}, \begin{pmatrix} -1/2 - j/2 \\ -1/2 - j/2 \end{pmatrix} \Bigg\}.
\end{aligned}
\tag{8.14}
$$

Note that using the above mappings, some of the elements of S^l will not be QPSK constellation points. For example, if we pick the initial vector $s_1^0 = s_2^0 = 1/\sqrt{2}$, the s_k^l symbols will be from a 9-QAM constellation. In other words, the elements of S^l are members of the set

$$
\left\{ 0, \frac{1}{\sqrt{2}}, \frac{j}{\sqrt{2}}, \frac{-1}{\sqrt{2}}, \frac{-j}{\sqrt{2}}, \frac{1}{\sqrt{2}} + \frac{j}{\sqrt{2}}, \frac{1}{\sqrt{2}} - \frac{j}{\sqrt{2}}, \frac{-1}{\sqrt{2}} + \frac{j}{\sqrt{2}}, \frac{-1}{\sqrt{2}} - \frac{j}{\sqrt{2}} \right\}.
\tag{8.15}
$$

The last example is not just an isolated example. Picking two symbols from a constellation to build vector P^l can be utilized for any bit rate. In what follows, we consider such an encoder in which the mapping $\beta(\cdot)$ picks two symbols from a given constellation.

8.2.1 A special case of differential encoding

The above differential encoding scheme is general and works for any set $\mathcal{V}$ and mapping $\beta(\cdot)$ as long as the vectors in $\mathcal{V}$ have the same length. One special case is if we use the input bits to pick signals from a unit-length constellation, for example PSK. In other words, let us assume that the $2b$ input bits for block l pick z_1^l and z_2^l and we set

$$
P_1^l = z_1^l,
\tag{8.16}
$$
$$
P_2^l = -z_2^l.
\tag{8.17}
$$

Then, using (8.8), we have

$$
S^l = z_1^l \cdot \begin{pmatrix} s_1^{l-1} \\ s_2^{l-1} \end{pmatrix} - z_2^l \cdot \begin{pmatrix} (s_2^{l-1})^* \\ -(s_1^{l-1})^* \end{pmatrix} = \begin{pmatrix} z_1^l s_1^{l-1} - z_2^l (s_2^{l-1})^* \\ z_1^l s_2^{l-1} + z_2^l (s_1^{l-1})^* \end{pmatrix}.
\tag{8.18}
$$

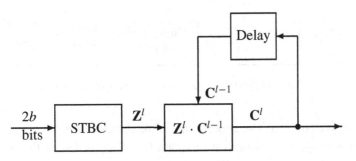

Fig. 8.4. Differential space-time encoder block diagram.

Then, using S^l in the STBC, as in Figure 8.3, results in transmitting the following codeword

$$\mathbf{C}(S^l) = \begin{pmatrix} z_1^l s_1^{l-1} - z_2^l (s_2^{l-1})^* & z_1^l s_2^{l-1} + z_2^l (s_1^{l-1})^* \\ -(z_1^l)^* (s_2^{l-1})^* - (z_2^l)^* s_1^{l-1} & (z_1^l)^* (s_1^{l-1})^* - (z_2^l)^* s_2^{l-1} \end{pmatrix}. \quad (8.19)$$

Note that picking

$$V_2(S^l) = \begin{pmatrix} -(s_2^l)^* \\ (s_1^l)^* \end{pmatrix}, \quad (8.20)$$

instead of the $V_2(S^l)$ in (8.6) and $P_2^l = z_2^l$ instead of the P_2^l in (8.17) also results in the same codeword. This is equivalent to the differential-encoding block diagram in Figure 8.4. The input codeword $\mathbf{Z}^l$ is constructed using the $2b$ input bits for the lth block and the corresponding input symbols z_1^l and z_2^l

$$\mathbf{Z}^l = \begin{pmatrix} z_1^l & z_2^l \\ -(z_2^l)^* & (z_1^l)^* \end{pmatrix}. \quad (8.21)$$

Then, $\mathbf{C}^l$ is calculated based on $\mathbf{C}^{l-1}$ as follows

$$\mathbf{C}^l = \mathbf{Z}^l \cdot \mathbf{C}^{l-1} = \begin{pmatrix} z_1^l s_1^{l-1} - z_2^l (s_2^{l-1})^* & z_1^l s_2^{l-1} + z_2^l (s_1^{l-1})^* \\ -(z_2^l)^* s_1^{l-1} - (z_1^l)^* (s_2^{l-1})^* & -(z_2^l)^* s_2^{l-1} + (z_1^l)^* (s_1^{l-1})^* \end{pmatrix}. \quad (8.22)$$

The codewords in (8.19) and (8.22) are the same which means the differential encoder in Figure 8.4 is a special case of the encoder in Figure 8.3.

8.3 Differential decoding

In this section, we consider the decoding of the differential space-time modulation for two transmit antennas. For the sake of simplicity of presentation, we consider

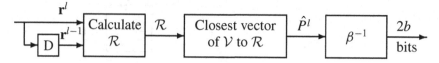

Fig. 8.5.   Differential space-time decoder block diagram.

only one receive antenna. The general case of M receive antennas is easily handled by maximum ratio combining as we explain later. Let us denote the two received signals for block l by r_1^l and r_2^l. We have

$$\begin{cases} r_1^l = \alpha_1 s_1^l + \alpha_2 s_2^l + \eta_1^l \\ r_2^l = -\alpha_1 (s_2^l)^* + \alpha_2 (s_1^l)^* + \eta_2^l. \end{cases} \tag{8.23}$$

where η_1^l and η_2^l are the noise samples for block l. We define a vector $\mathcal{R}$ to reconstruct a noisy version of the scaled P^l as follows:

$$\mathcal{R} = \begin{pmatrix} (r_1^{l-1})^* r_1^l + r_2^{l-1}(r_2^l)^* \\ (r_2^l)^* r_1^{l-1} - r_1^l (r_2^{l-1})^* \end{pmatrix} = (|\alpha_1|^2 + |\alpha_2|^2) P^l + \mathcal{N}. \tag{8.24}$$

Then, similar to the argument for the DPSK decoding, the scaling does not change the geometry of the detection regions. Therefore, the decoder finds the closest vector P^l in $\mathcal{V}$ and declares it as the best estimate of the transmitted vector. The inverse mapping $\beta^{-1}(\cdot)$ provides the decoded bits as depicted in Figure 8.5. Similar to the case of DPSK for one transmitter, the result of the differential decoding is a 3 dB loss in performance due to the doubling of the effective noise.

To find the closest vector of $\mathcal{V}$ to $\mathcal{R}$ in (8.24), denoted by $\hat{P}^l$, the most complex method is to perform a full search over all vectors in set $\mathcal{V}$. In general, this may need 2^{2b} comparisons. However, depending on the structure of the vectors in $\mathcal{V}$, simpler methods are possible. For example, the decoding regions may have a geometric property such that a simple lattice decoding is possible. In the case of DPSK, the detection regions are bordered by lines passing the origin. Therefore, the angle of the decision variable in a polar coordinate is good enough to find the closest constellation point. For many examples of set $\mathcal{V}$, a similar simple approach in a higher dimensional space is possible. For the special case of the encoder in Section 8.2.1, a simple separate ML decoding is possible [47]. We discuss the separate decoding of space-time modulation schemes for two transmit antennas in the next section. Later by extending to more than two antennas, we demonstrate that similar to coherent detection of STBCs, the decoding complexity of the differential space-time modulation grows linearly by rate and number of antennas.

8.3.1 A simple decoding for the special case

In this section, we consider a differential decoding for the encoding in Figure 8.4. This is a special case of the encoder in Figure 8.3 as we discussed in Section 8.2.1. In what follows, we show that a simple separate decoding of the symbols is possible for this special case.

The data matrix $\mathbf{Z}^l$ in (8.21) is an orthogonal design with the following property:

$$(\mathbf{Z}^l)^H \cdot \mathbf{Z}^l = \left(|z_1^l|^2 + |z_2^l|^2\right) I_2. \tag{8.25}$$

If all symbols have the same energy, for example PSK with $|z_k^l| = 1/\sqrt{2}$, the codewords are unitary, that is

$$(\mathbf{Z}^l)^H \cdot \mathbf{Z}^l = I_2. \tag{8.26}$$

The encoding for unitary differential space-time modulation starts with the transmission of an arbitrary unitary codeword $\mathbf{C}^0$. As in Figure 8.4, the codeword $\mathbf{Z}^l$ is utilized to calculate the transmitted matrix $\mathbf{C}^l$ by

$$\mathbf{C}^l = \mathbf{Z}^l \cdot \mathbf{C}^{l-1}. \tag{8.27}$$

Due to the orthogonality of the codewords, the transmitted matrix $\mathbf{C}^l$ is also unitary. In other words,

$$(\mathbf{C}^l)^H \cdot \mathbf{C}^l = (\mathbf{C}^{l-1})^H \cdot (\mathbf{Z}^l)^H \cdot \mathbf{Z}^l \cdot \mathbf{C}^{l-1} = (\mathbf{C}^{l-1})^H \cdot \mathbf{C}^{l-1}, \tag{8.28}$$

and using an induction argument, we have $(\mathbf{C}^l)^H \cdot \mathbf{C}^l = I_2$.

The codeword $\mathbf{C}^l$ is transmitted over the fading channel with path gain matrix $\mathbf{H}$ and noise matrix $\mathcal{N}^l$. Assuming that the path gain matrix does not change during consecutive blocks $l-1$ and l, the received vector for block l is calculated as

$$\mathbf{r}^l = \mathbf{C}^l \cdot \mathbf{H} + \mathcal{N}^l = \mathbf{Z}^l \cdot \mathbf{C}^{l-1} \cdot \mathbf{H} + \mathcal{N}^l = \mathbf{Z}^l \cdot (\mathbf{r}^{l-1} - \mathcal{N}^{l-1}) + \mathcal{N}^l$$
$$= \mathbf{Z}^l \cdot \mathbf{r}^{l-1} - \mathbf{Z}^l \cdot \mathcal{N}^{l-1} + \mathcal{N}^l. \tag{8.29}$$

Equation (8.29) represents the data matrix $\mathbf{Z}^l$ in terms of the received vectors and noise. Therefore, the best estimate of the data matrix, denoted by $\hat{\mathbf{Z}}^l$, is the matrix that minimizes $||\mathbf{r}^l - \mathbf{Z}^l \cdot \mathbf{r}^{l-1}||_F$. In other words, the ML decoding amounts to the following minimization problem

$$\hat{\mathbf{Z}}^l = \underset{\mathbf{Z}^l}{\arg\min} ||\mathbf{r}^l - \mathbf{Z}^l \cdot \mathbf{r}^{l-1}||_F^2. \tag{8.30}$$

Replacing the Frobenius norm with its trace definition results in

$$\hat{\mathbf{Z}}^l = \underset{\mathbf{Z}^l}{\arg\min} \operatorname{Tr}\left[(\mathbf{r}^l - \mathbf{Z}^l \cdot \mathbf{r}^{l-1})^H \cdot (\mathbf{r}^l - \mathbf{Z}^l \cdot \mathbf{r}^{l-1})\right]. \tag{8.31}$$

Expanding the terms in (8.31) and ignoring the constant terms results in

$$\hat{\mathbf{Z}}^l = \underset{\mathbf{Z}^l}{\arg\min} \ \mathrm{Tr}\left[-(\mathbf{r}^{l-1})^H \cdot (\mathbf{Z}^l)^H \cdot \mathbf{r}^l - (\mathbf{r}^l)^H \cdot \mathbf{Z}^l \cdot \mathbf{r}^{l-1}\right]$$

$$= \underset{\mathbf{Z}^l}{\arg\max} \ \mathrm{Tr}\left[\Re\left\{(\mathbf{r}^l)^H \cdot \mathbf{Z}^l \cdot \mathbf{r}^{l-1}\right\}\right]. \tag{8.32}$$

The data matrix $\mathbf{Z}^l$ in (8.21) can be written as the linear combination of the real and imaginary parts of the symbols, $\Re\{z_k^l\}$ and $\Im\{z_k^l\}$:

$$\mathbf{Z}^l = \Re\{z_1^l\} \begin{pmatrix} 1 & 0 \\ 0 & 1 \end{pmatrix} + \Re\{z_2^l\} \begin{pmatrix} 0 & 1 \\ -1 & 0 \end{pmatrix}$$

$$+ j\Im\{z_1^l\} \begin{pmatrix} 1 & 0 \\ 0 & -1 \end{pmatrix} + j\Im\{z_2^l\} \begin{pmatrix} 0 & 1 \\ 1 & 0 \end{pmatrix}. \tag{8.33}$$

Replacing (8.33) in (8.32), we can separately decode different data symbols z_1^l and z_2^l for every block l using the received vectors at blocks $l - 1$ and l. The ML decoder finds data symbols z_1^l and z_2^l independently from

$$\hat{z}_1^l = \underset{z_1^l}{\arg\max} \left[\Re\left\{\mathrm{Tr}\left[(\mathbf{r}^l)^H \begin{pmatrix} 1 & 0 \\ 0 & 1 \end{pmatrix} \mathbf{r}^{l-1}\right]\right\} \Re\{z_1^l\} \right.$$

$$\left. + \Re\left\{\mathrm{Tr}\left[(\mathbf{r}^l)^H j \begin{pmatrix} 1 & 0 \\ 0 & -1 \end{pmatrix} \mathbf{r}^{l-1}\right]\right\} \Im\{z_1^l\}\right],$$

$$\hat{z}_2^l = \underset{z_2^l}{\arg\max} \left[\Re\left\{\mathrm{Tr}\left[(\mathbf{r}^l)^H \begin{pmatrix} 0 & 1 \\ -1 & 0 \end{pmatrix} \mathbf{r}^{l-1}\right]\right\} \Re\{z_2^l\} \right.$$

$$\left. + \Re\left\{\mathrm{Tr}\left[(\mathbf{r}^l)^H j \begin{pmatrix} 0 & 1 \\ 1 & 0 \end{pmatrix} \mathbf{r}^{l-1}\right]\right\} \Im\{z_2^l\}\right]. \tag{8.34}$$

Assuming only one receive antenna results in

$$\hat{z}_1^l = \underset{z_1^l}{\arg\max}\left[\Re\left\{(r_1^l)^* r_1^{l-1} + (r_2^l)^* r_2^{l-1}\right\} \Re\{z_1^l\} \right.$$

$$\left. + \Re\left\{j\left[(r_1^l)^* r_1^{l-1} - (r_2^l)^* r_2^{l-1}\right]\right\} \Im\{z_1^l\}\right],$$

$$\hat{z}_2^l = \underset{z_2^l}{\arg\max}\left[\Re\left\{-(r_2^l)^* r_1^{l-1} + (r_1^l)^* r_2^{l-1}\right\} \Re\{z_2^l\} \right.$$

$$\left. + \Re\left\{j\left[(r_2^l)^* r_1^{l-1} + (r_1^l)^* r_2^{l-1}\right]\right\} \Im\{z_2^l\}\right]. \tag{8.35}$$

Using the fact that $\Re\{ja\} = -\Im\{a\}$, we have

$$\hat{z}_1^l = \underset{z_1^l}{\operatorname{argmax}}\big[\Re\{(r_1^l)^*r_1^{l-1} + (r_2^l)^*r_2^{l-1}\}\Re\{z_1^l\}$$

$$+ \Im\{-(r_1^l)^*r_1^{l-1} + (r_2^l)^*r_2^{l-1}\}\Im\{z_1^l\}\big],$$

$$\hat{z}_2^l = \underset{z_2^l}{\operatorname{argmax}}\big[\Re\{-(r_2^l)^*r_1^{l-1} + (r_1^l)^*r_2^{l-1}\}\Re\{z_2^l\}$$

$$- \Im\{(r_2^l)^*r_1^{l-1} + (r_1^l)^*r_2^{l-1}\}\Im\{z_2^l\}\big]. \qquad (8.36)$$

One can rewrite (8.36) as

$$\hat{z}_1^l = \underset{z_1^l}{\operatorname{argmax}}\ \Re\big\{\big[(r_1^l)^*r_1^{l-1} + r_2^l(r_2^{l-1})^*\big]z_1^l\big\},$$

$$\hat{z}_2^l = \underset{z_2^l}{\operatorname{argmax}}\ \Re\big\{\big[-(r_1^{l-1})^*r_2^l + r_2^{l-1}(r_1^l)^*\big]z_2^l\big\}. \qquad (8.37)$$

Obviously, (8.37) is a special case of finding the closet vector $\mathcal{R}$ in (8.24) to vectors $P^l = \begin{pmatrix} z_1^l \\ -z_2^l \end{pmatrix}$ in $\mathcal{V}$. In fact, the ML decoding in (8.37) can be derived directly from (8.24) by calculating $\Re\{\mathcal{R}^H \cdot P^l\}$.

8.4 Extension to more than two antennas

In this section, we show how to design differential space-time modulation schemes for more than two transmit antennas. We consider the extension for both the general case and the special case.

For the general case, we focus on the full-diversity and full-rate codes. Full-rate orthogonal STBCs only exist for real constellations with four or eight transmit antennas. We consider the case of four transmit antennas while emphasizing that a similar code can be designed for eight transmit antennas. Also, a differential space-time modulation scheme for three antennas is possible by removing the columns of the scheme for four antennas. Similarly, codes for five, six, and seven transmit antennas can be designed from the scheme for eight antennas. Half-rate codes for complex constellations are also possible [73, 69].

For the special case, we consider a universal formulation that provides differential modulation for any orthogonal STBC.

8.4.1 Differential modulation for four transmit antennas

The block diagram of differential encoding for more than two transmit antennas is similar to that of Figure 8.3. The number of input bits is Kb, instead of $2b$, for transmitting K symbols per block. For a full-rate orthogonal STBC with four transmit antennas, the number of symbols is $K = 4$ as in (4.38). Vectors S^l and P^l

are 4×1 vectors such that

$$
S^l = \begin{pmatrix} s_1^l \\ s_2^l \\ s_3^l \\ s_4^l \end{pmatrix},
\tag{8.38}
$$

and

$$
S^l = \sum_{k=1}^{4} P_k^l V_k(S^{l-1}),
\tag{8.39}
$$

where P_k^l is the kth element of the vector P^l and

$$
V_1(S^l) = \begin{pmatrix} s_1^l \\ s_2^l \\ s_3^l \\ s_4^l \end{pmatrix}, \qquad
V_2(S^l) = \begin{pmatrix} s_2^l \\ -s_1^l \\ s_4^l \\ -s_3^l \end{pmatrix},
$$

$$
V_3(S^l) = \begin{pmatrix} s_3^l \\ -s_4^l \\ -s_1^l \\ s_2^l \end{pmatrix}, \qquad
V_4(S^l) = \begin{pmatrix} s_4^l \\ s_3^l \\ -s_2^l \\ -s_1^l \end{pmatrix}.
\tag{8.40}
$$

For each specific constellation symbol vector S^l, vectors $V_1(S^l)$, $V_2(S^l)$, $V_3(S^l)$, and $V_4(S^l)$ are orthogonal to each other. They create a basis for the four-dimensional sub-space of any arbitrary four-dimensional constellation vector. Similar to the case of two transmit antennas, we define set $\mathcal{V}$ which consists of 2^{4b} unit-length distinct vectors $P_1, P_2, \ldots, P_{2^{4b}}$. Each vector P_v is a 4×1 vector of real numbers, $P_v = (P_{v1}, P_{v2}, P_{v3}, P_{v4})^T$. The arbitrary one-to-one mapping $\beta(\cdot)$ maps the $4b$ input bits onto $\mathcal{V}$ as in Figure 8.3. The choice of the real constellation, the set $\mathcal{V}$, and mapping $\beta(\cdot)$ completely define the encoder. Note that since $V_1(S^{l-1})$, $V_2(S^{l-1})$, $V_3(S^{l-1})$, and $V_4(S^{l-1})$ create an orthogonal basis, P_k^l can be derived from the inner product of S^l and $V_k(S^{l-1})$.

Example 8.4.1 *In this example, we consider the encoding of a differential space-time modulation for four transmit antennas and $b = 1$ bit/(s Hz). A real orthogonal*

STBC from the orthogonal design in (4.38) is utilized as repeated below:

$$\begin{pmatrix} x_1 & x_2 & x_3 & x_4 \\ -x_2 & x_1 & -x_4 & x_3 \\ -x_3 & x_4 & x_1 & -x_2 \\ -x_4 & -x_3 & x_2 & x_1 \end{pmatrix}. \tag{8.41}$$

Now, we need to specify the one-to-one mapping $\beta(\cdot)$. As an example, let us consider the following one-to-one mapping $\beta(\cdot)$:

$$\beta(0000) = \begin{pmatrix} 1 \\ 0 \\ 0 \\ 0 \end{pmatrix}, \quad \beta(0001) = \begin{pmatrix} 0.5 \\ 0.5 \\ -0.5 \\ 0.5 \end{pmatrix}, \quad \beta(0010) = \begin{pmatrix} 0.5 \\ -0.5 \\ 0.5 \\ 0.5 \end{pmatrix},$$

$$\beta(0011) = \begin{pmatrix} 0 \\ 0 \\ 0 \\ 1 \end{pmatrix}, \quad \beta(0100) = \begin{pmatrix} 0.5 \\ 0.5 \\ 0.5 \\ -0.5 \end{pmatrix}, \quad \beta(0101) = \begin{pmatrix} 0 \\ 1 \\ 0 \\ 0 \end{pmatrix},$$

$$\beta(0110) = \begin{pmatrix} 0 \\ 0 \\ 1 \\ 0 \end{pmatrix}, \quad \beta(0111) = \begin{pmatrix} -0.5 \\ 0.5 \\ 0.5 \\ 0.5 \end{pmatrix}, \quad \beta(1000) = \begin{pmatrix} 0.5 \\ -0.5 \\ -0.5 \\ -0.5 \end{pmatrix},$$

$$\beta(1001) = \begin{pmatrix} 0 \\ 0 \\ -1 \\ 0 \end{pmatrix}, \quad \beta(1010) = \begin{pmatrix} 0 \\ -1 \\ 0 \\ 0 \end{pmatrix}, \quad \beta(1011) = \begin{pmatrix} -0.5 \\ -0.5 \\ -0.5 \\ 0.5 \end{pmatrix},$$

$$\beta(1100) = \begin{pmatrix} 0 \\ 0 \\ 0 \\ -1 \end{pmatrix}, \quad \beta(1101) = \begin{pmatrix} -0.5 \\ 0.5 \\ -0.5 \\ -0.5 \end{pmatrix}, \quad \beta(1110) = \begin{pmatrix} -0.5 \\ -0.5 \\ 0.5 \\ -0.5 \end{pmatrix},$$

$$\beta(1111) = \begin{pmatrix} -1 \\ 0 \\ 0 \\ 0 \end{pmatrix}. \tag{8.42}$$

The corresponding set $\mathcal{V}$ includes the 16 vectors on the right-hand side of (8.42). All these vectors have unit length. The one-to-one mapping $\beta(\cdot)$ maps four input bits onto $\mathcal{V}$. In fact, the above example of mapping $\beta(\cdot)$ can also be defined by calculation. For example, the four input bits at block l select four BPSK symbols, that is symbol $z_k^l = 1/2$ if the corresponding bit is zero and $z_k^l = -1/2$ if the bit is one. Then, the elements of P^l are defined as the projection of the vector $(z_1^l, z_2^l, z_3^l, z_4^l)^T$ onto $V_1(S), V_2(S), V_3(S), V_4(S)$, where $S = (1/2, 1/2, 1/2, 1/2)^T$.

In fact, the elements of P^l can be calculated using the following equations:

$$P_1^l = \frac{1}{2}(z_1^l + z_2^l + z_3^l + z_4^l),$$

$$P_2^l = \frac{1}{2}(z_1^l - z_2^l + z_3^l - z_4^l),$$

$$P_3^l = \frac{1}{2}(z_1^l - z_2^l - z_3^l + z_4^l),$$

$$P_4^l = \frac{1}{2}(z_1^l + z_2^l - z_3^l - z_4^l). \tag{8.43}$$

The result is exactly the same as the one-to-one mapping in (8.42).

To study the decoding, we consider only one receive antenna for the sake of simplicity. The received signals for block l are denoted by r_1^l, r_2^l, r_3^l, and r_4^l. Then, we have

$$(r_1^l, \ r_2^l, \ r_3^l, \ r_4^l) = (s_1^l, \ s_2^l, \ s_3^l, \ s_4^l)\Omega + (\eta_1^l, \ \eta_2^l, \ \eta_3^l, \ \eta_4^l), \tag{8.44}$$

where

$$\Omega = \begin{pmatrix} \alpha_1 & \alpha_2 & \alpha_3 & \alpha_4 \\ \alpha_2 & -\alpha_1 & -\alpha_4 & \alpha_3 \\ \alpha_3 & \alpha_4 & -\alpha_1 & -\alpha_2 \\ \alpha_4 & -\alpha_3 & \alpha_2 & -\alpha_1 \end{pmatrix}. \tag{8.45}$$

From the orthogonality of the STBC, we have

$$\Omega \cdot \Omega^T = \sum_{n=1}^{4} \alpha_n^2 I. \tag{8.46}$$

From (8.44), we can write the following equivalent equations:

$$\begin{aligned}
(-r_2^l, \ r_1^l, \ r_4^l, \ -r_3^l) &= (s_2^l, \ -s_1^l, \ s_4^l, \ -s_3^l)\Omega + (-\eta_2^l, \ \eta_1^l, \ \eta_4^l, \ -\eta_3^l), \\
(-r_3^l, \ -r_4^l, \ r_1^l, \ r_2^l) &= (s_3^l, \ -s_4^l, \ -s_1^l, \ s_2^l)\Omega + (-\eta_3^l, \ -\eta_4^l, \ \eta_1^l, \ \eta_2^l), \quad (8.47) \\
(-r_4^l, \ r_3^l, \ -r_2^l, \ r_1^l) &= (s_4^l, \ s_3^l, \ -s_2^l, \ -s_1^l)\Omega + (-\eta_4^l, \ \eta_3^l, \ -\eta_2^l, \ \eta_1^l).
\end{aligned}$$

To write these equations in a compact vector way, we use the definition of $V_k(S^l)$ in (8.40) and define the following vectors:

$$\begin{aligned}
R_1^l &= (r_1^l, \ r_2^l, \ r_3^l, \ r_4^l) &&= V_1(S^l)^T \cdot \Omega + (\eta_1^l, \ \eta_2^l, \ \eta_3^l, \ \eta_4^l), \\
R_2^l &= (-r_2^l, \ r_1^l, \ r_4^l, \ -r_3^l) &&= V_2(S^l)^T \cdot \Omega + (-\eta_2^l, \ \eta_1^l, \ \eta_4^l, \ -\eta_3^l), \\
R_3^l &= (-r_3^l, \ -r_4^l, \ r_1^l, \ r_2^l) &&= V_3(S^l)^T \cdot \Omega + (-\eta_3^l, \ -\eta_4^l, \ \eta_1^l, \ \eta_2^l), \\
R_4^l &= (-r_4^l, \ r_3^l, \ -r_2^l, \ r_1^l) &&= V_4(S^l)^T \cdot \Omega + (-\eta_4^l, \ \eta_3^l, \ -\eta_2^l, \ \eta_1^l).
\end{aligned} \tag{8.48}$$

Then, using (8.46), we have

$$
\begin{aligned}
R_1^l \cdot (R_k^{l-1})^T &= (S_1^l)^T \cdot \Omega \cdot \Omega^T \cdot V_k(S^{l-1}) + \mathcal{N}_k \\
&= \left(\sum_{n=1}^{4} \alpha_n^2 \right) (S_1^l)^T \cdot V_k(S^{l-1}) + \mathcal{N}_k \\
&= \left(\sum_{n=1}^{4} \alpha_n^2 \right) P_k^l + \mathcal{N}_k, \quad k = 1, 2, 3, 4,
\end{aligned}
\tag{8.49}
$$

where $\mathcal{N}_k$ is a noise term. We rewrite the above equations in a vector equation as follows:

$$
\mathcal{R} = \begin{pmatrix} R_1^l \cdot (R_1^{l-1})^T \\ R_1^l \cdot (R_2^{l-1})^T \\ R_1^l \cdot (R_3^{l-1})^T \\ R_1^l \cdot (R_4^{l-1})^T \end{pmatrix} = \left(\sum_{n=1}^{4} \alpha_n^2 \right) P^l + \mathcal{N}.
\tag{8.50}
$$

Because the elements of V have equal lengths, to compute P^l, the receiver can compute the closest vector of V to $\mathcal{R}$. Then, the inverse mapping $\beta^{-1}(\cdot)$ is applied to recover the transmitted bits. The block diagram of the decoder is similar to Figure 8.5 although the output provides Kb bits instead of $2b$ bits. As it is evident from (8.50), the decision factor $\mathcal{R}$ is a scaled version of the vector P^l. Similar to the argument for the DPSK decoding, the scaling does not change the geometry of the detection regions and the result of the differential decoding is a 3 dB loss in performance due to the doubling of the effective noise.

It can be proved that the above detection method provides N-level diversity assuming N transmit and one receive antennas. Instead of a mathematical proof that is very similar to the analysis provided in Chapter 4, we provide a physical argument. The coefficient of P^l in (8.50) is small only if all path gains are small. In other words, all sub-channels from the N transmit antennas to the receive antenna must undergo fading to damage the signal. This means that the fading has a detrimental effect only if all of the N sub-channels have small path gains. This results in an N-level diversity.

If there is more than one receive antenna, a maximum ratio combining (MRC) method similar to the MRC of Chapter 4 can be utilized. In the case of differential space-time modulation, first assuming that only the receive antenna m exists, we compute $\mathcal{R}_m$ as follows:

$$\mathcal{R}_m = \begin{pmatrix} R_{1,m}^l \cdot (R_{1,m}^{l-1})^T \\ R_{1,m}^l \cdot (R_{2,m}^{l-1})^T \\ R_{1,m}^l \cdot (R_{3,m}^{l-1})^T \\ R_{1,m}^l \cdot (R_{4,m}^{l-1})^T \end{pmatrix}. \tag{8.51}$$

Note that we should replace r_n^l with $r_{n,m}^l$ in (8.48) to calculate $R_{n,m}^l$ instead of R_n^l. Then, after calculating M vectors $\mathcal{R}_m$, $m = 1, 2, \ldots, M$, the closest vector of $\mathcal{V}$ to $\sum_{m=1}^M \mathcal{R}_m$ is computed. The inverse mapping of $\beta^{-1}(\cdot)$ is applied to the closest vector to detect the transmitted bits. It is easy to show that NM-level diversity is achieved by using this method.

8.4.2 The extension of the simple special case

The differential space-time modulation in Figure 8.3 and its extension put the STBC block as the last stage of the encoder. In fact, the STBC codewords are outside the differential loop. Instead of the codewords, the symbols that define these codewords generate the loop. Another approach to differential space-time modulation is to put the STBC block as the first stage of the encoder. Then, the STBC codewords generate the differential loop as shown in Figure 8.4. This is a special case of the encoder in Figure 8.3 as we discussed in Section 8.2.1. For this special case, the extension to more than two transmit antennas is straightforward. In fact, most equations are identical to the equations in Sections 8.2.1 and 8.3.1.

Let us consider a complex orthogonal STBC that transmits K symbols s_k, $k = 1, 2, \ldots, K$ from N transmit antennas. We consider square $N \times N$ orthogonal designs as defined in Chapter 4. Examples of square orthogonal designs include real rate one designs for $N = 2, 4, 8$, a complex rate one design for $N = 2$, a complex rate 3/4 design for $N = 4$, and so on. In general, a data matrix $\mathbf{Z}^l$ from an orthogonal design with elements z_k^l has the following property as discussed in Chapter 4:

$$(\mathbf{Z}^l)^H \cdot \mathbf{Z}^l = \kappa \left(\sum_{k=1}^K |z_k^l|^2 \right) I_N. \tag{8.52}$$

If all symbols have the same energy, for example PSK, with an appropriate normalization, the codewords are unitary, that is

$$(\mathbf{Z}^l)^H \cdot \mathbf{Z}^l = I_N. \tag{8.53}$$

Following the encoding for two transmit antennas, the transmitter first sends an arbitrary unitary codeword $\mathbf{C}^0$. For block l, we use the Kb input bits and the orthogonal STBC to pick $\mathbf{Z}^l$. As in Figure 8.4, the codeword $\mathbf{Z}^l$ is utilized to calculate $\mathbf{C}^l$ by

$$\mathbf{C}^l = \mathbf{Z}^l \cdot \mathbf{C}^{l-1}. \tag{8.54}$$

Then, the codeword $\mathbf{C}^l$ is transmitted over the channel. Using an induction argument similar to that of the two transmit antennas, we have

$$(\mathbf{C}^l)^H \cdot \mathbf{C}^l = (\mathbf{C}^{l-1})^H \cdot \mathbf{C}^{l-1} = I_N. \tag{8.55}$$

Assuming that the path gain matrix $\mathbf{H}$ does not change during consecutive blocks $l-1$ and l, the received vector for block l is calculated as

$$\begin{aligned}
\mathbf{r}^l &= \mathbf{C}^l \cdot \mathbf{H} + \mathcal{N}^l \\
&= \mathbf{Z}^l \cdot \mathbf{C}^{l-1} \cdot \mathbf{H} + \mathcal{N}^l \\
&= \mathbf{Z}^l \cdot (\mathbf{r}^{l-1} - \mathcal{N}^{l-1}) + \mathcal{N}^l \\
&= \mathbf{Z}^l \cdot \mathbf{r}^{l-1} - \mathbf{Z}^l \cdot \mathcal{N}^{l-1} + \mathcal{N}^l.
\end{aligned} \tag{8.56}$$

Equation (8.56) represents the data matrix $\mathbf{Z}^l$ in terms of the received vectors and noise. The ML decoding results in the following minimization problem

$$\begin{aligned}
\hat{\mathbf{Z}}^l &= \underset{\mathbf{Z}^l}{\operatorname{argmin}} \| \mathbf{r}^l - \mathbf{Z}^l \cdot \mathbf{r}^{l-1} \|_F^2 \\
&= \underset{\mathbf{Z}^l}{\operatorname{argmin}} \operatorname{Tr}\left[(\mathbf{r}^l - \mathbf{Z}^l \cdot \mathbf{r}^{l-1})^H \cdot (\mathbf{r}^l - \mathbf{Z}^l \cdot \mathbf{r}^{l-1}) \right].
\end{aligned} \tag{8.57}$$

Expanding the terms in (8.57) and ignoring the constant terms results in

$$\begin{aligned}
\hat{\mathbf{Z}}^l &= \underset{\mathbf{Z}^l}{\operatorname{argmin}} \operatorname{Tr}\left[-(\mathbf{r}^{l-1})^H \cdot (\mathbf{Z}^l)^H \cdot \mathbf{r}^l - (\mathbf{r}^l)^H \cdot \mathbf{Z}^l \cdot \mathbf{r}^{l-1} \right] \\
&= \underset{\mathbf{Z}^l}{\operatorname{argmax}} \operatorname{Tr}\left[\Re\{ (\mathbf{r}^l)^H \cdot \mathbf{Z}^l \cdot \mathbf{r}^{l-1} \} \right].
\end{aligned} \tag{8.58}$$

The orthogonal STBCs considered in Chapter 4 are linear codes in terms of the real and imaginary parts of the symbols s_k, $k = 1, 2, \ldots, K$. Therefore, the data matrix $\mathbf{Z}^l$ can be written as

$$\mathbf{Z}^l = \sum_{k=1}^{K} \Re\{z_k^l\} X_k + j \sum_{k=1}^{K} \Im\{z_k^l\} Y_k, \tag{8.59}$$

where X_k and Y_k are $N \times N$ real matrices. For $N = K = 2$, the OSTBC in (4.1) and (8.21), we have

$$X_1 = \begin{pmatrix} 1 & 0 \\ 0 & 1 \end{pmatrix}, \, X_2 = \begin{pmatrix} 0 & 1 \\ -1 & 0 \end{pmatrix}, \, Y_1 = \begin{pmatrix} 1 & 0 \\ 0 & -1 \end{pmatrix}, \, Y_2 = \begin{pmatrix} 0 & 1 \\ 1 & 0 \end{pmatrix}. \tag{8.60}$$

In this case, (8.59) becomes (8.33) and the ML decoding formulas in Section 8.3.1 follow. Note that in general a normalization factor of $1/\sqrt{K}$ or an appropriate normalized choice of the constellation is necessary to validate (8.53). Replacing (8.59) in (8.58), we can separately decode different data symbols z_k^l for every block l using the received vectors at blocks $l-1$ and l. The ML decoder finds different data symbols z_k^l independently from

$$\hat{z}_k^l = \arg \max_{z_k^l} \left[\Re\left\{\mathrm{Tr}[(\mathbf{r}^l)^H X_k \mathbf{r}^{l-1}]\right\}\Re\{z_k^l\} + \Re\left\{\mathrm{Tr}[(\mathbf{r}^l)^H j Y_k \mathbf{r}^{l-1}]\right\}\Im\{z_k^l\}\right].$$

(8.61)

Equivalently, for the kth symbol in the lth block, the ML decoding results in the constellation point that maximizes the following cost function:

$$\Re\left\{\mathrm{Tr}[(\mathbf{r}^{l-1})^H X_k^H \mathbf{r}^l]\right\}\Re\{z_k^l\} - \Re\left\{\mathrm{Tr}[(\mathbf{r}^{l-1})^H j Y_k^H \mathbf{r}^l]\right\}\Im\{z_k^l\},$$

(8.62)

or equivalently,

$$\Re\left\{\mathrm{Tr}[(\mathbf{r}^{l-1})^H X_k^H \mathbf{r}^l]\right\}\Re\{z_k^l\} + \Im\left\{\mathrm{Tr}[(\mathbf{r}^{l-1})^H Y_k^H \mathbf{r}^l]\right\}\Im\{z_k^l\},$$

(8.63)

8.5 Simulation results

In this section, we provide simulation results for the above differential space-time modulation schemes using one receive antenna. We assume a quasi-static flat Rayleigh fading model for the channel. Therefore, the path gains are independent complex Gaussian random variables. The channel is fixed during the transmission of one frame. The receiver does not know the channel path gains and utilizes maximum-likelihood decoding to estimate the transmitted symbols and bits. The transmission bit rate is a function of the constellation and the code rate of the underlying STBC. We use Monte Carlo simulations to derive the bit error probability versus the received SNR. We also provide the simulation results for a coherent detection scheme for comparison. The coherent scheme has perfect knowledge of the channel path gains at the receiver.

Figure 8.6 provides simulation results for a transmission bit rate of 1 bit/(s Hz) using two transmit antennas. In this chapter, a perfect knowledge of the channel path gains at the receiver is denoted by "coherent." On the other hand, we use "non-coherent" to denote the case that neither the encoder nor the decoder knows the channel path gains. As evident from Figure 8.6, the performance of the differential space-time modulation is about 3 dB worse than that of the coherent orthogonal STBC. As we discussed before, the effect of non-coherent decoding is equivalent to doubling the noise. Therefore, the 3 dB difference in Figure 8.6 is expected. This is in fact an extension of DPSK to multiple

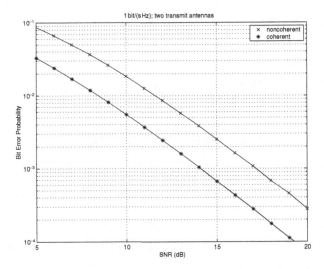

Fig. 8.6. Performance results for coherent and non-coherent detection schemes; two transmit antennas, one receive antenna, $b = 1$ bit/(s Hz).

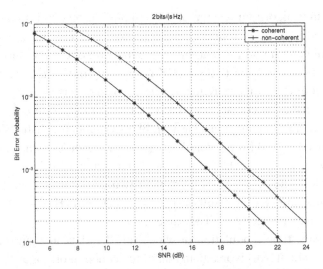

Fig. 8.7. Performance results for coherent and non-coherent detection schemes; two transmit antennas, one receive antenna, $b = 2$ bits/(s Hz).

transmit antennas. Figure 8.7 depicts similar results for a rate of 2 bits/(s Hz). Again there is only about a 3 dB difference between the performance of the coherent and non-coherent schemes. Note that the differential schemes in Examples 8.2.2 and 8.2.3 both provide the same performance. Using the simpler special decoding from Section 8.3.1 only reduces the decoding complexity without degrading the performance.

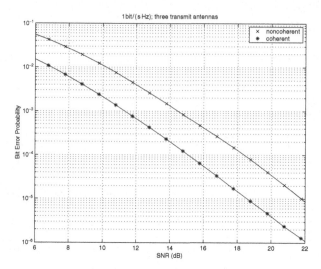

Fig. 8.8.   Performance results for coherent and non-coherent detection schemes; three transmit antennas, one receive antenna, $b = 1$ bit/(s Hz).

The decoding algorithm for the above simulation results utilizes two blocks of the received signals, $\mathbf{r}^{l-1}$ and $\mathbf{r}^l$ to detect the differentially transmitted data. The underlying assumption is that the path gains do not change during those two transmission blocks. Similar to the DPSK case, using more than two transmission blocks in decoding results in reducing the 3 dB gap between the coherent and non-coherent cases. Of course, utilizing L blocks in detection only works if the assumption of constant path gains during L blocks is not unrealistic. The interested reader is referred to [48] for the details of decoding formulas using L transmission blocks. Utilizing $L = 8$ blocks reduces the gap to about 1.5 dB.

Similar results are derived for more than two transmit antennas. Figures 8.8 and 8.9 provide simulation results for a rate of 1 bit/(s Hz) for three and four transmit antennas, respectively. The same trend is observed in these figures. The differential space-time modulation provides full diversity and full rate. Also, its performance is about 3 dB worse than that of the coherent orthogonal STBC under similar conditions.

Similar to the case of coherent detection, full diversity and full rate complex orthogonal differential space-time modulation schemes do not exist for more than two transmit antennas. Rate 1/2 and 3/4 full diversity orthogonal STBCs exist for three and four transmit antennas. Corresponding differential space-time modulation schemes can be designed using the encoding and decoding in Sections 8.2.1 and 8.3.1, respectively. The resulting differential schemes suffer from rate deficiency. For example, to transmit 2 bits/(s Hz), instead of using QPSK one needs to use a constellation with more than four points. This translates to a lower coding gain

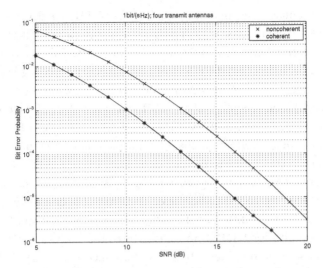

Fig. 8.9. Performance results for coherent and non-coherent detection schemes; four transmit antennas, one receive antenna, $b = 1$ bit/(s Hz).

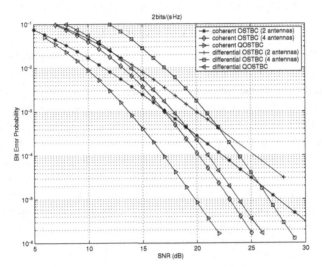

Fig. 8.10. Performance results for coherent and non-coherent detection schemes; one receive antenna, $b = 2$ bits/(s Hz).

compared to full rate codes. A quasi-orthogonal STBC provides full diversity and full rate for three and four transmit antennas as we discussed in Chapter 5. Differential modulation based on quasi-orthogonal STBCs is possible as explained in [161]. The result is a full diversity and full rate complex differential space-time modulation scheme. Figure 8.10 provides simulation results at rate 2 bits/(s Hz) for different differential space-time modulation schemes. It depicts the results for two and four transmit antennas using an orthogonal STBC (OSTBC) and a quasi-orthogonal STBC (QOSTBC). For the case of OSTBCs, there is about a 3 dB difference between the performance of the coherent and non-coherent codes. This is expected since in the case of orthogonal codes, the effect of non-coherent decoding is equivalent to doubling the noise. However, the performance of the quasi-orthogonal differential space-time modulation scheme is about 4 dB worse than that of the corresponding coherent code from Chapter 5. Its performance is about 2.8–3.2 dB, depending on the bit error probability, better than that of the orthogonal differential code for four transmit antennas. This is due to the rate deficiency of the OSTBCs for more than two transmit antennas. In fact, Figure 8.10 demonstrates how this rate deficiency translates into a coding gain disadvantage. Therefore, the comparison of orthogonal and quasi-orthogonal differential modulation schemes are similar to the case of coherent decoding for STBCs. In other words, the quasi-orthogonal differential modulation scheme provides a better performance in a trade-off for pairwise decoding instead of separate symbol decoding. Both schemes provide full diversity and very low encoding complexity.

8.6 Summary of important results

- Differential detection schemes for multiple transmit antennas are possible.
- Differential space-time modulation provides full spatial diversity and requires no channel state side information at the receiver or transmitter.
- The complexity of the differential detection schemes is very low. Similar to coherent orthogonal STBCs, separate ML decoding with linear complexity is possible.
- Like a regular DPSK scheme, the performance of the orthogonal differential detection schemes is about 3 dB worse than those of the corresponding coherent schemes.
- The 3 dB performance gap can be decreased by using more than two blocks for decoding.
- Full rate and full diversity quasi-orthogonal differential space-time modulation schemes exist for four transmit antennas. The quasi-orthogonal codes outperform their orthogonal counterparts while using a slightly more complex pairwise decoding.

8.7 Problems

1 Consider the differential modulation scheme in Example 8.2.2. What are the transmitted symbols from each antenna for the following input bit stream: 01110010?

2 Consider the differential modulation scheme in Example 8.2.3. What are the transmitted symbols from each antenna for the following input bit stream: 01110010? Compare the transmitted symbols with those in the previous problem.

3 Consider a differential modulation scheme for two transmit antennas and one receive antenna that uses the data matrix $\mathbf{Z}^l$ in (8.59) and the matrices X_1, X_2, Y_1, and Y_2 in (8.60). We consider a differential scheme that uses a QAM constellation instead of the

PSK constellation. In the first block, the transmitter sends $\mathbf{C}^0 = I_2$. Let us define $a^0 = 1$ and $a^l = \sqrt{|z_1^l|^2 + |z_2^l|^2}$, where l is the block number. Also, let us assume that the transmitted codeword for block l is $\mathbf{C}^l = (1/a^{l-1})\mathbf{Z}^l \cdot \mathbf{C}^{l-1}$ and satisfies the energy constraint: $E\left[\sum_{t=1}^{T}\sum_{n=1}^{N}|\mathbf{C}_{t,n}^l|^2\right] = T$.

(a) Show that $(\mathbf{C}^l)^H \cdot \mathbf{C}^l = (a^l)^2 I_2$.
(b) Assuming that the path gain matrix $\mathbf{H}$ does not change during consecutive blocks $l-1$ and l, show that $\mathbf{r}^l = (1/a^{l-1})\mathbf{Z}^l \cdot \mathbf{r}^{l-1} - (1/a^{l-1})\mathbf{Z}^l \cdot \mathcal{N}^{l-1} + \mathcal{N}^l$.
(c) Defining the decoding as $\hat{\mathbf{Z}}^l = \underset{\mathbf{Z}^l}{\operatorname{\mathbf{argmin}}}\, \|\mathbf{r}^l - (1/a^{l-1})\mathbf{Z}^l \cdot \mathbf{r}^{l-1}\|_F^2$, derive the decoding formulas for the real and imaginary parts of the kth data symbol in the lth block for a rectangular QAM constellation.

4 As discussed in Section 8.5, the performance of a differential detection scheme can be improved by considering more than two consecutive received blocks. Consider a differential modulation scheme for two transmit antennas and one receive antenna. Let us consider three consecutive blocks $l-2$, $l-1$ and l for decoding. Following the differential encoding in Section 8.2, we define

$$P^l = \begin{pmatrix} P_1^l \\ P_2^l \end{pmatrix}, \quad V_1(S^l) = \begin{pmatrix} s_1^l \\ s_2^l \end{pmatrix}, \quad V_2(S^l) = \begin{pmatrix} (s_2^l)^* \\ -(s_1^l)^* \end{pmatrix},$$

$$P^{l,l-1} = \begin{pmatrix} P_1^l P_1^{l-1} - P_2^l (P_2^{l-1})^* \\ P_1^l P_2^{l-1} + P_2^l (P_1^{l-1})^* \end{pmatrix}, \quad \mathcal{R}^{i,j} = \begin{pmatrix} (r_1^j)^* r_1^i + r_2^j (r_2^i)^* \\ (r_2^j)^* r_1^i - r_1^i (r_2^j)^* \end{pmatrix}.$$

Show that the decoder should find the vectors P^l, $P^{l-1} \in \mathcal{V}$ that maximize

$$\Re\{(\mathcal{R}^{l,l-1})^H \cdot P^l + (\mathcal{R}^{l-1,l-2})^H \cdot P^{l-1} + (\mathcal{R}^{l,l-2})^H \cdot P^{l,l-1}\}.$$

5 Instead of using STBCs, one can use unitary group codes as the codewords of a differential modulation scheme. Consider a differential modulation scheme for two transmit antennas and one receive antenna. Consider the following group under matrix multiplication:

$$\mathcal{G} = \left\{ \pm \begin{pmatrix} 1 & 0 \\ 0 & 1 \end{pmatrix}, \pm \begin{pmatrix} j & 0 \\ 0 & -j \end{pmatrix}, \pm \begin{pmatrix} 0 & -1 \\ 1 & 0 \end{pmatrix}, \pm \begin{pmatrix} 0 & j \\ j & 0 \end{pmatrix} \right\}.$$

All members of $\mathcal{G}$ are unitary. Also, let us define

$$D = \begin{pmatrix} 1 & 1 \\ -1 & 1 \end{pmatrix}.$$

The pair $\mathcal{G}$ and D construct a group code over the QPSK constellation $\{1, j, -1, -j\}$. The encoder, initially, sends $\mathbf{C}^0 = D$. At block l, the transmitter sends $\mathbf{C}^l = G^l \cdot \mathbf{C}^{l-1}$, where $G^l \in \mathcal{G}$. The group structure of $\mathcal{G}$ guarantees that $\mathbf{C}^l \in \mathcal{G} \cdot D = \{G \cdot D | G \in \mathcal{G}\}$ whenever $\mathbf{C}^{l-1} \in \mathcal{G} \cdot D$.

(a) Derive the ML decoder for the above differential scheme assuming the path gain matrix does not change during blocks $l-1$ and l.

(b) Compare the complexity of the encoding of the above differential scheme with that of the differential encoding in Section 8.2.1.

(c) Compare the complexity of the decoding of the above differential scheme with that of the differential decoding in Section 8.3.1.

9

Spatial multiplexing and receiver design

9.1 Introduction

Using multiple antennas can result in a smaller probability of error for the same throughput because of the diversity gain. The main objective of space-time codes is to achieve the maximum possible diversity. As we have shown in previous chapters, space-time codes provide a diversity gain equal to the product of the number of transmit and receive antennas NM. Also, we have demonstrated that the maximum throughput of the space-time codes is one symbol per channel use for any number of transmit antennas. The use of multiple antennas results in increasing the capacity of MIMO channels as shown in Chapter 2. Therefore, one may transmit at a higher throughput, compared to SISO channels, for a given probability of error. The capacity analysis of Chapter 2 shows that when the number of transmit and receive antennas are the same, the capacity grows at least linearly by the number of antennas. Instead of utilizing the multiple antennas to achieve the maximum possible diversity gain, one can use multiple antennas to increase the transmission rate. In fact, as we discussed in Chapter , there is a trade-off between these two gains from multiple antennas.

One approach to achieve the higher possible throughput is spatial multiplexing (SM). One simple example of spatial multiplexing is when the input is demultiplexed into N separate streams, using a serial-to-parallel converter, and each stream is transmitted from an independent antenna. As a result, the throughput is N symbols per channel use for a MIMO channel with N transmit antennas. This N-fold increase in throughput will generally come at the cost of a lower diversity gain compared to space-time coding. Therefore, spatial multiplexing is a better choice for high rate systems operating at relatively high SNRs while space-time coding is more appropriate for transmitting at relatively low rates and low SNRs.

In this chapter, we study space-time architectures that achieve such a high throughput. Foschini proposed the first example of such a system in [43]. Since

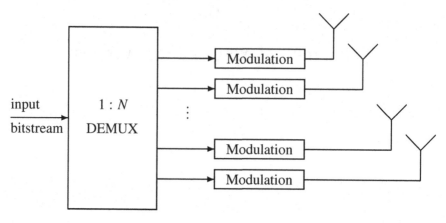

Fig. 9.1. A simple example of spatial multiplexing (V-BLAST).

then, different flavors of such a space-time architecture have been proposed under the general framework of Bell Labs Layered Space-Time (BLAST) architectures [44]. In this chapter, we study different versions of BLAST in addition to different possible methods for designing their decoders. We start with a review of different decoder structures that work for any block code. Later, we present specific formulas for different BLAST architectures.

9.2 Spatial multiplexing

Let us start with a simple example of spatial multiplexing as shown in Figure 9.1. A serial to parallel demultiplexer generates N separate substreams from the input. Each substream is processed separately and is transmitted from a different antenna. The processing may include temporal coding in addition to the depicted modulation.

Denoting the transmitted $1 \times N$ vector at each time slot by $\mathbf{C}$, the $1 \times M$ output vector $\mathbf{r}$ is

$$\mathbf{r} = \mathbf{C} \cdot \mathbf{H} + \mathcal{N}, \tag{9.1}$$

where $\mathbf{H}$ is the $N \times M$ channel matrix and $\mathcal{N}$ is the $1 \times M$ noise matrix. Let us assume for the sake of simplicity that there is no temporal coding. The maximum-likelihood decoding finds the codeword $\mathbf{C}$ that minimizes the Frobenius norm $||\mathbf{r} - \mathbf{C} \cdot \mathbf{H}||_F$. Using a full search to find the optimal codeword is computationally very demanding. If the modulation utilizes a constellation with 2^b points to transmit b bits, the number of possibilities for $\mathbf{C}$ is 2^{bN}. For four transmit antennas and 16-QAM, $b = 4$, there are 65 536 possibilities, which is impractical in most cases. The space-time coding structures that we have considered so far make it possible to implement a simpler ML decoding. For example, symbols can be decoded

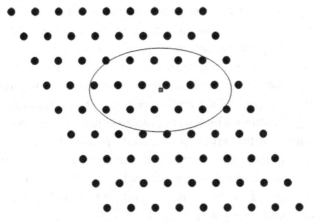

Fig. 9.2. Sphere decoding.

separately for an orthogonal STBC resulting in a total of $N2^b$ possibilities to check. Another example is the use of Viterbi algorithm to reduce the complexity of the ML decoding for STTCs. When a simpler ML decoding does not exist, we have to use suboptimal decoding methods to reduce the complexity of the receiver. Such systems generally operate at high bit rates with sometimes more than four transmit antennas. Therefore, the complexity of full search is even more critical than that of the space-time coding methods. A lower complexity alternative for the decoding of spatial multiplexing systems is essential in practice. We start with some general receiver architectures that work for any system that operates on input blocks. Later, we provide more efficient methods for specific spatial multiplexing systems.

9.3 Sphere decoding

The main idea behind sphere decoding is to limit the number of possible codewords by considering only those codewords that are within a sphere centered at the received signal vector [146]. The concept is depicted in Figure 9.2, in which the received signal is represented by a small square. To have the transmitted signal far from the received signal, the instantaneous power of the noise should be large. A larger instantaneous power of the noise is less likely than a smaller one because of the white Gaussian nature of the noise. Therefore, intuitively, it is more likely to have the most likely codeword in a neighborhood close to the received signal. The main question is how to separate the codewords inside a sphere from the set of all possible codewords. This should be done for any arbitrary received signal which is the center of the sphere. The complexity of separating these signals should be small enough such that the overall complexity of the sphere decoding is lower than that of the full search. This can be done since the ML decoding candidates are the elements of a lattice. A lattice, generated by a basis, consists of all linear combinations of the basis elements with integer coefficients. In fact, the problem of finding the closest point of a lattice to a given point appears in many other areas as well [26]. For example, lattice codes are used in vector quantization and source coding. Also, finding the closest point in a lattice has applications in cryptography. In general, it can be an NP-hard problem for some lattice structures. Therefore, finding suboptimal lattice decoding algorithms with lower complexity has been an important problem. Different solutions for this problem have been proposed in the literature. For a survey of closest point search methods, the interested reader is referred to [2]. The idea of limiting the search in a given region that includes the optimal lattice point has been proposed for the shortest vector problem in [28, 36]. The "limiting region" can be a rectangular parallelepiped [58, 77] or a hyper-sphere [98]. In what follows, we consider the latter option, which is called sphere decoding, and explain the details of the algorithm.

In general, one has to follow different steps in a sphere-decoding algorithm. First, the complex equations are converted to real equations. Second, the lattice structure of the noiseless vectors are defined. Third, the lattice points inside the sphere are enumerated. Finally, the closest point among the vectors inside the sphere to the received vector is found. We explain the details of each step in what follows.

First, let us separate the real and imaginary parts of the received vector in (9.1) to define the following $1 \times 2M$ signal

$$(\Re\{\mathbf{r}\}\Im\{\mathbf{r}\}) = (\Re\{\mathbf{C}\}\Im\{\mathbf{C}\}) \cdot \begin{pmatrix} \Re\{\mathbf{H}\} & \Im\{\mathbf{H}\} \\ -\Im\{\mathbf{H}\} & \Re\{\mathbf{H}\} \end{pmatrix} + (\Re\{\mathcal{N}\}\Im\{\mathcal{N}\}). \qquad (9.2)$$

The above equation includes only real numbers and can be written in terms of real matrices as

$$\mathbf{r}' = \mathbf{C}' \cdot \mathbf{H}' + \mathcal{N}', \qquad (9.3)$$

where $\mathbf{r}' = (\Re\{\mathbf{r}\}\Im\{\mathbf{r}\})$, $\mathbf{C}' = (\Re\{\mathbf{C}\}\Im\{\mathbf{C}\})$, $\mathcal{N}' = (\Re\{\mathcal{N}\}\Im\{\mathcal{N}\})$, and

$$\mathbf{H}' = \begin{pmatrix} \Re\{\mathbf{H}\} & \Im\{\mathbf{H}\} \\ -\Im\{\mathbf{H}\} & \Re\{\mathbf{H}\} \end{pmatrix}. \qquad (9.4)$$

Note that it is very unlikely to have a singular $\mathbf{H}'$. If the transmitted constellation is QAM, the elements of the $1 \times 2N$ vector $\mathbf{C}'$ are from the set of integers $\mathbb{Z}$, that is $\mathbf{C}' \in \mathbb{Z}^{2N}$. Then, the received vectors without noise are represented by a lattice [35]. Using the discrete nature of the noiseless received vectors that belong to a lattice, it is easy to enumerate them. Let us define the following set of points that creates a lattice

$$\Lambda = \{\mathbf{x} | \mathbf{x} = \mathbf{C}' \cdot \mathbf{H}'\}, \qquad (9.5)$$

and assume $N = M$ for the sake of simplicity. Then, the lattice Λ is defined by its generator matrix $\mathbf{H}'$. Since $\mathbf{C}'$ belongs to a cubic lattice in $\mathbb{Z}^{2N}$, the lattice Λ is a linear transform defined by the $2N \times 2N$ generator matrix $\mathbf{H}'$ in the $2N$-dimensional space $\mathbb{R}^{2N}$. Since the vector $\mathbf{r}'$ is a row vector, the Frobenius norm $||\mathbf{r}' - \mathbf{C}' \cdot \mathbf{H}'||_F$ is the same as the Euclidean norm $||\mathbf{r}' - \mathbf{C}' \cdot \mathbf{H}'||$. Therefore, the ML decoding is to solve the following minimization problem:

$$\min_{\mathbf{x} \in \Lambda} ||\mathbf{r} - \mathbf{x}|| = \min_{\mathbf{y} \in \mathbf{r} - \Lambda} ||\mathbf{y}||, \tag{9.6}$$

where $\mathbf{r} - \Lambda$ is the transformed lattice in the $2N$-dimensional space. Since the generator matrix $\mathbf{H}'$ is not singular, using its inverse, any vector in $\mathbb{R}^{2N}$ can be represented as the product of a $1 \times 2N$ vector and $\mathbf{H}'$. For example, $\mathbf{r} = \mathbf{a} \cdot \mathbf{H}'$ and $\mathbf{y} = \mathbf{b} \cdot \mathbf{H}'$, where $\mathbf{a} = \mathbf{r} \cdot (\mathbf{H}')^{-1}$ and $\mathbf{b} = \mathbf{y} \cdot (\mathbf{H}')^{-1}$ are $1 \times 2N$ vectors. Then, we have $\mathbf{b} = \mathbf{a} - \mathbf{C}'$ and $\mathbf{b}$ defines a new coordinate system centered at the received vector. The idea of sphere decoding is to consider the lattice points inside the sphere $||\mathbf{y}||^2 \leq D$, where $\sqrt{D}$ is the arbitrary radius of the sphere. Although the radius is an arbitrary number, it plays an important role in the performance and the speed of the sphere decoding algorithm as we discuss later. Using the new coordinate system defined by $\mathbf{b}$, such a sphere is transformed to an ellipsoid centered at the origin. Let us define the Gram matrix of $\mathbf{H}'$ by $G = \mathbf{H}' \cdot \mathbf{H}'^T$, then we have

$$||\mathbf{y}||^2 = \mathbf{b} \cdot \mathbf{H}' \cdot \mathbf{H}'^T \cdot \mathbf{b}^T = \mathbf{b} \cdot G \cdot \mathbf{b}^T \leq D. \tag{9.7}$$

The Gram matrix G is a positive definite matrix and therefore using Cholesky factorization it can be written as [62]

$$G = Q \cdot Q^T, \tag{9.8}$$

where Q is a $2N \times 2N$ real lower triangular matrix with non-negative diagonal entries. Note that a matrix Q is lower triangular if $Q_{ij} = 0$ whenever $i < j$ and

$$\mathbf{b} \cdot Q = \begin{pmatrix} \sum_{i=1}^{2N} \mathbf{b}_i Q_{i1} \\ \vdots \\ \mathbf{b}_{2N} Q_{2N,2N-1} + \mathbf{b}_{2N-1} Q_{2N-1,2N-1} \\ \mathbf{b}_{2N} Q_{2N,2N} \end{pmatrix}^T. \tag{9.9}$$

Then, replacing (9.8) in (9.7) results in

$$||\mathbf{y}||^2 = \mathbf{b} \cdot Q \cdot Q^T \cdot \mathbf{b}^T = ||\mathbf{b} \cdot Q||^2 \leq D. \tag{9.10}$$

Now, the problem is to find the limits of $\mathbf{b}_{2N}, \mathbf{b}_{2N-1}, \ldots, \mathbf{b}_1$ from (9.10). Then, using the limits of $\mathbf{b}_n$ and $\mathbf{b}_n = \mathbf{a}_n - \mathbf{C}'_n$, we find the limits of the possible integer values for $\mathbf{C}'_n$. To reduce the complexity of the search, we usually use the lower triangular nature of Q to sequentially and recursively find the limits of $\mathbf{b}_{2N}, \mathbf{b}_{2N-1}, \ldots, \mathbf{b}_1$.

For example, considering only the last column of Q in (9.10), $\mathbf{b}_{2N}$ is bounded by

$$\frac{-\sqrt{D}}{Q_{2N,2N}} \leq \mathbf{b}_{2N} \leq \frac{\sqrt{D}}{Q_{2N,2N}}. \tag{9.11}$$

Then, for $\mathbf{C}'_{2N}$ we have

$$\left\lceil \mathbf{a}_{2N} - \frac{\sqrt{D}}{Q_{2N,2N}} \right\rceil \leq \mathbf{C}'_{2N} \leq \left\lfloor \mathbf{a}_{2N} + \frac{\sqrt{D}}{Q_{2N,2N}} \right\rfloor, \tag{9.12}$$

where $\lceil a \rceil$ is the smallest integer greater than a and $\lfloor a \rfloor$ is the greatest integer smaller than a. Finding the limited number of indices for $\mathbf{C}'_{2N}$ and the corresponding possible values of $\mathbf{b}_{2N}$, we need to find the limits of the next symbol $\mathbf{C}'_{2N-1}$. Then, the limits of $\mathbf{C}'_{2N}$ and $\mathbf{C}'_{2N-1}$ are utilized to find the limits of $\mathbf{C}'_{2N-2}$. Sequentially, we can limit the possibilities for all symbols in the codeword. The main difference between different decoding algorithms is in the method that they limit the nth symbol using all available limits of the symbols from the previous steps. For example, to find the limits of $\mathbf{C}'_{2N-1}$ from those of $\mathbf{C}'_{2N}$, one can upper-bound the sum of the squares of the $2N - 1$st and $2N$th elements of (9.9) to satisfy the upper bound in (9.10). Using this method, we have

$$(\mathbf{b}_{2N} Q_{2N,2N-1} + \mathbf{b}_{2N-1} Q_{2N-1,2N-1})^2 + (\mathbf{b}_{2N} Q_{2N,2N})^2 \leq D. \tag{9.13}$$

Then, using the limits of $\mathbf{b}_{2N}$ from (9.11), we find the limits of $\mathbf{b}_{2N-1}$ from (9.13) and the corresponding limits for $\mathbf{C}'_{2N-1}$. The sequence continues recursively as we find the bounds of $\mathbf{C}'_n$ for all values of n. For example, using the last $2N - n + 1$ elements of (9.9) in (9.10) results in

$$\sum_{j=n}^{2N} \left(\sum_{i=j}^{2N} \mathbf{b}_i Q_{ij} \right)^2 \leq D. \tag{9.14}$$

The bounds for $\mathbf{b}_n$, and consequently $\mathbf{C}'_n$, are found using the above equation. Let us denote these bounds by L_n and U_n, that is

$$L_n \leq \mathbf{C}'_n \leq U_n, \quad n = 2N, 2N - 1, \ldots, 1. \tag{9.15}$$

Using (9.15), the sphere decoding algorithm calculates the limits of the possible indices inside the sphere for each received vector. Then, among all possible codewords inside the sphere, it finds the closest codeword to the received vector. Of course, the limits L_n and U_n for the nth element of the codeword depend on the choice of the radius $\sqrt{D}$. For a very small value of D, it is possible that the sphere does not contain any lattice point. In this case, either the decoder reports an error or it increases the value of D and repeats the process. On the other hand, a large value of D results in a huge number of possibilities and a slow decoding process. In practice, the value of D may be selected according to the SNR or may be adjusted adaptively. The complexity of finding the above limits is polynomial with the dimension of the lattice [41]. After finding the limits, instead of a full search ML decoding that requires an exponential complexity, we may only search within the calculated limits.

While the above equations show how to find the limits of each symbol to bound the number of possibilities, in practice we follow a sequence of bounds as explained above. For example, (9.12) provides a finite number of possibilities for $\mathbf{C}'_{2N}$. Then, we should try all possible values of $\mathbf{C}'_{2N}$ within the limits in a given order. The order that we pick may affect the complexity of the decoding. This is because of the fact that the limits of $\mathbf{C}'_{2N-1}$, derived from the limits of $\mathbf{b}_{2N-1}$, depend on the value of $\mathbf{C}'_{2N}$, equivalently the value of $\mathbf{b}_{2N}$, that we put in (9.13). We may try the limited values of $\mathbf{b}_{2N}$ in any order that we want. A good ordering strategy is to start from the middle point and use a sequence with a nondecreasing order from the middle point [107]. Defining $\lfloor a \rceil$ as the closest integer to a, the corresponding ordering for $\mathbf{C}'_{2N}$ in (9.12) could be

$$\lfloor \mathbf{a}_{2N} \rceil, \lfloor \mathbf{a}_{2N} \rceil + 1, \lfloor \mathbf{a}_{2N} \rceil - 1, \ldots \qquad (9.16)$$

With such an ordering the chance of finding the closest point in earlier recursions is higher. Also, at each stage, we may decrease the boundary of the search to the best point that has been found up to that stage. This will eliminate some of the choices and even further limit the number of search points. A complete presentation of this closest point search algorithm with a step-by-step implementation detail is provided in [2].

Before we finish our discussion on sphere decoding, we briefly mention a related decoding algorithm called the Babai nearest plane algorithm [6]. In this algorithm, we try to find only one candidate for the closest point by picking only the middle point at each stage. In other words, for $\mathbf{C}'_{2N}$, we select the first candidate in (9.16), that is $\mathbf{C}'_{2N} = \lfloor \mathbf{a}_{2N} \rceil$. Then, using the value of $\mathbf{C}'_{2N}$, and the corresponding value of $\mathbf{b}_{2N}$, we pick the middle point for $\mathbf{C}'_{2N-1}$ using the corresponding inequality for $\mathbf{C}'_{2N-1}$ derived from (9.13). The algorithm continues iteratively until we select all the elements of the codeword. In fact, the resulting lattice point is the first generated candidate in the last sphere decoding algorithm that we discussed, that is the algorithm using the ordering in [107].

9.4 Using equalization techniques in receiver design

One general approach to designing receivers with a complexity lower than that of ML decoding is to use equalization techniques to separate different symbols. In fact, this class of techniques first tries to find the best signal that represents each of the symbols and then decodes the symbol using the detected signal. In detecting the best representation of each symbol, the effects of other symbols are considered as interference. Therefore, the equalization ideas to remove intersymbol interference (ISI) can be used. A linear equalizer usually tries to separate symbols without enhancing the noise. In most cases, the two goals are contradictory. Therefore, the objective is to separate the symbols with a minimum enhancement of the noise. The two popular equalization methods that we consider in the design of a MIMO receiver are the zero-forcing (ZF) equalizer and the minimum mean-squared error (MMSE) equalizer.

9.4.1 Zero-forcing

An ISI channel may be modeled by an equivalent finite-impulse response (FIR) filter plus noise. A zero-forcing equalizer uses an inverse filter to compensate for the channel response function. In other words, at the output of the equalizer, it has an overall response function equal to one for the symbol that is being detected and an overall zero response for other symbols. If possible, this results in the removal of the interference from all other symbols in the absence of the noise. Zero forcing is a linear equalization method that does not consider the effects of noise. In fact, the noise may be enhanced in the process of eliminating the interference.

To show the main idea and its application to the system represented by (9.1), let us assume the case that $N = M$ and $\mathbf{H}$ is a full rank square matrix. In this case, the inverse of the channel matrix $\mathbf{H}$ exists and if we multiply both sides of (9.1) by $\mathbf{H}^{-1}$, we have

$$\mathbf{r} \cdot \mathbf{H}^{-1} = \mathbf{C} + \mathcal{N} \cdot \mathbf{H}^{-1}. \tag{9.17}$$

As can be seen from (9.17), the symbols are separated from each other. The noise is still Gaussian and the nth symbol can be decoded by finding the closest constellation point to the nth element of $\mathbf{r} \cdot \mathbf{H}^{-1}$. However, the power of the effective noise $\mathcal{N} \cdot \mathbf{H}^{-1}$ may be more than the power of the original noise $\mathcal{N}$. In fact, the covariance matrix of the effective noise for a given channel matrix $\mathbf{H} = H$ can be calculated as

$$E[(\mathcal{N} \cdot H^{-1})^H \cdot \mathcal{N} \cdot H^{-1}] = (H^{-1})^H \cdot E[\mathcal{N}^H \cdot \mathcal{N}] \cdot H^{-1}$$
$$= N_0(H \cdot H^H)^{-1}. \tag{9.18}$$

Clearly, the noise power may increase because of the factor $(H \cdot H^H)^{-1}$.

In general if the number of transmit and receive antennas are not the same, we may multiply (9.1) by the Moore–Penrose generalized inverse, pseudo-inverse

[62] of **H** to achieve a similar zero-forcing result. To define the Moore–Penrose generalized inverse, pseudo-inverse, of a matrix, we use the singular value decomposition theorem [62]. There are unitary matrices U and V, respectively $N \times N$ and $M \times M$, such that

$$\mathbf{H} = U \cdot \Sigma \cdot V^H, \tag{9.19}$$

where Σ is an $N \times M$ matrix with elements $\sigma_{j,j} = \sqrt{\lambda_j}$, $j = 1, 2, \dots, N$, where λ_j is the jth eigenvalue of $\mathbf{H} \cdot \mathbf{H}^H$. The columns of U are the eigenvectors of $\mathbf{H} \cdot \mathbf{H}^H$ while the columns of V are the eigenvectors of $\mathbf{H}^H \cdot \mathbf{H}$. Let us assume that r is the rank of $\mathbf{H}$, equivalently the rank of $\mathbf{H} \cdot \mathbf{H}^H$. Without loss of generality, we also assume that the eigenvalues are in a non-increasing order. Then, the Moore–Penrose generalized inverse, pseudo-inverse, of **H** is defined by [62]

$$\mathbf{H}^+ = V \Sigma^+ U^H, \tag{9.20}$$

where Σ^+ is the transpose of Σ in which $\sigma_{j,j}$, $j = 1, 2, \dots, r$ is replaced by $1/\sigma_{j,j}$. Note that if **H** is square and non-singular, we have $\mathbf{H}^+ = \mathbf{H}^{-1}$ and we achieve the same results as before. Also, if $M > N$ and **H** is full rank, we have $\mathbf{H}^+ = \mathbf{H}^H \cdot (\mathbf{H} \cdot \mathbf{H}^H)^{-1}$. Therefore, multiplying (9.1) by $\mathbf{H}^+$ results in

$$\mathbf{r} \cdot \mathbf{H}^+ = \mathbf{C} \cdot \mathbf{H} \cdot \mathbf{H}^H \cdot (\mathbf{H} \cdot \mathbf{H}^H)^{-1} + \mathcal{N} \cdot \mathbf{H}^+ = \mathbf{C} + \mathcal{N} \cdot \mathbf{H}^+. \tag{9.21}$$

Again, separate decoding of the symbols is possible by finding the closest constellation point to the nth element of $\mathbf{r} \cdot \mathbf{H}^+$. The noise enhancing factor for a given channel matrix $\mathbf{H} = H$ is

$$\begin{aligned}(H^+)^H \cdot H^+ &= [H^H \cdot (H \cdot H^H)^{-1}]^H \cdot [H^H \cdot (H \cdot H^H)^{-1}] \\ &= (H \cdot H^H)^{-1}.\end{aligned} \tag{9.22}$$

9.4.2 Minimum mean-squared error

As we discussed before, the ZF equalization does not consider the effects of the equalization in enhancing the noise. To address this problem, a mean-square error criterion is minimized. The goal of the linear MMSE equalizer is to multiply (9.1) by a matrix such that the resulting effective noise is minimized. Equivalently, MMSE equalizer maximizes the effective SNR.

Let us assume the following normalization

$$\begin{aligned}E[\mathbf{C}^H \cdot \mathbf{C}] &= \gamma I_N, \\ E[\mathcal{N}^H \cdot \mathcal{N}] &= I_M,\end{aligned} \tag{9.23}$$

where as usual γ is the received SNR. The received vector **r** in (9.1) is the noisy observation of the receiver for the input **C**. Using the MMSE criterion, the linear least-mean-squares estimate of $\mathbf{C}$ is

$$\mathbf{r} \cdot \mathbf{H}^H \cdot \left(\frac{I_N}{\gamma} + \mathbf{H} \cdot \mathbf{H}^H \right)^{-1}. \tag{9.24}$$

Unlike the ZF method, the received vector is multiplied by a matrix that is a function of SNR. When noise is negligible, that is $\gamma \to \infty$, the MMSE detection matrix $\mathbf{H}^H \cdot [(I_N/\gamma) + \mathbf{H} \cdot \mathbf{H}^H]^{-1}$ converges to $\mathbf{H}^+$ which is the detection matrix for the ZF method.

The above linear equalization methods are based on multiplying the received vector by a detection matrix and then decoding the symbols separately. Another equalization approach is decision feedback equalization (DFE) [100]. In the context of MIMO channels, DFE is usually combined with ZF or MMSE. Since such a combination for MIMO channels has been originally proposed as a receiver for the BLAST architectures, we discuss it in detail for BLAST in the next section.

9.5 V-BLAST

In this section, we study an architecture called vertical BLAST (V-BLAST) [155, 45]. The encoder of V-BLAST is depicted in Figure 9.1. First, the input bitstream is demultiplexed into N parallel substreams. Then, each substream is modulated and transmitted from the corresponding transmit antenna. It is also possible to use coding for each substream to improve the performance in a trade-off with the bandwidth [44]. Here, for the sake of brevity, we only consider uncoded substreams. Note that since each substream is treated separately, mathematically the structure of the interference of different substreams to each other is similar to that of a synchronized multi-user system. Therefore, the decoding methods that we discuss are also applicable to multi-user interference cancellation algorithms and vice versa. In fact, some of the methods were first introduced for multi-user detection; the interested reader is referred to [158, 64] for examples.

Representing the nth symbol of the codeword $\mathbf{C}$ by c_n and the nth row of $\mathbf{H}$ by H_n, (9.1) can be written as

$$\mathbf{r} = \sum_{n=1}^{N} c_n H_n + \mathcal{N}. \tag{9.25}$$

One approach to a lower complexity design of the receiver is to use a "divide-and-conquer" strategy instead of decoding all symbols jointly. First, the algorithm decodes the strongest symbol. Then, canceling the effects of this strongest symbol from all received signals, the algorithm detects the next strongest symbol. The algorithm continues by canceling the effects of the detected symbol and the decoding of the next strongest symbol until all symbols are detected. The optimal detection

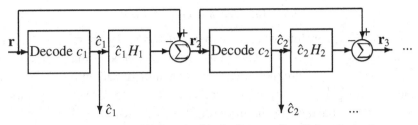

Fig. 9.3.   V-BLAST decoder block diagram (implicit ordering).

order is from the strongest symbol to the weakest one. This is the original decoding algorithm of V-BLAST presented in [155, 45]. It only works if the number of receive antennas is more than the number of transmit antennas, that is $M \geq N$. In this decoding algorithm, the decoder detects symbols one by one as in Figure 9.3. In decoding the first symbol, the interference from all other symbols are considered as noise. After finding the best candidate for the first symbol, the effects of this symbol in all of the receiver equations are canceled. Then, the second symbol is detected from the new sets of equations. The effects of the second detected symbol is canceled next to derive a new set of equations. The process continues until all symbols are detected. Of course, the order in which the symbols are detected will impact the final solution.

The algorithm includes three steps:

- ordering;
- interference cancellation;
- interference nulling.

Symbols are detected one by one. The purpose of the ordering step is to decide which transmitted symbol to detect at each stage of the decoding. The symbol with highest SNR is the best pick in this step. The goal of the interference cancelation is to remove the interference from the already detected symbols in decoding the next symbol. Finally, interference nulling finds the best estimate of a symbol from the updated equations. This step is called interference nulling since it can be considered as removing the interference effects of undetected symbols from the one that is being decoded. In what follows, we describe these steps in detail. Without loss of generality, we assume that $c_1, c_2, \ldots, c_N$ are in the optimal order. Note that changing the order of the symbols in codeword $\mathbf{C}$, that is $c_1, c_2, \ldots, c_N$, will not change (9.25) if the order of the rows of channel matrix $\mathbf{H}$, that is $H_1, H_2, \ldots, H_N$, changes accordingly. If the interference cancelation step is removed, the algorithm is modified to one of the equalization methods in the last section. In this case, the codeword is detected using ZF or MMSE for interference nulling. If, for example, we use ZF for interference nulling, its combination with interference cancelation is called ZF-DFE. Then, ZF-DFE usually outperforms ZF as we discuss later.

9.5.1 Interference cancelation

At stage n of the algorithm, when c_n is being detected, symbols $c_1, c_2, \ldots, c_{n-1}$ have been already detected. Let us assume a perfect decoder, that is the decoded symbols $\hat{c}_1, \hat{c}_2, \ldots, \hat{c}_{n-1}$ are the same as the transmitted symbols $c_1, c_2, \ldots, c_{n-1}$. One can subtract $\sum_{i=1}^{n-1} c_i H_i$ from the received vector $\mathbf{r}$ to derive an equation that relates remaining undetected symbols to the received vector:

$$\mathbf{r}_n = \mathbf{r} - \sum_{i=1}^{n-1} c_i H_i = \sum_{i=n}^{N} c_i H_i + \mathcal{N}, \quad n = 2, 3, \ldots, N - 1. \qquad (9.26)$$

In fact, by using induction in addition to the convention $\mathbf{r}_1 = \mathbf{r}$, one can show that

$$\mathbf{r}_{n+1} = \mathbf{r}_n - c_n H_n, \quad n = 1, 2, \ldots, N - 2. \qquad (9.27)$$

Therefore, at the nth stage of the algorithm after detecting the nth symbol as $\hat{c}_n$, its effect is canceled from the equations by

$$\mathbf{r}_{n+1} = \mathbf{r}_n - \hat{c}_n H_n. \qquad (9.28)$$

This interference cancelation is conceptually similar to DFE.

9.5.2 Interference nulling

Interference nulling is the process of detecting c_n from $\mathbf{r}_n$ by first removing the effects of undetected symbols. Basically, in this step the nth symbol is detected by nulling the interference caused by symbols $c_{n+1}, c_{n+2}, \ldots, c_N$. Like any other interference suppression problem, there are many different methods to detect a symbol in the presence of interference. Two examples of these methods are MMSE and ZF. We describe these two methods separately.

9.5.2.1 Zero-forcing nulling

Using zero-forcing for interference nulling is common in practice. First, let us assume perfect detection of symbols as in (9.26). We would like to separate the term $c_n H_n$ from $\mathbf{r}_n$. This can be done through multiplying $\mathbf{r}_n$ by an $M \times 1$ vector W_n that is orthogonal to interference vectors $H_{n+1}, H_{n+2}, \ldots, H_N$ but not orthogonal to H_n. In other words, W_n should be such that

$$\begin{aligned}
H_i \cdot W_n &= 0, \quad i = n + 1, n + 2, \ldots, N, \\
H_n \cdot W_n &= 1.
\end{aligned} \qquad (9.29)$$

In fact, W_n is called the zero-forcing nulling vector with the minimum norm. Such a vector is uniquely calculated from the channel matrix $\mathbf{H}$. To calculate W_n from $\mathbf{H}$, for $M \geq N$, first we should replace the rows $1, 2, \ldots, n - 1$ of $\mathbf{H}$ by zero.

Let us denote the resulting matrix by Z. Then, W_n is the nth column of Z^+, the Moore–Penrose generalized inverse, pseudo-inverse, of Z.

Using the error-free detection formula for $\mathbf{r}_n$ in (9.26) and W_n in (9.29), we have

$$\mathbf{r}_n \cdot W_n = c_n + \mathcal{N} \cdot W_n. \tag{9.30}$$

The noise in (9.30) is still Gaussian and the symbol c_n can be easily decoded. The decoded symbol $\hat{c}_n$ is the closest constellation point to $\mathbf{r}_n \cdot W_n$. The noise enhancing factor using (9.30) is

$$\begin{aligned} E[(\mathcal{N} \cdot W_n)^H \cdot \mathcal{N} \cdot W_n] &= W_n^H \cdot E[\mathcal{N}^H \cdot \mathcal{N}] \cdot W_n \\ &= N_0 ||W_n||^2. \end{aligned} \tag{9.31}$$

Comparing (9.31) with (9.22) demonstrates why adding an interference cancelation step improves the performance. Using the combination of canceling and nulling in a ZF-DFE structure enhances the noise by a factor of $||W_n||^2$. Vector W_n is orthogonal to $N - n$ rows of the channel matrix $\mathbf{H}$. On the other hand, using a pure interference nulling method like ZF, the corresponding vector that detects the nth symbol, the nth column of the pseudo-inverse, is orthogonal to $N - 1$ rows of the channel matrix $\mathbf{H}$. Using the Cauchy–Schwartz inequality, it can be shown that the norm of a vector is larger if it has to be orthogonal to a greater number of rows. Therefore, the enhancing factor for the case of nulling alone, ZF, is more than that of the canceling and nulling, ZF-DFE. For the first vector, $n = 1$, the two cases are identical.

Another practical way of nulling the interference is to remove the projection of $\mathbf{r}_n$ on the subspace created by $H_{n+1}, H_{n+2}, \ldots, H_N$. Let us assume that the set $\{A_{n+1}, A_{n+2}, \ldots, A_N\}$ is an orthonormal basis for the subspace created from $H_{n+1}, H_{n+2}, \ldots, H_N$. Then, one can subtract the projection of $\mathbf{r}_n$ on each of these vectors to remove the interference caused by undetected symbols. Note that an orthonormal basis can be created from $H_{n+1}, H_{n+2}, \ldots, H_N$ using the Gram–Schmidt orthonormalization process [62]. To describe the details of the Gram–Schmidt process, let us define the inner product of two $1 \times M$ vectors x and y by

$$< x, y > = x \cdot y^H. \tag{9.32}$$

Then, to find the first vector of the orthonormal basis, we have

$$A_{n+1} = \frac{H_{n+1}}{< H_{n+1}, H_{n+1} >^{1/2}}. \tag{9.33}$$

To find the second vector, we calculate the projection of H_{n+2} on A_{n+1} and subtract it from H_{n+2} as

$$x_2 = H_{n+2} - < H_{n+2}, A_{n+1} > A_{n+1}. \tag{9.34}$$

The vector x_2 is orthogonal to A_{n+1} and normalizing x_2 provides the second vector of the orthonormal basis

$$A_{n+2} = \frac{x_2}{< x_2, x_2 >^{1/2}}. \tag{9.35}$$

The process continues in a similar fashion. The ith vector A_{n+i} is derived first by calculating x_i as

$$x_i = H_{n+i} - < H_{n+i}, A_{n+i-1} > A_{n+i-1} - < H_{n+i}, A_{n+i-2} > A_{n+i-2} \\ - \ldots - < H_{n+i}, A_1 > A_1. \tag{9.36}$$

Then, A_{n+i} is the normalized version of x_i as

$$A_{n+i} = \frac{x_i}{< x_i, x_i >^{1/2}}. \tag{9.37}$$

To use the orthonormal basis in removing the interference, we calculate the interference free vector u_n by

$$u_n = \mathbf{r}_n - < \mathbf{r}_n, A_{n+1} > A_{n+1} - < \mathbf{r}_n, A_{n+2} > A_{n+2} \\ - \ldots - < \mathbf{r}_n, A_N > A_N. \tag{9.38}$$

The vector u_n is utilized to detect the nth symbol c_n. Each of the M elements of u_n consists of a noise component and a known multiple of c_n. Therefore, a standard maximum ratio combining algorithm can be utilized to decode the symbol c_n [45].

9.5.2.2 Minimum mean-squared error nulling

Another approach for interference nulling is MMSE. Let us assume that the transmitted vector is a zero-mean random vector that is uncorrelated to the noise. Considering the received vector $\mathbf{r}$ in (9.1) as a noisy observation of the input $\mathbf{C}$, the linear least-mean-squares estimator of $\mathbf{C}$ is

$$\mathbf{M} = \mathbf{H}^H \cdot \left(\frac{I_N}{\gamma} + \mathbf{H} \cdot \mathbf{H}^H \right)^{-1}, \tag{9.39}$$

as we discussed before.

Note that in the nth stage of the algorithm, the effects of $c_1, c_2, \ldots, c_{n-1}$ have been canceled. Therefore, similar to the ZF nulling, to calculate c_n, first we should replace the rows $1, 2, \ldots, n-1$ of $\mathbf{H}$ by zero. Let us denote the resulting matrix by Z as we did in the ZF case. Now, to find the best estimate of the nth symbol, that is $\hat{c}_n$, we replace $\mathbf{H}$ with Z in (9.39) to calculate the best linear MMSE estimator at stage n as

$$\mathbf{M} = Z^H \cdot \left(\frac{I_N}{\gamma} + Z \cdot Z^H \right)^{-1}. \tag{9.40}$$

Then, the nth column of $\mathbf{M}$, denoted by $\mathbf{M}_n$ is utilized as the MMSE nulling vector for the nth symbol. In other words, the decoded symbol $\hat{c}_n$ is the closest constellation point to $\mathbf{r}_n \cdot \mathbf{M}_n$. A lower complexity implementation of the MMSE nulling is presented in [54].

9.5.3 Optimal ordering

For the above V-BLAST algorithm to work, the receiver needs at least N receive antennas, that is $M \geq N$. At each stage of the algorithm, we detect one of the symbols using interference nulling. In fact, in the nth stage, when we detect the nth symbol, there are $N - n$ rows of $\mathbf{H}$ that are utilized to construct the nulling vector, for example, based on (9.29). The extra $(M - N + n)$ receive antennas can provide diversity gain. Therefore, the diversity for the nth transmit symbol, $n = 1, 2, \ldots, N$, is $(M - N + n)$, which is an increasing function of n. For the last symbol, the diversity gain is maximum and is equal to M. As a result, the receiver should decode a more reliable symbol earlier than a less reliable one.

Let us concentrate on the case of ZF nulling for the sake of brevity. The noise enhancing factor for the nth symbol is $||W_n||^2$ as we calculated in (9.31). Therefore, the order in which we detect the symbols will effect the noise enhancing factors. Since each symbol has its own enhancing factor we need to define a global criterion to be optimized for the overall system. A popular criterion is to optimize the worst-case scenario. In other words, we would like to minimize the maximum enhancing factor or equivalently maximize the minimum SNR. To solve such a problem, one needs to consider all possible ordering of symbols and find the maximum enhancing factor in each case. Then, the ordering set that includes the lowest maximum enhancing factor is the best solution for the given criterion. Obviously, the complexity of such a method is very high. However, an equivalent alternative is the following method. At each stage, among all remaining symbols, we pick the symbol that provides the lowest enhancing factor, maximum SNR, for that stage. It has been proved in [155, 45] that the sequence of these locally optimized symbols is identical to the globally optimized sequence for the given criterion. Of course, the complexity of this sequential ordering is much lower than that of its equivalent global ordering.

9.6 D-BLAST

In this section, we study the transmitter and receiver of diagonal BLAST (D-BLAST) [43, 44]. The encoder of D-BLAST is very similar to that of V-BLAST as depicted in Figure 9.4. The main difference is in the way that the signals are transmitted from different antennas. In V-BLAST, all signals in each layer are

Table 9.1. *Layering in D-BLAST*

	Antenna 1	Antenna 2	Antenna 3	Antenna 4
Block 1	Layer 1			
Block 2	Layer 4	Layer 1		
Block 3	Layer 3	Layer 4	Layer 1	
Block 4	Layer 2	Layer 3	Layer 4	Layer 1
Block 5	Layer 1	Layer 2	Layer 3	Layer 4
Block 6	Layer 4	Layer 1	Layer 2	Layer 3
Block 7	Layer 3	Layer 4	Layer 1	Layer 2
Block 8	Layer 2	Layer 3	Layer 4	Layer 1
$\vdots$	$\vdots$	$\vdots$	$\vdots$	$\vdots$
Block $T-2$		Layer 2	Layer 3	Layer 4
Block $T-1$			Layer 2	Layer 3
Block T				Layer 2

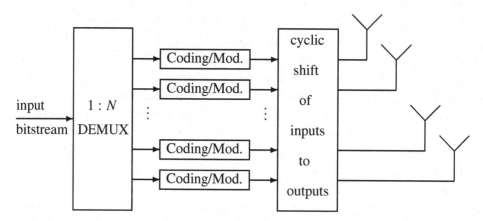

Fig. 9.4. D-BLAST encoder.

transmitted from the same antenna. However, in D-BLAST, the signals are shifted before transmission. Let us enumerate the output of the demultiplexing as layers 1 to N. For the first block of T time slots, Layer n is sent by the nth transmit antenna. Note that the block size T can be as small as $T = 1$. In the second block, Layer n is transmitted by the $(n + 1)$th antenna, for $n = 1, 2, \ldots, N - 1$, while Layer N is transmitted by the first antenna. The shift continues as a cyclic shift after every block of T time slots. Table 9.1 demonstrates how different layers are transmitted diagonally because of the cyclic shift for a system with $N = 4$ transmit antennas. As shown, in the first block of T time slots, we only transmit Layer 1 from the first antenna. In the second block, Layers 1 and 4 are transmitted from the second and first antennas, respectively. Cyclic shifting is the reason for transmitting Layer 1 from the second antenna instead of the first antenna in the second block. The transmission of different layers continues as tabulated. Note that during the first $N - 1 = 3$ blocks, similarly the last three blocks, we do not transmit any data to simplify the decoding. This is called the boundary effect and reduces the throughput of the system a little bit.

The main reason for transmitting layers from different antennas is to avoid deep fades through diversity. Let us assume that the path gains from one of the transmit antennas are in deep fade. In this case, without the cyclic shift, all symbols of the corresponding layer will be affected. However, using the cyclic shift, only one out of N blocks of each layer is affected by the deep fade. Therefore, it will be easier to overcome the fading through the transmit diversity. In fact, the role of this cyclic shifting in combating the fading is similar to the job of interleaving to overcome burst errors.

The receiver of a D-BLAST architecture is similar to that of a V-BLAST system although the shifting creates a higher complexity. To decode a symbol, the main steps are interference nulling and interference cancelation to detect the best estimate of the symbol. Layers are detected one by one following the diagonal pattern of the transmitter. There is a boundary effect that may result in some waste of bandwidth. Let us assume $N = 4$ transmit antennas as in Table 9.1 to explain the decoding. First, we should detect the best signals to decode the symbols in Layer 1 during the first $N = 4$ blocks. Since we only transmit Layer 1 during the first block, there is no interference and a maximum ratio combining provides the best candidate for decoding. During the second block for Layer 1, there is interference from Layer 4. Considering the interference from Layer 4 as noise, the receiver uses ZF or MMSE nulling to remove its effect to the symbols in Layer 1. Similarly, the interference of Layers 3 and 4 to Layer 1 are nulled during the third block. Finally, during the fourth block, the interference of Layers 2, 3, and 4 is removed from Layer 1. We call the outcome of the interference nulling process of this first diagonal the detected Layer 1. Then, the detected Layer 1 is utilized to decode the corresponding symbols. The decoded symbols of Layer 1 are used to cancel their interference on the undetected layers. After canceling the interference of the decoded Layer 1, a similar method is applied to detect Layer 4 in the second diagonal starting from Block 2. A combination of interference nulling and cancelation is sequentially applied to every diagonal to detect all layers. At the time of detection of each layer, since the effects of all layers in the upper right side are canceled, a similar process is applied. The details of interference nulling and canceling is the same as that of the V-BLAST in Section 9.5. However, there is no ordering in the decoding of D-BLAST. The diagonal nature of the encoding results in a natural order that cannot be changed easily.

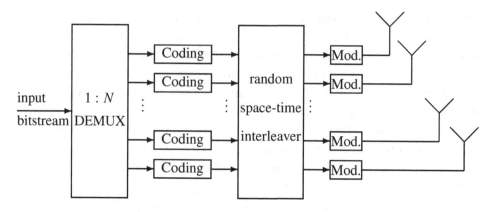

Fig. 9.5. Turbo-BLAST encoder.

9.7 Turbo-BLAST

So far, most of our discussion on BLAST architectures has not included the coding of different layers. One may add coding to the modulation of each layer. Also, it is possible to combine coding and modulation using a TCM for each layer. The main goal of the coding, when applied to different layers, is to improve the performance of the system. Different coding methods provide different coding gains resulting in distinctive performance. On the other hand, using coding creates additional decoding complexity and delay. The decoding complexity and the delay of different coding methods are also different. Usually, there is a trade-off between the performance and the decoding complexity of the coding algorithms. One class of codes that provides performances very close to the Shannon theoretical limits for AWGN channels is turbo code [11]. Usually, the encoder of a turbo code consists of an interleaver. The goal of the interleaver is the creation of "random codewords" to achieve the capacity as promised by the proof of the Shannon channel coding theorem [27]. On the other hand, the receiver is usually an iterative decoder. In this section, we study the combination of a BLAST architecture with that of a turbo code to improve its performance. Such a combination is called turbo-BLAST [108, 109].

Figure 9.5 shows the block diagram of the turbo-BLAST encoder. First, the input bit stream is demultiplexed into N equal rate substreams. Then, each substream is independently encoded using a linear block code. Usually, all bitstreams use the same block code with a given rate. The output bits of the block codes are bit-interleaved using a random interleaver. The interleaver is independent of the input bitstream and is designed offline. It is important that each substream uses different paths equally. One example of such an interleaver is a diagonal space interleaver followed by random time interleavers for the bitstream of each antenna. The diagonal space interleaver is depicted in Table 9.2. The main difference between

Table 9.2. *Diagonal interleaving*

	Antenna 1	Antenna 2	Antenna 3	Antenna 4
Block 1	Layer 1	Layer 2	Layer 3	Layer 4
Block 2	Layer 4	Layer 1	Layer 2	Layer 3
Block 3	Layer 3	Layer 4	Layer 1	Layer 2
Block 4	Layer 2	Layer 3	Layer 4	Layer 1
Block 5	Layer 1	Layer 2	Layer 3	Layer 4
Block 6	Layer 4	Layer 1	Layer 2	Layer 3
Block 7	Layer 3	Layer 4	Layer 1	Layer 2
Block 8	Layer 2	Layer 3	Layer 4	Layer 1
⋮	⋮	⋮	⋮	⋮

the diagonal space interleaving for turbo-BLAST in Table 9.2 and the diagonal layering for D-BLAST in Table 9.1 is in the boundaries. The original diagonal layering of D-BLAST suffers from boundary effects that result in some throughput waste as we discussed before. Finally, the output of the space-time interleaver is modulated separately for each antenna and is transmitted.

The combination of the independent block coding of the substreams and the random space-time interleaving is equivalent to a random layered space-time (RLST) architecture. The effects of the diagonal interleaving of Table 9.2 on the quasi-static fading channel is to create an "effective" time-varying channel. Using this effective time-varying channel, the encoder can be represented as serially concatenated turbo code. Figure 9.6 depicts such a representation of the encoder. The "outer code" consists of block encoders for different substreams. The "inner code" is the effective channel resulting from diagonal interleaving. Similar to a serially concatenated turbo code, the inner and outer codes are separated by interleavers. This representation allows the use of iterative decoding as we discuss next.

The complexity of an optimal decoder for turbo codes grows exponentially with N, constellation size, and the block size. Therefore, it is not practical to use an optimal decoder because of its high complexity. Instead, similar to serially concatenated turbo codes, we can use an iterative decoding algorithm for turbo-BLAST as suggested in [108, 109]. The main idea behind an iterative decoding algorithm is first to divide the problem into two stages. Then, we solve the problem for each stage optimally and iteratively while exchanging the information between the two stages. Of course, the resulting decoder is not optimal; however, it provides a very good performance with a manageable complexity. The two stages of iterative decoding for turbo-BLAST are the decoding of the inner code and the decoding of the

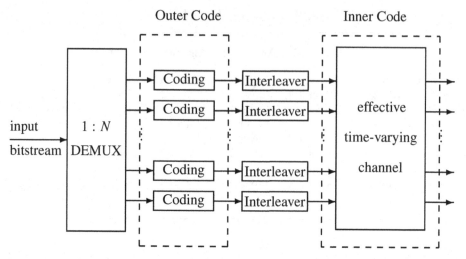

Fig. 9.6.   Turbo-BLAST encoder represented by serially concatenated codes.

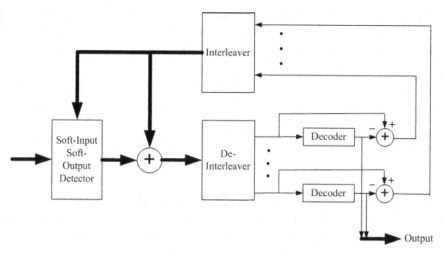

Fig. 9.7.   Turbo-BLAST decoder block diagram.

outer block code. The former is also called "channel detector" since it consists of a soft-input soft-output detector to address the ISI created by diagonal interleaving and fading. The inner decoder and outer decoder are separated by space-time interleavers and de-interleavers similar to a regular turbo decoder. The role of the space-time interleavers and de-interleavers is to reverse the effects of the space-time interleaving at the encoder. This is done such that the input of each decoder is properly aligned with the output of the corresponding encoder. The block diagram of the above iterative decoder is depicted in Figure 9.7.

9.8 Combined spatial multiplexing and space-time coding

So far in this chapter, we have studied space-time architectures that transmit up to one symbol per transmit antenna per time slot. This is equal to N symbols per time slot for a system with N transmit antennas. Consequently, using spatial multiplexing results in a huge increase in throughput. Most decoding algorithms with reasonable complexity only work when the number of receive antennas is at least as many as the number of transmit antennas. Another drawback of these systems is that they do not achieve the maximum diversity of MN for N transmit and M receive antennas. In fact, for some symbols they only achieve a diversity of M. On the other hand, space-time coding is designed to achieve the maximum diversity with very low complexity ML decoding. The drawback of space-time coding compared to BLAST architectures is its maximum possible transmission rate. As shown in Chapter 4, to achieve full diversity, one cannot transmit more than one symbol per time slot. Therefore, the number of transmitted symbols for BLAST architectures may be up to N times that of space-time coding systems. This extra throughput is achieved in a trade-off with BLAST's lower diversity and much higher decoding complexity in addition to the need for higher number of receive antennas.

In this section, we study systems that provide a trade-off between the number of transmitted symbols and the achieved diversity [137]. The minimum required number of receive antennas for these systems can be less than the number of transmit antennas.

9.8.1 General framework

In this section, we present a general framework for combining spatial multiplexing and space-time coding. Let us divide the N transmit antennas to J groups of $N_1, N_2, \ldots, N_J$ transmit antennas such that $N_1 + N_2 + \cdots + N_J = N$. Then, we utilize a space-time code to achieve the maximum diversity of each group. If a rate-one code exists for each group, the total throughput of the system will be J symbols per time slot. Let us denote these J space-time codes by $C_1, C_2, \ldots, C_J$. At each time slot, the encoder separates the B input bits into J groups of $B_1, B_2, \ldots, B_J$ bits. The jth substream is used as the input of the jth space-time code C_j. The output of the jth space-time code C_j is transmitted from the N_j antennas in group j as depicted in Figure 9.8. The resulting code is called a product space-time code in [137]. All transmissions are done simultaneously, therefore a sum of their faded versions plus Gaussian noise is received at each receive antenna. Then, the receiver should decode each symbol while keeping the effects of the other simultaneously transmitted symbols as small as possible. So far, since there is no collaboration among different space-time codes, the mathematical model of the system is

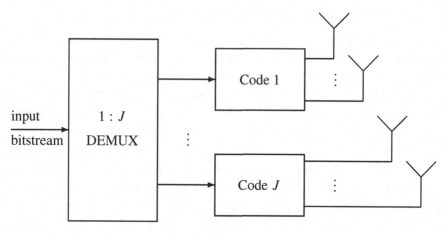

Fig. 9.8. Combined spatial multiplexing and space-time coding.

equivalent to that of a multi-user system. In other words, the J groups may be considered as J different users transmitting J independent data streams. Therefore, the problem of decoding is similar to that of interference suppression as discussed in [20]. One can use the existing interference-suppression methods at the receiver. On the other hand, the schemes that we develop in this section can be utilized in a multi-user system. While we mostly present the system as a space-time architecture, it is described as a synchronous multi-user system in many related literature [94].

The main idea behind the decoding is similar to the idea of V-BLAST decoding that was discussed in Section 9.5. In fact, the V-BLAST architecture can be considered as a special case of the above product space-time code when every group consists of only one transmit antenna, that is $J = N$ and $N_1 = N_2 = \cdots = N_N = 1$. Therefore, the decoder in this section can be considered as the generalization of the decoder of V-BLAST.

The decoding starts with nulling or suppressing the effects of the symbols from space-time codes $C_2, C_3, \ldots, C_J$ to the symbols from C_1. Then, a space-time decoder for C_1 is utilized to decode the first group of symbols and consequently the first B_1 bits. In the next stage, the effects of the decoded symbols of C_1 are canceled from the remaining symbols. To detect the symbols from C_2, first, the suppression of the symbols from $C_3, \ldots, C_J$ is done. Then, using a space-time decoder for C_2, the second group of B_2 bits are decoded. The decoding process continues as we decode the bits of different groups one after the other. Of course, the details of interference nulling and cancelation steps depend on the structure of the particular space-time codes $C_1, C_2, \ldots, C_J$. However, the general principles behind them are similar for any choice of space-time codes. In general, the receiver needs at least

$N - N_1 + 1$ receive antennas, that is $M \geq N - N_1 + 1$. Then, the diversity gain for the first group is $N_1(M - N + N_1)$ as proved in [137]. The diversity for the jth group, $j = 1, 2, \ldots, J$, is $N_j(M - N + \sum_{i=1}^{j} N_i)$. Note that since $N = \sum_{j=1}^{J} N_j$, a diversity gain of $N_J M$ is provided for the last group of bits. If all groups contain the same number of antennas, the sequence of diversity gains is an increasing sequence. Therefore, the receiver should decode a more reliable group earlier than a less reliable group. In fact, since the diversity gain of different component codes is different, the transmitter may allocate its power unequally among different groups. Such an unequal power allocation improves the overall performance of the system as suggested in [137]. In fact, denoting the average transmitted power from antenna n by E_n, one can optimize these power values given a constraint on the total power [78].

9.8.2 General decoding

In this section, we explain the details of the decoding for the above general framework. Similar to the case of BLAST architecture, first we separate the temporal processing from spatial processing. For each group, we detect the best signals after canceling the interference from other symbols. Then, we decode the transmitted bits, through temporal processing, using the decoder of the corresponding space-time code. Let us concentrate on the spatial processing first. At each time slot, the $1 \times N$ transmitted vector $\mathbf{C}$ in (9.1) can be divided into J parts, one for each group. We denote the resulting J vectors by $\mathbf{C}_1, \mathbf{C}_2, \ldots, \mathbf{C}_J$, where $\mathbf{C}_j$ is a $1 \times N_j$ codeword that is transmitted from the N_j antennas of group j. Representing the first N_1 rows of $\mathbf{H}$ by $\mathbf{H}_1$, the next N_2 rows by $\mathbf{H}_2$ and so on, we divide $\mathbf{H}$ into J parts. The channel matrix $\mathbf{H}_j$ is an $N_j \times M$ matrix of path gains from the transmit antennas in group j to M receive antennas. Therefore, the input–output relationship in (9.1) can be rewritten as

$$\mathbf{r} = \sum_{j=1}^{J} \mathbf{C}_j \cdot \mathbf{H}_j + \mathcal{N}. \qquad (9.41)$$

Mathematically, (9.41) also models a synchronous multi-user system in which each group is a separate user. In other words, $\mathbf{C}_j$ and $\mathbf{H}_j$ are the transmitted codeword and the channel matrix for user j, respectively. Therefore, any decoding method for the proposed space-time architecture is also applicable for interference suppression in a synchronous multi-user system. Also, note that (9.25) is a special case of (9.41) when every group contains only one transmit antenna. Therefore, a generalization of the 'divide-and-conquer' strategy proposed in Section 9.5 can be utilized to detect the codewords in (9.41). The main steps are interference cancelation and interference nulling. At stage j of the algorithm $\mathbf{C}_j$ is being detected, while codewords $\mathbf{C}_1, \mathbf{C}_2, \ldots, \mathbf{C}_{j-1}$ have been already detected. Subtracting $\sum_{i=1}^{j-1} \mathbf{C}_i \cdot \mathbf{H}_i$ from the received vector $\mathbf{r}$ derives an equation that relates undetected codewords to the received vector by

$$\mathbf{r}_j = \mathbf{r} - \sum_{i=1}^{j-1} \mathbf{C}_i \cdot \mathbf{H}_i = \sum_{i=j}^{J} \mathbf{C}_i \cdot \mathbf{H}_i + \mathcal{N}, \quad j = 2, 3, \ldots, J-1. \quad (9.42)$$

As a result, we have

$$\mathbf{r}_{j+1} = \mathbf{r}_j - \mathbf{C}_j \cdot \mathbf{H}_j, \quad j = 1, 2, \ldots, J-2, \quad (9.43)$$

where $\mathbf{r}_1 = \mathbf{r}$. Therefore, at the jth stage of the algorithm after detecting the jth codeword as $\hat{\mathbf{C}}_j$, its effect is canceled from the received vector in a DFE fashion by

$$\mathbf{r}_{j+1} = \mathbf{r}_j - \hat{\mathbf{C}}_j \mathbf{H}_j. \quad (9.44)$$

Then, to remove the effects of undetected codewords through interference nulling, which is called interference suppression for multi-user systems, we may use zero-forcing or MMSE. The main ideas are very similar to those of the interference nulling methods for V-BLAST in Section 9.5. Here, we deal with different groups consisting of space-time codewords instead of symbols. Therefore, the interference suppression is sometimes called group interference suppression. For example, to separate the term $\mathbf{C}_1 \cdot \mathbf{H}_1$ from $\mathbf{r} = \mathbf{r}_1$ using zero-forcing, we find an $M \times (M - N + N_1)$ matrix $\mathbf{W}_1$ such that

$$\mathbf{H}_i \cdot \mathbf{W}_1 = 0, \quad i = 2, 3, \ldots, J. \quad (9.45)$$

Then, $\mathbf{C}_1$ is detected by calculating $\mathbf{r} \cdot \mathbf{W}_1$ as follows:

$$\mathbf{r} \cdot \mathbf{W}_1 = \mathbf{C}_1 \cdot \mathbf{H}_1' + \mathcal{N} \cdot \mathbf{W}_1. \quad (9.46)$$

Decoding $\mathbf{C}_1$ is easily done using the decoder of the first space-time code assuming $\mathbf{r} \cdot \mathbf{W}_1$ and $\mathbf{H}_1' = \mathbf{H}_1 \cdot \mathbf{W}_1$ as the received vector and the channel matrix, respectively. The next stage is to remove the effects of $\mathbf{C}_1$ from other codewords using (9.44). In general, $\mathbf{W}_j$ is an $M \times (M - N + \sum_{i=1}^{j} N_i)$ matrix such that

$$\mathbf{H}_i \cdot \mathbf{W}_j = 0, \quad i = j+1, j+2, \ldots, J, \quad (9.47)$$

and we use

$$\mathbf{r}_j \cdot \mathbf{W}_j = \mathbf{C}_j \cdot \mathbf{H}_j' + \mathcal{N} \cdot \mathbf{W}_j, \quad (9.48)$$

to detect $\mathbf{C}_j$. There are many methods to calculate $\mathbf{W}_j$ as explained in [137].

9.8.3 Joint array processing and space-time block coding

So far, among different groups, we have used spatial processing and temporal processing separately to decode the codewords of different space-time codes. In the special case of orthogonal STBCs, one can use the orthogonality of the code to combine spatial processing and temporal processing among different groups. As a result, the number of required receive antennas is less than $N - N_1 + 1$. For example, when all groups have the same number of antennas, that is $N_1 = N_2 = \cdots = N_J$, interference suppression can be done by only J receive antennas. This is possible by combining spatial and temporal processing in addition to utilizing the structure of orthogonal STBCs. For the sake of simplicity, we show how to modify the decoding algorithm in the case of orthogonal STBCs through specific examples [94].

Example 9.8.1 *Let us start with an architecture that uses 2×2 Alamouti STBCs as building blocks. For example, we concentrate on a system that includes four transmit antennas in two groups of two antennas and two receive antennas. In other words, we have $N = 4$, $J = 2$, $N_1 = N_2 = 2$, and $M = 2$. Since the space-time coding building block is an orthogonal STBC, the interference suppression needs at least $J = 2$ receive antennas instead of at least $N - N_1 + 1 = 3$ antennas.*

Considering the channel as a MIMO channel with four transmit antennas and two receive antennas, the input–output relationship is

$$\mathbf{r} = \mathbf{C} \cdot \mathbf{H} + \mathcal{N}, \tag{9.49}$$

where $\mathbf{C}$ is a 4×2 codeword transmitting four symbols in two time slots as

$$\mathbf{C} = \begin{pmatrix} s_1 & s_2 & s_3 & s_4 \\ -s_2^* & s_1^* & -s_4^* & s_3^* \end{pmatrix}, \tag{9.50}$$

where s_1, s_2 are the symbols of the first group and s_3, s_4 are the symbols of the second group. Similar to (4.17), we may rearrange the equations to write the following matrix equation.

$$(r_{1,1}, r_{2,1}^*, r_{1,2}, r_{2,2}^*) = (s_1, s_2, s_3, s_4) \cdot \Omega + (\eta_{1,1}, \eta_{2,1}^*, \eta_{1,2}, \eta_{2,2}^*), \tag{9.51}$$

where

$$\Omega = \begin{pmatrix} \alpha_{1,1} & \alpha_{2,1}^* & \alpha_{1,2} & \alpha_{2,2}^* \\ \alpha_{2,1} & -\alpha_{1,1}^* & \alpha_{2,2} & -\alpha_{1,2}^* \\ \alpha_{3,1} & \alpha_{4,1}^* & \alpha_{3,2} & \alpha_{4,2}^* \\ \alpha_{4,1} & -\alpha_{3,1}^* & \alpha_{4,2} & -\alpha_{3,2}^* \end{pmatrix} = \begin{pmatrix} \Omega_{1,1} & \Omega_{1,2} \\ \Omega_{2,1} & \Omega_{2,2} \end{pmatrix}. \tag{9.52}$$

Note that $\Omega_{1,1}$, $\Omega_{1,2}$, $\Omega_{2,1}$, and $\Omega_{2,2}$ all have the same format as Ω in (4.17), that is the format of (4.18). Separating the received signals at each receive antenna *results in*

$$\begin{cases} (r_{1,1}, \quad r_{2,1}^*) = (s_1, \quad s_2) \cdot \Omega_{1,1} + (s_3, \quad s_4) \cdot \Omega_{2,1} + (\eta_{1,1}, \quad \eta_{2,1}^*) \\ (r_{1,2}, \quad r_{2,2}^*) = (s_1, \quad s_2) \cdot \Omega_{1,2} + (s_3, \quad s_4) \cdot \Omega_{2,2} + (\eta_{1,2}, \quad \eta_{2,2}^*) . \end{cases} \quad (9.53)$$

Using zero-forcing for interference suppression, we multiply both sides of (9.51) by the following linear combiner matrix

$$\mathbf{W} = \begin{pmatrix} I_2 & -\Omega_{1,1}^{-1} \cdot \Omega_{1,2} \\ -\Omega_{2,2}^{-1} \cdot \Omega_{2,1} & I_2 \end{pmatrix}. \quad (9.54)$$

Since, we have

$$\Omega \cdot \mathbf{W} = \begin{pmatrix} \Omega_{1,1} - \Omega_{1,2} \cdot \Omega_{2,2}^{-1} \cdot \Omega_{2,1} & 0 \\ 0 & \Omega_{2,2} - \Omega_{2,1} \cdot \Omega_{1,1}^{-1} \cdot \Omega_{1,2} \end{pmatrix}, \quad (9.55)$$

the symbols of the two groups can be decoded separately after multiplying both sides of (9.51) by $\mathbf{W}$. Also, note that both matrices $\Omega_{1,1} - \Omega_{1,2} \cdot \Omega_{2,2}^{-1} \cdot \Omega_{2,1}$ and $\Omega_{2,2} - \Omega_{2,1} \cdot \Omega_{1,1}^{-1} \cdot \Omega_{1,2}$ have formats similar to the format of Ω in (4.18). Therefore, each group can have a separate decoding after interference suppression similar to that of an orthogonal STBC.

The zero-forcing solution is independent of SNR and does not consider the noise enhancement. Instead, an MMSE criterion can be utilized to minimize the effects of noise enhancement. To detect symbols of the first group using MMSE, we calculate the best linear estimator as

$$\mathbf{M} = \left(\frac{I_4}{\gamma} + \Omega^H \cdot \Omega \right)^{-1} \cdot \Omega^H. \quad (9.56)$$

Then, the first and second columns of $\mathbf{M}$, denoted by $\mathbf{M}_1$ and $\mathbf{M}_2$ are utilized as the MMSE nulling vectors for the first and second symbols, respectively. In other words, the decoded symbol $\hat{s}_k$ is the closest constellation point to $(r_{1,1}, \quad r_{2,1}^*, \quad r_{1,2}, \quad r_{2,2}^*) \cdot \mathbf{M}_k$, for $k = 1, 2$. Similarly after canceling the effects of s_1 and s_2, we can detect the symbols of the second group, s_3 and s_4. Therefore, again the symbols are decoded separately after finding the best MMSE estimator. Note that as $\gamma \to \infty$, the MMSE nulling converges to zero-forcing nulling.

9.9 Simulation results

In this section, we provide simulation results using spatial multiplexing and different decoding methods presented in the previous sections. We assume a quasi-static flat Rayleigh fading model for the channel. Figure 9.9 shows the simulation results for transmitting 4 bits/(s Hz) over two transmit and two receive antennas

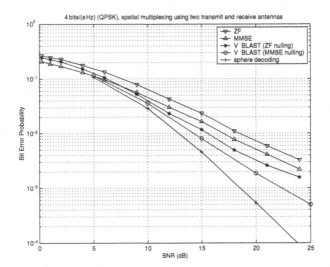

Fig. 9.9.   Bit error probability plotted against SNR for spatial multiplexing using QPSK, 4 bits/(s Hz); two transmit and receive antennas. (Courtesy of Ashish Bhargave and Mohamed O. Damen.)

using QPSK. The results are reported for the decoding using the ZF and MMSE techniques in Section 9.4, interference cancelation and nulling as explained in Section 9.5, and sphere decoding as explained in Section 9.3. In the case of interference cancelation and nulling, we report results for both ZF nulling and MMSE nulling. In both cases, a greedy ordering method is utilized without optimizing the ordering. As expected, linear decoding methods like ZF and MMSE perform worse than other methods while requiring a lower complexity. The interference cancelation and nulling methods, denoted as V-BLAST in the figure, outperform the linear decoding methods; however, their complexity is higher due to the iterative nature of the algorithm. Sphere decoding provides the best performance and the highest decoding complexity among the three methods. Note that MMSE outperforms ZF at all SNRs and in both linear and iterative, V-BLAST, methods. Similar results and behavior are seen if the number of transmit and receive antennas is four or eight. The results for four and eight antennas using QPSK are demonstrated in Figures 9.10 and 9.11, respectively. By increasing the number of antennas, the transmission bit rate of the spatial multiplexing system increases. The slope of the sphere decoding curve is steeper than that of the other methods. In other words, sphere decoding provides a higher diversity gain. In general, for $M \geq N$, it can be shown that a maximum-likelihood decoder, which is non-linear, can achieve a diversity of M and the linear decoders like ZF and MMSE achieve a diversity of $M - N + 1$ [152]. For V-BLAST, the diversity gain for different symbols is different as we

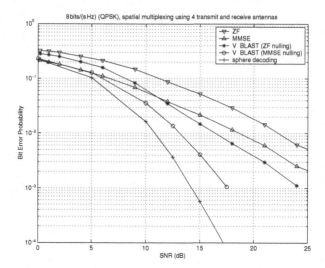

Fig. 9.10.   Bit error probability plotted against SNR for spatial multiplexing using QPSK, 8 bits/(s Hz); four transmit and receive antennas. (Courtesy of Ashish Bhargave and Mohamed O. Damen.)

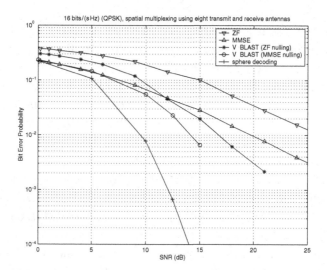

Fig. 9.11.   Bit error probability plotted against SNR for spatial multiplexing using QPSK, 16 bits/(s Hz); eight transmit and receive antennas. (Courtesy of Ashish Bhargave and Mohamed O. Damen.)

explained before. The diversity gain for the nth symbol is $M - N + n$. Therefore, the slope of the average error curve is between those of the linear decoders and the maximum-likelihood decoder. This is also evident from Figures 9.9, 9.10, and 9.11. There is a trade-off between the decoding complexity and the diversity gain, or equivalently the performance.

9.10 Summary of important results

- Spatial multiplexing is utilized to increase the throughput of communication systems with multiple antennas.
- If the structure of space-time architecture does not allow the simplification of the receiver, in general, the complexity of ML decoding grows exponentially.
- In sphere decoding only the codewords inside a sphere centered at the received vector are considered as decoding candidates.
- For a given channel matrix $\mathbf{H} = H$, zero-forcing enhances the noise by a factor of $(H \cdot H^H)^{-1}$.
- V-BLAST is a practical approach to achieve spatial multiplexiing. Its decoder consists of ordering, interference cancelation, and interference nulling.
- In V-BLAST, all signals in each layer are transmitted from the same antenna. However, the signals are shifted before the transmission in D-BLAST.
- Adding coding to different BLAST architectures improves the performance.
- Combining spatial multiplexing and space-time coding is possible resulting in a range of architectures providing different rates, diversities, and performance.
- When array processing and space-time block coding are combined, the system may use a smaller number of receive antennas compared to that of V-BLAST.
- Transmitting N symbols per time slot from N transmit antennas, linear decoding methods like ZF and MMSE provide a diversity gain of $M - N + 1$ for $M \geq N$ receive antennas. For the same case, ML decoding, which is a non-linear decoding method, provides a diversity gain of M.
- There is a trade-off between the complexity of the decoding and the achieved diversity. Going from linear decoding methods to V-BLAST and to ML decoding, the diversity gain and the decoding complexity increase simultaneously.

9.11 Problems

1 Derive the optimum detection criteria for the spatial multiplexing system in (9.1).

2 Consider the zero-forcing approach in (9.21).

(a) What is the statistics of the noise after applying the zero-forcing detection?
(b) What is the disadvantage of the zero-forcing detection?

3 Consider a spatial multiplexing system with $N = 4$ transmit and $M = 4$ receive antennas using QPSK. Also, consider a receiver using interference cancelation and zero-forcing nulling without any ordering. Use Monte Carlo simulations to derive the bit error probability versus the received SNR simulation results. We may call such a system V-BLAST with no ordering. How does it compare with the V-BLAST (ZF nulling) curve in Figure 9.10? How important is ordering in V-BLAST?

4 For the V-BLAST system in this chapter, consider the following modifications.

(a) If the first detected symbol is incorrectly detected, it results in error propagation. To avoid such an error propagation, use the correct first symbol during the symbol cancelation step. Compare the results with V-BLAST and observe the difference in performance.

(b) Measure the distance of each detected symbol from the statistic in the constellation space. Add these distances for all the symbols in the transmitted vector to obtain the distance metric for the entire detected vector. We call such a distance D_{vector}. Now, for the first detected symbol, in addition to the best symbol, which was closest to the statistic, use the second best. Continue the detection process, with symbol cancelation using this second best symbol and obtain D_{vector} for this new vector. Finally, choose the best vector amongst the two which has the minimum distance metric. How much improvement do you achieve? Is it possible to improve the results by considering the first three best symbols?

5 Use Monte Carlo simulations to derive the bit error probability versus the received SNR simulation results for the system in Example 9.8.1.

10

Non-orthogonal space-time block codes

10.1 Introduction

We introduced space-time block coding in Chapter 4. The space-time block codes in Chapters 4 and 5 have interesting structures that can be utilized to simplify the ML decoding. The orthogonality of OSTBCs makes it possible to decouple the decoding of different symbols. Similarly, a pairwise decoding is possible for QOSTBCs. Such simple ML decoding methods restrict the cardinality of transmitted symbols. It may also impose a limit on the performance of the code. On the other hand, there is no restriction on the number of receive antennas and even one receive antenna is enough. In addition to simple ML decoding and full diversity, the nice structure of these codes makes it possible to combine them with other blocks of the system. In fact, the transmission of an orthogonal STBC over a MIMO channel usually translates into N parallel SISO channels. As a result many wireless communication methods and building blocks that have been developed for SISO channels can be easily generalized using an orthogonal STBC. In many wireless communication systems the above advantages and disadvantages makes OSTBC a very attractive candidate, especially at low transmission bit rates.

As discussed in Chapter 4, OSTBCs are more restrictive for channels with higher number of antennas. When decoding complexity is not an issue, one may design codes that provide higher rates or better performance compared to OSTBCs. There are different schemes that provide a higher rate in a trade-off with a higher complexity. Usually to transmit more than one symbol per time slot, more than one receive antenna is needed. We studied a class of such non-orthogonal codes, based on spatial multiplexing, in Chapter 9. In this chapter, we study other classes of codes that are not orthogonal.

10.2 Linear dispersion space-time block codes

In the original STBC proposed in [136], every element of a codeword matrix is a linear combination of the input symbols. The number of symbols is selected such that an orthogonal STBC is feasible. Such a limit on the number of symbols is not necessary if the orthogonality condition of the STBC is relaxed. Instead, the maximum number of symbols is dictated by the availability of a decoder. By keeping the linearity of the STBC, usually the maximum possible number of symbols per time slot is upper bounded by the number of receive antennas. In other words, decoding is possible if $K \leq MT$. Therefore, designing codes with rates up to M symbols per time slot are possible. Of course, such codes are designed specifically for the given number of receive antennas. This is in contrast with orthogonal STBCs that are designed independently of the number of receive antennas.

In this section, we consider a class of STBCs that has the structure of orthogonal STBCs but relaxes the orthogonality condition. The encoder block diagram of these codes is the same as that of the orthogonal STBCs in Figure 4.4. Assuming a constellation with 2^b elements, Kb input bits select constellation signals $s_1, s_2, \ldots, s_K$ at the encoder. For an orthogonal STBC, these constellation symbols replace the indeterminant variables $x_1, x_2, \ldots, x_K$ in Definition 4.7.1 to construct the codeword. The definition of a linear dispersion STBC (LDSTBC) is similar to that of an orthogonal STBC after relaxing the condition in (4.76). In other words, a linear dispersion codeword is a $T \times N$ matrix $\mathbf{C}$ whose entries are linear combinations of $s_1, s_2, \ldots, s_K$ and their conjugates. This is equivalent to the following definition from [55].

Definition 10.2.1 *A linear dispersion codeword is constructed from input constellation signals $s_1, s_2, \ldots, s_K$ and their conjugates by*

$$\mathbf{C} = \sum_{k=1}^{K} \left(s_k E_k + s_k^* F_k \right), \tag{10.1}$$

where $\mathbf{C}$, E_k, and F_k are $T \times N$ matrices.

Since Definition 10.2.1, unlike Definition 4.7.1, has no restriction on the dispersion matrices E_k and F_k, codes with rates higher than one are possible. This is in a trade-off with the higher complexity of the decoding. Different decoding methods may put different restrictions on the rate of the code. For example, using a decoding algorithm similar to that of the V-BLAST interference cancelation and nulling, the number of transmit symbols should be less than MT. Otherwise, the system of equations at the receiver is under-determined and the V-BLAST decoding does not work. Due to the linearity of the system, a sphere decoding method, presented in

Section 9.3, is also possible. In both cases, the decoder is much more complex than that of the orthogonal STBC.

Rewriting the linear dispersion codeword in (10.1) in terms of the real and imaginary parts of the input constellation signals, we have

$$\mathbf{C} = \sum_{k=1}^{K} (\Re\{s_k\}X_k + j\Im\{s_k\}Y_k), \qquad (10.2)$$

where $X_k = E_k + F_k$ and $Y_k = E_k - F_k$. Therefore, X_k and Y_k equivalently define the LDSTBC. This is similar to the separation of real and imaginary parts of the orthogonal STBCs in (8.59) that has been used for differential decoding. Note that X_k and Y_k may be complex matrices and we do not restrict them to be real.

10.2.1 Decoding

Due to the linearity of (10.2) in variables $\Re\{s_k\}$ and $\Im\{s_k\}$, both V-BLAST interference cancellation and nulling decoding and sphere decoding are possible. To use the sphere decoding equations in Section 9.3, we need to represent the input–output relations as in (9.3). Towards that goal, we separate the real and imaginary parts of the input–output equation $\mathbf{r} = \mathbf{C} \cdot \mathbf{H} + \mathcal{N}$. We use the uppercase R and I to denote the real and imaginary parts of a matrix, respectively. Then, we have

$$\mathbf{r} = \left(\sum_{k=1}^{K} (s_k^R(X_k^R + jX_k^I) + js_k^I(Y_k^R + jY_k^I)) \right) \cdot (\mathbf{H}^R + j\mathbf{H}^I) + \mathcal{N}. \quad (10.3)$$

Equivalently, we can write the following two equations, containing only real numbers and matrices

$$\mathbf{r}^R = \sum_{k=1}^{K} \left[s_k^R \left(X_k^R \cdot \mathbf{H}^R - X_k^I \cdot \mathbf{H}^I \right) - s_k^I \left(Y_k^R \cdot \mathbf{H}^I + Y_k^I \cdot \mathbf{H}^R \right) \right] + \mathcal{N}^R, \quad (10.4)$$

$$\mathbf{r}^I = \sum_{k=1}^{K} \left[s_k^R \left(X_k^I \cdot \mathbf{H}^R + X_k^R \cdot \mathbf{H}^I \right) + s_k^I \left(Y_k^R \cdot \mathbf{H}^R - Y_k^I \cdot \mathbf{H}^I \right) \right] + \mathcal{N}^I. \quad (10.5)$$

Then, to form a real input–output relationship that contains row vectors as inputs and outputs similar to those of (9.3), we need to rearrange the equations. Let us denote the ith column of a matrix by a subscript. First, we construct the $1 \times 2K$

input vector and the $1 \times 2MT$ output vector, respectively, as

$$
\underline{s}^T = \begin{pmatrix} s_1^R \\ s_1^I \\ \vdots \\ s_K^R \\ s_K^I \end{pmatrix}, \quad \underline{r}^T = \begin{pmatrix} \mathbf{r}_1^R \\ \mathbf{r}_1^I \\ \vdots \\ \mathbf{r}_M^R \\ \mathbf{r}_M^I \end{pmatrix}. \tag{10.6}
$$

Defining the following intermediate matrices:

$$
A_k = \begin{pmatrix} X_k^R & -X_k^I \\ X_k^I & X_k^R \end{pmatrix}, \quad B_k = \begin{pmatrix} -Y_k^I & -Y_k^R \\ Y_k^R & -Y_k^I \end{pmatrix}, \quad \underline{h}_i = \begin{pmatrix} \mathbf{H}_i^R \\ \mathbf{H}_i^I \end{pmatrix}, \tag{10.7}
$$

the transpose of the equivalent channel will be

$$
H^T = \begin{pmatrix} A_1\underline{h}_1 & B_1\underline{h}_1 & \cdots & A_K\underline{h}_1 & B_K\underline{h}_1 \\ \vdots & \vdots & \ddots & \vdots & \vdots \\ A_1\underline{h}_M & B_1\underline{h}_M & \cdots & A_K\underline{h}_M & B_K\underline{h}_M \end{pmatrix}. \tag{10.8}
$$

Finally, the equivalent of (9.3) is

$$
\underline{r} = \underline{s} \cdot H + \underline{N}, \tag{10.9}
$$

where the transpose of the $1 \times 2M$ noise matrix $\underline{N}$ is

$$
\underline{N}^T = \begin{pmatrix} \mathcal{N}_1^R \\ \mathcal{N}_1^I \\ \vdots \\ \mathcal{N}_M^R \\ \mathcal{N}_M^I \end{pmatrix}. \tag{10.10}
$$

Therefore, sphere decoding can be used for decoding after calculating the $2K \times 2MT$ matrix H from (10.8) at the receiver. For $K \leq \min\{N, M\}T$, a "divide-and-conquer" receiver similar to that of V-BLAST works as well.

10.2.2 Design of LDSTBCs

For a given number of transmit and receive antennas, designing an LDSTBC amounts to defining matrices X_k and Y_k, $k = 1, 2, \ldots, K$, or equivalently matrices A_K and B_k. Like any other code design, the main question is how to pick these matrices to optimize the performance of the code. The choice of matrices X_k and Y_k changes the elements of the equivalent channel matrix in (10.8). As a result the mutual information between the equivalent input and output matrices changes too. The design criterion that is usually used in designing LDSTBCs is to maximize the mutual information between the input and output signals. The rank, determinant, and trace criteria, are derived from minimizing the maximum pairwise error probability. Such design criteria usually translate to a condition on one of the columns in (10.8) and is not very useful. For example, designs based on determinant criterion may lead to a near singular H matrix that results in other forms of errors.

Following the mutual information formulas for an ergodic channel, since the $2K \times 2MT$ matrix H is real, its average mutual information is

$$\frac{1}{2} E\left[\log_2\left(\det\{I_{2MT} + (\gamma/N)H^H \cdot H\}\right)\right]. \tag{10.11}$$

Since the LDSTBC transmits over T time slots, the design criterion is to maximize the following mutual information per time slot

$$\frac{1}{2T} E\left[\log_2\left(\det\{I_{2MT} + (\gamma/N)H^H \cdot H\}\right)\right]. \tag{10.12}$$

Note that the average mutual information in (10.12) is a function of matrices X_k and Y_k, $k = 1, 2, \ldots, K$ and the received signal to noise ratio γ. Of course, increasing the transmission power while keeping a constant noise power results in higher average mutual information. Therefore, the maximization of (10.12) should be a constraint maximization with a constraint over the transmission power. There are different ways to impose a constant average transmission power. In [55], three constraints with different levels of restrictions have been proposed. All three constraints results in $E[\text{Tr}(\mathbf{C}^H \cdot \mathbf{C})] = NT$. We list these three constraints with an increasing level of restriction.

- The first constraint is the power constraint that ensures $E[\text{Tr}(\mathbf{C}^H \cdot \mathbf{C})] = NT$ in terms of matrices X_k and Y_k:

$$\sum_{k=1}^{K} \text{Tr}\left(X_k^H \cdot X_k + Y_k^H \cdot Y_k\right) = 2NT. \tag{10.13}$$

- The second constraint restricts the real and imaginary parts of each transmitted signal, $\Re\{s_k\}$ and $\Im\{s_k\}$, to have the same total power in T time slots and from N transmit antennas:

$$\text{Tr}\left(X_k^H \cdot X_k\right) = \text{Tr}\left(Y_k^H \cdot Y_k\right) = \frac{NT}{K}, \quad k = 1, 2, \ldots, K. \tag{10.14}$$

- The third constraint restricts $\Re\{s_k\}$ and $\Im\{s_k\}$ to have the same power in every use of the channel from every transmit antenna:

$$X_k^H \cdot X_k = Y_k^H \cdot Y_k = \frac{T}{K} I_N, \quad k = 1, 2, \ldots, K. \tag{10.15}$$

It is important to note that maximizing the mutual information in (10.12) with any of the above constraints may result in a different code. Also, the optimality of the resulting code is only valid for the given SNR and the number of transmit and receive antennas. Usually, one fixes all parameters, that is K, T, N, M, and γ, and solves the constraint optimization problem for the specific parameters. Using Lagrangian theory, one can convert the constraint optimization problem to an optimization problem without constraints. Then, a gradient-based or steepest descent numerical optimization method [12] can be utilized to solve the problem. Since the mutual information formula is neither convex nor concave, the numerical solution may end up in local maxima.

The fact that the solution depends on the SNR and the number of receive antennas is limiting. Nevertheless, a code that is designed for M receive antennas can be decoded for any $M' > M$ receive antennas similar to the case of OSTBCs. Also, one may design a code for a given SNR and use the resulting code for all SNRs. Such a code may not be optimal at all SNRs; however, the encoding and decoding work for any SNR.

The above design method is a numerical method. As we discussed before, the solution may be a local maxima and is not unique. Reordering the matrices X_k and Y_k results in another solution. Also, multiplying all matrices by a unitary matrix provides another solution. Multiplying the input vector $\underline{s}$ in (10.9) by a $2K \times 2K$ orthogonal matrix Φ does not change the input–output relation. In fact, defining the new input vector $\underline{s}' = \underline{s} \cdot \Phi$ results in the equivalent channel matrix $H' = \Phi^T \cdot H$ and (10.9) can be rewritten as

$$\underline{r} = \underline{s}' \cdot \Phi^T \cdot H + \underline{N} = \underline{s}' \cdot H' + \underline{N}. \tag{10.16}$$

Therefore, such a transform preserves the mutual information while changing the dispersion matrices X_k and Y_k. Such a transform may change the performance of the code. Finding the best orthogonal matrix Φ can be considered as the second phase of the design. For example, V-BLAST provides the maximum mutual information. The performance of V-BLAST can be improved by multiplying the corresponding symbols by an appropriate matrix Φ. In fact, some of the good LDSTBCs are designed by starting from a V-BLAST code and picking the right Φ matrix as we show in the following example.

Example 10.2.1 *In this example, we consider a class of LDSTBCs for $T = N$ and $M \geq N$ to transmit N symbols per time slot, that is $K = NT$. Note that this is the same rate as that of V-BLAST and in fact the codes can be designed by transforming V-BLAST. To derive the dispersion matrices for this class of LDSTBCs, first we define* *the following two matrices:*

$$\Pi = \begin{pmatrix} 0 & 0 & \cdots & 0 & 1 \\ 1 & 0 & \cdots & 0 & 0 \\ 0 & 1 & \ddots & 0 & 0 \\ \vdots & & \ddots & & \vdots \\ 0 & 0 & \cdots & 1 & 0 \end{pmatrix}, \tag{10.17}$$

$$D = diag\left(1, e^{j\frac{2\pi}{N}}, \ldots, e^{j\frac{2\pi(N-1)}{N}}\right), \tag{10.18}$$

where diag(...) is a diagonal matrix that contains its arguments as the diagonal elements. Then, the code is defined by the following matrices

$$X'_{N(i-1)+j} = Y'_{N(i-1)+j} = \frac{1}{\sqrt{N}} D^{i-1} \cdot \Pi^{j-1}, \ i, j = 1, 2, \ldots, N. \tag{10.19}$$

For two transmit and receive antennas, we have

$$X'_1 = Y'_1 = \frac{1}{\sqrt{2}}\begin{pmatrix} 1 & 0 \\ 0 & 1 \end{pmatrix}, \qquad X'_2 = Y'_2 = \frac{1}{\sqrt{2}}\begin{pmatrix} 0 & 1 \\ 1 & 0 \end{pmatrix},$$

$$X'_3 = Y'_3 = \frac{1}{\sqrt{2}}\begin{pmatrix} 1 & 0 \\ 0 & -1 \end{pmatrix}, \qquad X'_4 = Y'_4 = \frac{1}{\sqrt{2}}\begin{pmatrix} 0 & 1 \\ -1 & 0 \end{pmatrix}. \tag{10.20}$$

The resulting codeword will be

$$\mathbf{C} = \frac{1}{\sqrt{2}}\begin{pmatrix} s_1 + s_3 & s_2 + s_4 \\ s_2 - s_4 & s_1 - s_3 \end{pmatrix}. \tag{10.21}$$

Note that for V-BLAST, we have

$$X_1 = Y_1 = \begin{pmatrix} 1 & 0 \\ 0 & 0 \end{pmatrix}, \qquad X_2 = Y_2 = \begin{pmatrix} 0 & 1 \\ 0 & 0 \end{pmatrix},$$

$$X_3 = Y_3 = \begin{pmatrix} 0 & 0 \\ 1 & 0 \end{pmatrix}, \qquad X_4 = Y_4 = \begin{pmatrix} 0 & 0 \\ 0 & 1 \end{pmatrix}. \tag{10.22}$$

Therefore, the proposed LDSTBC is just a transform of V-BLAST matrices by

$$\begin{array}{ll} X'_1 = \frac{1}{\sqrt{2}}(X_1 + X_4), & Y'_1 = \frac{1}{\sqrt{2}}(Y_1 + Y_4), \\ X'_2 = \frac{1}{\sqrt{2}}(X_2 + X_3), & Y'_2 = \frac{1}{\sqrt{2}}(Y_2 + Y_3), \\ X'_3 = \frac{1}{\sqrt{2}}(X_1 - X_4), & Y'_3 = \frac{1}{\sqrt{2}}(Y_1 - Y_4), \\ X'_4 = \frac{1}{\sqrt{2}}(X_2 - X_3), & Y'_4 = \frac{1}{\sqrt{2}}(Y_2 - Y_3). \end{array} \tag{10.23}$$

Note that both V-BLAST and the LDSTBC in (10.19) provide the maximum mutual information and satisfy the first two constraints. However, V-BLAST does not satisfy the third constraint while the LDSTBC in (10.19) satisfies it. As a result, the LDSTBC provides a higher diversity gain.

For three transmit antennas, the OSTBC in (4.154) is actually a local maximum of the mutual information in (10.12). At $\gamma = 20$ dB, it provides a mutual information of 5.13 bits/(s Hz) for one receive antenna which is less than the channel capacity of 6.41 bits/(s Hz). The OSTBC provides the maximum diversity and a very simple separate ML decoding. However, its rate is $R = 3/4$ which does not allow it to achieve the capacity. In Chapter 5, we discussed QOSTBCs that provide rate-one and maximum diversity with pairwise ML decoding. In the following example, we provide an example from [55] that provides mutual information of 6.25 bits/(s Hz).

Example 10.2.2 *The number of symbols should satisfy $K \leq MT$. With one receive antenna, to design a rate one LDSTBC, we pick $K = T = 4$. In this example, we consider the following LDSTBC that utilizes three transmit antennas and one receive antenna.*

$$
\mathbf{C} = \begin{pmatrix}
s_1^R + s_3^R + j\left(\frac{s_2^I + s_4^I}{\sqrt{2}} + s_4^I\right) & \frac{s_2^R - s_4^R}{\sqrt{2}} - j\left(\frac{s_1^I}{\sqrt{2}} + \frac{s_2^I - s_4^I}{2}\right) & 0 \\
\frac{s_4^R - s_2^R}{\sqrt{2}} - j\left(\frac{s_1^I}{\sqrt{2}} + \frac{s_2^I - s_4^I}{2}\right) & s_1^R - j\frac{s_2^I + s_4^I}{\sqrt{2}} & -\frac{s_2^R + s_4^R}{\sqrt{2}} + j\left(\frac{s_1^I}{\sqrt{2}} - \frac{s_2^I - s_4^I}{2}\right) \\
0 & \frac{s_4^R + s_2^R}{\sqrt{2}} + j\left(\frac{s_1^I}{\sqrt{2}} - \frac{s_2^I - s_4^I}{2}\right) & s_1^R - s_3^R + j\left(\frac{s_2^I + s_4^I}{\sqrt{2}} - s_4^I\right) \\
\frac{s_2^R - s_4^R}{\sqrt{2}} + j\left(\frac{s_1^I}{\sqrt{2}} + \frac{s_2^I - s_4^I}{2}\right) & -s_3^R + js_4^I & -\frac{s_2^R + s_4^R}{\sqrt{2}} + j\left(\frac{s_1^I}{\sqrt{2}} - \frac{s_2^I - s_4^I}{2}\right)
\end{pmatrix}.
$$

$$(10.24)$$

The above code provides a mutual information of 6.25 bits/(s Hz) at $\gamma = 20$ dB.

In general by relaxing the orthogonality condition on STBCs, the LDSTBCs provide additional degrees of freedom to improve the performance. The good codes are usually designed by numerical methods and sometimes exhaustive search. Like any other non-orthogonal STBC, the drawback is the increased decoding complexity and the high peak-to-average power ratio (PAPR). While the average transmitted energy may be preserved, the dynamic range of each transmitted signal is increased. Therefore, the ratio of the pick value to the average value increases. This may cause a practical problem if the system amplifiers operate in their nonlinear region. To avoid the possible distortions due to the nonlinearity of the amplifier, one should either decrease the PAPR or use more expensive amplifiers with larger linear regions. This is a well-known problem in orthogonal frequency division multiplexing (OFDM) systems. We refer the interested reader to [24, 97, 134] and the references therein.

10.3 Space-time block codes using number theory

In [34], a class of STBCs is proposed that provides full diversity with no mutual information penalty. The code is linear but not orthogonal and therefore separate ML decoding is not possible. Instead, sphere decoding can be utilized to decode the symbols. The properties of this class of codes are investigated using number theory [34]. Here, we present the code as an example and provide its properties without proving them.

Example 10.3.1 *Let us consider $N = 2$ transmit antennas and a STBC over $T = 2$ time slots. To transmit one symbol per time slot from each antenna, the codeword should contain four symbols. Defining $\phi = \theta^2 = e^{j\lambda}$, where λ is a design parameter, the following STBC transmits two symbols per time slot over two transmit antennas:*

$$\mathbf{C} = \frac{1}{\sqrt{2}} \begin{pmatrix} s_1 + \phi s_2 & \theta(s_3 - \phi s_4) \\ \theta(s_3 + \phi s_4) & s_1 - \phi s_2 \end{pmatrix}. \tag{10.25}$$

The above STBC is information lossless, that is there is no mutual information penalty. With the right choice of the constellation and parameter λ, the code in (10.25) provides full diversity as well. For example, if $\lambda \neq 0$ is an algebraic number and the constellation is a subset of the ring of complex integers, the code provides a full diversity of $2M$. A complex number is called an algebraic number if it is the root of a polynomial with coefficients from the ring of rational integers.

The performance of the code depends on the choice of λ. For a given constellation, one can pick the value of λ to maximize the minimum coding gain. Such an optimization problem will have several local maximums. For example, $\lambda = \pi/4$ provides a local maximum for QPSK and 16-QAM constellations. The corresponding CGDs are 0.0074 and 0.0007, respectively. The global maximum can be achieved by an exhaustive search in the range of $0 \le \lambda \le \pi/2$ as done in [34]. The optimal values for QPSK and 16-QAM constellations are $\lambda = 0.5$ and $\lambda = 0.521$, respectively. The corresponding CGDs are 0.0561 for QPSK and 0.0035 for 16-QAM.

One drawback of the code in this example is its high PAPR. To decode the symbols, one may use sphere decoding. Therefore, the decoding complexity is much more than that of the orthogonal STBCs. Although the code is designed using number theory, due to its linearity it can be represented as an LDSTBC. We discuss the relationship between LDSTBCs and the code in this example in the sequel.

To show the relationship between the STBC in (10.25) and the LDSTBC in (10.21), we use the following equivalent LDSTBC instead of the one in (10.21)

$$\mathbf{C} = \frac{1}{\sqrt{2}} \begin{pmatrix} s_1 + s_2 & s_3 - s_4 \\ s_3 + s_4 & s_1 - s_2 \end{pmatrix}. \tag{10.26}$$

Note that (10.26) is achieved from (10.21) just by renumbering the input symbols and sign changes. Therefore, for symmetric constellations like PSK and QAM, the two LDSTBCs are identical. Now, if we multiply s_2, s_3, s_4 in (10.26) by ϕ, θ, $\phi\theta$, respectively, we derive the STBC in (10.25). Such a rotation of constellation symbols will not change the mutual information. However, it will affect the diversity gain of the STBC. The STBCs in (10.21) and (10.26) do not provide full diversity while the STBC in (10.25) does. In fact, the above rotations can be considered as a unitary transform on the input symbols. The corresponding transform matrix is

$$\begin{pmatrix} 1 & 0 & 0 & 0 \\ 0 & \phi & 0 & 0 \\ 0 & 0 & \theta & 0 \\ 0 & 0 & 0 & \phi\theta \end{pmatrix}. \tag{10.27}$$

Any such a unitary transform will preserve the mutual information while changing the diversity and coding gain. Therefore, the symbols of LDSTBCs can be multiplied by unitary transforms to improve the performance.

The idea of rotating the constellation symbols to increase the diversity and coding gain was used in Chapter 5 to design full diversity QOSTBCs. As mentioned in Section 5.2, QOSTBC does not provide full diversity when the same constellation is used for all symbols. However, rotating some of the constellations makes the QOSTBC full diversity. While the main idea is similar for the above LDSTBC example and QOSTBCs, there is a difference in the resulting transmitted signals. Let us assume a constellation with 2^b elements. At each time slot and from each transmit antenna, the transmitted signals for a rotated QOSTBC are still from a set of 2^b elements with the same PAPR. For example, if the constellation is PSK, we always use a rotated PSK with PAPR of one. On the other hand, the PAPR may increase dramatically for rotated LDSTBCs. This is because of the fact that the PAPR is not preserved under "linear combination." Also, the number of possible transmitted signals may change as well.

10.4 Threaded algebraic space-time codes

A threaded space-time code is a generalization of BLAST architecture. The layers are structured such that the code avoids spatial interference within a layer [40]. A layer is defined as a part of transmission in which each time slot is allocated to at most one transmit antenna. In other words, at each time slot, only one of the transmit antennas is utilized by each layer. The other transmit antennas, at that particular time slot, may be used by other layers. Therefore, the source of interference from different transmit antennas is from different layers. Then, given a layer, the spatial interference imposed on the layer is caused by other layers.

Table 10.1. *An example of layering in a threaded space-time code*

Time slot	Antenna 1	Antenna 2	Antenna 3	Antenna 4
1	Layer 1	Layer 2	Layer 3	Layer 4
2	Layer 4	Layer 1	Layer 2	Layer 3
3	Layer 3	Layer 4	Layer 1	Layer 2
4	Layer 2	Layer 3	Layer 4	Layer 1
5	Layer 1	Layer 2	Layer 3	Layer 4
6	Layer 4	Layer 1	Layer 2	Layer 3
7	Layer 3	Layer 4	Layer 1	Layer 2
8	Layer 2	Layer 3	Layer 4	Layer 1
$\vdots$	$\vdots$	$\vdots$	$\vdots$	$\vdots$
$T-2$	Layer 1	Layer 2	Layer 3	Layer 4
$T-1$	Layer 4	Layer 1	Layer 2	Layer 3
T	Layer 3	Layer 4	Layer 1	Layer 2

Such outside interference can be compensated for by methods used in iterative multiuser receivers [40]. Note that D-BLAST and other schemes in Chapter 9 are special cases of threaded space-time codes. However, such a generalization may not allow the use of some of the decoding methods studied in Chapter 9. For example, a simple extension of D-BLAST is shown in Table 10.1 for four transmit antennas. Unlike D-BLAST, it does not suffer from the boundary effects; however, the interference nulling and cancelation method is not applicable for decoding.

One property of the example in Table 10.1 is the fact that different layers use available transmit antennas equally. Also, each layer extends to all available time slots. Therefore, it is possible to provide maximum diversity. It is possible to use coding at each layer to improve the performance. In fact, since each layer includes transmission from different transmit antennas, a space-time code can be utilized at each layer. For example, the space-time codes in [53] that are based on convolutional codes with rate K/N can be used at each layer. Changing the component space-time codes affects the rate and diversity of the resulting threaded space-time code. Since our main focus in this section is on STBCs and not STTCs, we refer the interested reader to [40] for details. We just mention that the resulting codes are similar to turbo-BLAST codes with iterative decoders in Section 9.7. For decoding, each layer is first detected by a soft-input soft-output detector. The detected symbols are decoded by the corresponding channel decoder of the corresponding layer code. The soft outputs of the channel decoders are utilized to improve the soft-input soft-output detection of each layer. The decoding continues in an iterative fashion.

In what follows, we discuss the design and properties of threaded algebraic STBCs. First, we start with a special case of threaded algebraic space-time (TAST) block codes called diagonal algebraic STBCs.

10.4.1 Diagonal algebraic STBCs

In Example 4.8.1, we showed how to design a pseudo-orthogonal STBC using Hadamard matrices. As mentioned in the example, the code does not provide full diversity. To design a full-diversity code, one may replace each indeterminate variable by a linear combination of input symbols. In other words, the indeterminate variables $x_1, x_2, \ldots, x_N$ are derived from

$$(x_1, x_2, \ldots, x_N)^T = \mathbf{M}_N \cdot (s_1, s_2, \ldots, s_N)^T, \qquad (10.28)$$

where $\mathbf{M}_N$ is an $N \times N$ orthogonal matrix. Of course, by using such a transform, instead of just replacing the indeterminate variables with data symbols, we change the structure of the code. Because of such a transform, the resulting STBC may not be an orthogonal design and the separate ML decoding of symbols may not be possible. The decoding complexity increases since a sphere decoder should be used instead of simple separate decoding of the symbols. In addition, the transmitted constellation is expanded and its PAPR is increased. In fact, the transmitted constellation consists of all linear combinations of the symbols in the original constellation. This increases the number of transmitted symbols which can be related to rate and the PAPR. On the other hand, such a constellation expansion results in a higher diversity and better performance. Each transmitted signal contains information about all data symbols. As a result, each data symbol goes through different path gains, which results in full diversity. In [32], the following orthogonal rotation matrices, originally from [18], have been used for $N = 2, 4$ antennas:

$$\mathbf{M}_2 = \begin{pmatrix} 0.5257 & 0.8507 \\ -0.8507 & 0.5257 \end{pmatrix}, \qquad (10.29)$$

$$\mathbf{M}_4 = \begin{pmatrix} 0.2012 & 0.3255 & -0.4857 & -0.7859 \\ -0.3255 & 0.2012 & 0.7859 & -0.4857 \\ 0.4857 & 0.7859 & 0.2012 & 0.3255 \\ -0.7859 & 0.4857 & -0.3255 & 0.2012 \end{pmatrix}. \qquad (10.30)$$

With the above introduction, let us present the diagonal algebraic space-time (DAST) block code in [32]. We follow a notation that is slightly different from that of Example 4.8.1.

Definition 10.4.1 *An $N \times N$ DAST block code (DASTBC) is defined by the following generating matrix:*

$$
\mathcal{G}_N = \begin{pmatrix} x_1 & 0 & \cdots & 0 \\ 0 & x_2 & \cdots & 0 \\ \vdots & \vdots & \ddots & \vdots \\ 0 & 0 & \cdots & x_N \end{pmatrix} \cdot A_N, \tag{10.31}
$$

where $x_1, x_2, \ldots, x_N$ are defined in (10.28) and A_N is an $N \times N$ Hadamard matrix defined in (4.113).

To clarify the above definition, we provide specific examples for $N = 2$ and $N = 4$ transmit antennas.

Example 10.4.1 *For two transmit antennas, using $\mathbf{M}_2$ in (10.29) and A_2 in (4.114), we have*

$$
\begin{cases} x_1 = 0.5257s_1 + 0.8507s_2 \\ x_2 = -0.8507s_1 + 0.5257s_2. \end{cases} \tag{10.32}
$$

and

$$
\mathcal{G}_2 = \begin{pmatrix} x_1 & x_1 \\ x_2 & -x_2 \end{pmatrix}. \tag{10.33}
$$

Example 10.4.2 *For four transmit antennas, using $\mathbf{M}_4$ in (10.30), we have*

$$
\begin{cases} x_1 = 0.2012s_1 + 0.3255s_2 - 0.4857s_3 - 0.7859s_4 \\ x_2 = -0.3255s_1 + 0.2012s_2 + 0.7859s_3 - 0.4857s_4 \\ x_3 = 0.4857s_1 + 0.7859s_2 + 0.2012s_3 + 0.3255s_4 \\ x_4 = -0.7859s_1 + 0.4857s_2 - 0.3255s_3 + 0.2012s_4. \end{cases} \tag{10.34}
$$

and

$$
\mathcal{G}_4 = \begin{pmatrix} x_1 & x_1 & x_1 & x_1 \\ x_2 & -x_2 & x_2 & -x_2 \\ x_3 & x_3 & -x_3 & -x_3 \\ x_4 & -x_4 & -x_4 & x_4 \end{pmatrix}. \tag{10.35}
$$

To calculate the coding gain of a DASTBC, first we compute

$$
\mathcal{G}_N^H \cdot \mathcal{G}_N = A_N^T \cdot \begin{pmatrix} |x_1|^2 & 0 & \cdots & 0 \\ 0 & |x_2|^2 & \cdots & 0 \\ \vdots & \vdots & \ddots & \vdots \\ 0 & 0 & \cdots & |x_N|^2 \end{pmatrix} \cdot A_N. \tag{10.36}
$$

Then, since the DASTBC is a linear code, the CGD between two sets of data symbols $(s_1, s_2, \ldots, s_N)$ and $(s_1', s_2', \ldots, s_N')$ is

$$\det\left[\mathcal{G}_N(x_1 - x_1', x_2 - x_2', \ldots, x_N - x_N')^H \cdot \mathcal{G}_N(x_1 - x_1', x_2 - x_2', \ldots, x_N - x_N')\right],$$

(10.37)

where

$$(x_1', x_2', \ldots, x_N')^T = \mathbf{M}_N \cdot (s_1', s_2', \ldots, s_N')^T.$$

(10.38)

By replacing (10.36) in (10.37) and using (4.113), it is easy to show that the CGD is $N^N \prod_{n=1}^{N} |x_n - x_n'|^2$. Therefore, following the determinant design criterion, a good choice of $\mathbf{M}_N$ maximizes the following minimum product distance:

$$\min_{x_n \neq x_n'} \prod_{n=1}^{N} |x_n - x_n'|.$$

(10.39)

10.4.2 Design of threaded algebraic space-time block codes

To design a threaded algebraic space-time block code (TASTBC), one needs to select STBCs for different threads [39, 33]. One approach is to send scaled DAST symbols in different threads. In other words, denoting the symbols transmitted in the ith thread by $x_{i1}, x_{i2}, \ldots, x_{iN}$, we have

$$\mathbf{x}_i = (x_{i1}, x_{i2}, \ldots, x_{iN})^T = \mathbf{M}_N^i \cdot (s_{i1}, s_{i2}, \ldots, s_{iN})^T,$$

(10.40)

where $\mathbf{M}_N^i$ is an $N \times N$ matrix to design a full-diversity DASTBC for the ith thread and s_{iN} is the Nth data symbol for the ith thread. Then, in the ith thread, we transmit $\phi_i \mathbf{x}_i$, where ϕ_i is a constant. If $\mathbf{M}_N^i = \mathbf{M}_N$ is the same for all threads, the code is called a symmetric TASTBC. One can design full diversity TASTBCs with maximum coding gains at different rates by choosing the ϕ_i coefficients optimally. One design parameter is the number of threads denoted by I. Of course, a higher rate code can be designed by increasing the number of threads. For $I \leq \min(N, M)$, it is always possible to design a full-diversity TASTBC when using DASTBCs in different threads. Therefore, to maximize the rate, a good choice for the number of threads is $I = \min(N, M)$. Note that for one receive antenna, $M = 1$, we only have $I = 1$ thread. The TASTBC reduces to a DASTBC and the transmission rate is one symbol per channel use. Following the thread structure in Table 10.1, we provide a specific positioning of different threads for $N = M = I = 4$ in Table 10.2. If the number of threads is less than the number of transmit antennas, that is $I = M < N$, some of the threads in Table 10.2 may be empty. For example, if we have only two threads, the third and fourth threads in Table 10.2 may be replaced by zeros.

Table 10.2. *An example of thread structure in a*
TASTBC with $N = M = I = 4$

Time slot	Antenna 1	Antenna 2	Antenna 3	Antenna 4
1	Thread 1	Thread 2	Thread 3	Thread 4
2	Thread 4	Thread 1	Thread 2	Thread 3
3	Thread 3	Thread 4	Thread 1	Thread 2
4	Thread 2	Thread 3	Thread 4	Thread 1

In what follows, we provide examples of TASTBCs for more than one receive antenna with $I = \min(N, M)$ threads and using DASTBCs in different threads.

Example 10.4.3 *For two transmit antennas and more than one receive antenna, we have $I = 2$. Using the above guidelines to design a TASTBC results in the following code:*

$$\begin{pmatrix} \phi_1 x_{11} & \phi_2 x_{21} \\ \phi_2 x_{22} & \phi_1 x_{12} \end{pmatrix}. \tag{10.41}$$

Here, x_{11} and x_{12} belong to the first thread and relate to the first set of data symbols, as in (10.28), by

$$\begin{pmatrix} x_{11} \\ x_{12} \end{pmatrix} = \mathbf{M}_2 \cdot \begin{pmatrix} s_{11} \\ s_{12} \end{pmatrix}. \tag{10.42}$$

Similarly, the second set of data symbols create the second thread as

$$\begin{pmatrix} x_{21} \\ x_{22} \end{pmatrix} = \mathbf{M}_2 \cdot \begin{pmatrix} s_{21} \\ s_{22} \end{pmatrix}. \tag{10.43}$$

To define the TASTBC completely, we need to identify $\mathbf{M}_2$ and the ϕ_i coefficients. A good choice of the $\mathbf{M}_2$ for complex symbols is the following matrix from [50]:

$$\mathbf{M}_2 = \frac{1}{\sqrt{2}} \begin{pmatrix} 1 & e^{j\pi/4} \\ 1 & -e^{j\pi/4} \end{pmatrix}. \tag{10.44}$$

By choosing $\phi_1 = 1$ and $\phi_2 = \phi^{1/2}$, we can pick the best parameter ϕ to maximize the coding gain. In this case, the coding gain is

$$CGD = \min \left| (x_{11} - x'_{11})(x_{12} - x'_{12}) - \phi(x_{21} - x'_{21})(x_{22} - x'_{22}) \right|. \tag{10.45}$$

For QPSK, the optimal choice is $\phi = e^{j\pi/6}$. Sphere decoding is utilized for decoding the transmitted symbols. Note that the STBC in Example 10.3.1 can be represented as a TASTBC in this example.

For three transmit antennas, the number of threads depends on the number of receive antennas. For $M = 2$ receive antennas, we have $I = \min(N, M) = 2$ threads while for $M > 2$ receive antennas, we have $I = 3$ threads. We provide examples of such TASTBCs next.

Example 10.4.4 *For $N = 3$ transmit and $M \geq 2$ receive antennas, the following TASTBC is proposed in [39]:*

$$\begin{pmatrix} x_{11} & \phi^{1/3}x_{21} & \phi^{2/3}x_{31} \\ \phi^{2/3}x_{32} & x_{12} & \phi^{1/3}x_{22} \\ \phi^{1/3}x_{23} & \phi^{2/3}x_{33} & x_{13} \end{pmatrix}. \tag{10.46}$$

*Choosing $\phi = e^{j\pi/12}$ and the right **M** matrix guarantees full diversity. If we have only $M = 2$ receive antennas, only $I = 2$ threads are practical. In this case, the TASTBC is simply derived from the one in (10.46) by replacing the third thread, that is x_{31}, x_{32}, x_{33}, by zero. The resulting TASTBC, with a smaller rate compared to (10.46), is*

$$\begin{pmatrix} x_{11} & \phi^{1/3}x_{21} & 0 \\ 0 & x_{12} & \phi^{1/3}x_{22} \\ \phi^{1/3}x_{23} & 0 & x_{13} \end{pmatrix}. \tag{10.47}$$

In this case, $\phi = e^{j\pi/6}$ guarantees full diversity.

The decoding complexity of TASTBCs is more than those of the OSTBCs and QOSTBCs. Similar to DASTBC building blocks, the transmitted constellation for TASTBCs is an expanded version of the original symbol constellation. This usually increases the PAPR. In particular, if some of the threads are transmitting zeros, that is there are empty threads, the PAPR is very high. In many cases, the PAPR can be reduced as shown in the next example.

Example 10.4.5 *Let us assume $N = 4$ transmit antennas and $M = 2$ receive antennas. The TASTBC will have $I = 2$ threads as follows:*

$$\begin{pmatrix} x_{11} & \phi^{1/4}x_{21} & 0 & 0 \\ 0 & x_{12} & \phi^{1/4}x_{22} & 0 \\ 0 & 0 & x_{13} & \phi^{1/4}x_{23} \\ \phi^{1/4}x_{24} & 0 & 0 & x_{14} \end{pmatrix}. \tag{10.48}$$

For this code, $\phi = e^{j\pi/6}$ guarantees full diversity. Filling the empty threads 3 and

4 with a modified version of the first and second threads, respectively, results in

$$
\begin{pmatrix}
x_{11} & \phi^{1/4}x_{21} & -x_{13} & -\phi^{1/4}x_{23} \\
-\phi^{1/4}x_{24} & x_{12} & \phi^{1/4}x_{22} & -x_{14} \\
x_{11} & \phi^{1/4}x_{21} & x_{13} & \phi^{1/4}x_{23} \\
\phi^{1/4}x_{24} & x_{12} & \phi^{1/4}x_{22} & x_{14}
\end{pmatrix}.
\tag{10.49}
$$

For a given pair of data symbols, the coding gains of the codewords in (10.48) and (10.49) are identical. Therefore, while the two codes provide identical performances, the PAPR of the TASTBC in (10.49) is lower than that of the code in (10.48). This is due to the removal of the empty threads and the corresponding zeros and in a trade-off with a slightly higher complexity decoder.

As can be seen above, TASTBCs are designed for a given number of receive antennas. A different STBC is optimal for a different number of receive antennas. Note that orthogonal and quasi-orthogonal STBCs are designed independently of the number of receive antennas. For one receive antenna, usually there is not much room for improvement. In a multicast or broadcast scenario, different receivers may have different number of receive antennas. In such a system with different number of receive antennas, it is impossible to design a STBC that is optimal for all receivers.

10.5 Simulation results

In this section, we provide simulation results for some of the codes presented in this chapter and compare them with OSTBCs and V-BLAST. We present results for the LDSTBC in (10.21), from Example 10.2.1, and denote it by "LDSTBC" in the figures. Also, we present the results for the code in (10.25), from Example 10.3.1. We use the optimal values of λ, that is $\lambda = 0.5$ and $\lambda = 0.521$, for QPSK and 16-QAM, respectively. We denote these results by "NT5" to separate them from the LDSTBC results. Note that the code in (10.25) is also an LDSTBC and for $\lambda = 0$ this becomes identical to the code in (10.21).

We assume a quasi-static flat Rayleigh fading channel model and a perfect knowledge of the channel at the receiver. To have a fair comparison, we keep the total transmission bit rate identical for all simulated codes. First, we present the result for two transmit and receive antennas, because the codes in (10.21) and (10.25) are designed for this case. Bit error probability versus received SNR for transmission of 4 and 8 bits/(s Hz) are shown in Figures 10.1 and 10.2, respectively. Note that the number of symbols transmitted by an OSTBC is half of the number of symbols in the other codes. Therefore, to have the same transmission bit rate and a fair comparison, a larger constellation has been used for OSTBCs. It is clear from the

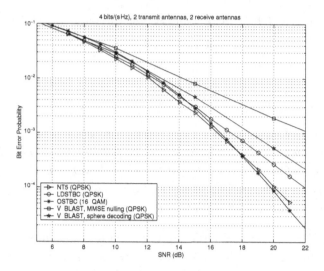

Fig. 10.1.   Bit error probability plotted against SNR at 4 bits/(s Hz); two transmit antennas, two receive antennas.

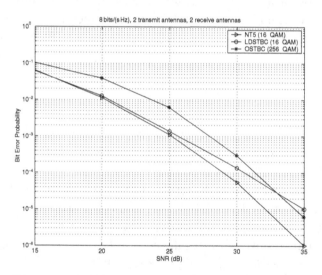

Fig. 10.2.   Bit error probability plotted against SNR at 8 bits/(s Hz); two transmit antennas, two receive antennas.

figures that NT5 and OSTBC provide full diversity and outperform the other codes at high SNRs. At 4 bits/(s Hz), the performance of OSTBC is as good as that of NT5 and better than the performance of LDSTBC and V-BLAST. For higher bit rates, for example 8 bits/(s Hz), NT5 starts to demonstrate a superior performance. For example, the performance of NT5 is about 2 dB better than that of the OSTBC at high SNRs. There are two factors that affect the performance of OSTBCs. First,

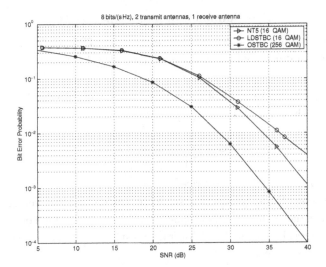

Fig. 10.3. Bit error probability plotted against SNR at 8 bits/(s Hz); two transmit antennas, one receive antenna. (Courtesy of Yun Zhu.)

OSTBCs do not provide the maximum mutual information for more than one receive antenna. Second, since OSTBCs transmit a smaller number of symbols compared to other codes in the figures, to have the same bit rate, a larger and less efficient constellation is utilized. On the other hand, one advantage of OSTBCs is the fact that the same code works for any number of receive antennas. For other considered systems, a minimum number of receive antennas may be needed. For example, if the system contains only one receive antenna, $K = 4$ and $\min\{N, M\}T = 2$, that is the condition $K \leq \min\{N, M\}T$, is violated. Therefore, the system equations may be under-determined and, for example, a V-BLAST decoding is impossible. Nevertheless, one can use an exhaustive search for ML decoding of LDSTBCs. The result of the ML decoding for two transmit antennas and one receive antenna at 8 bits/(s Hz) is shown in Figure 10.3. As can be seen from the figure, the OSTBC outperforms the other two schemes. The performance of the OSTBC is about 6 dB better than that of the NT5.

In general, OSTBCs are very attractive for low data rates and smaller number of antennas. They provide a very low complex ML decoding and can manage different number of receive antennas using MRC. Their main deficiencies are the lack of rate one codes for more than two transmit antennas and that they do not provide the maximum mutual information with the exception of the two transmit, one receive antenna case. For a high data rate and a large number of receive antennas, it is beneficial to use non-orthogonal STBCs if the complexity of the decoding is still practical.

10.6 Summary of important results

- Sphere decoding is utilized to decode a non-orthogonal STBC. The complexity of sphere decoding for non-orthogonal STBCs is more than that of the separate or pairwise ML decoding.
- A linear dispersion codeword is constructed from input constellation signals $s_1, s_2, \ldots, s_K$ and their conjugates by

$$\mathbf{C} = \sum_{k=1}^{K} \left(s_k E_k + s_k^* F_k \right),$$

where $\mathbf{C}$, E_k, and F_k are $T \times N$ matrices.

- A linear dispersion STBC is usually designed by maximizing the average mutual information between the input and output signals given a power constraint on the input power.
- An $N \times N$ diagonal algebraic STBC over data symbols $s_1, s_2, \ldots, s_N$ is defined by the following generating matrix:

$$\mathcal{G}_N = \begin{pmatrix} x_1 & 0 & \cdots & 0 \\ 0 & x_2 & \cdots & 0 \\ \vdots & \vdots & \ddots & \vdots \\ 0 & 0 & \cdots & x_N \end{pmatrix} \cdot A_N, \tag{10.50}$$

where $x_1, x_2, \ldots, x_N$ are defined by

$$(x_1, x_2, \ldots, x_N)^T = \mathbf{M}_N \cdot (s_1, s_2, \ldots, s_N)^T, \tag{10.51}$$

$\mathbf{M}_N$ is an orthogonal matrix and A_N is an $N \times N$ Hadamard matrix.

- In a threaded algebraic space-time code, a layer is defined as a part of transmission in which each time slot is allocated to, at most, one transmit antenna. At each time slot, each layer uses a different transmit antenna.
- A non-orthogonal STBC is usually designed specifically for the given number of receive antennas. This is in contrast with orthogonal and quasi-orthogonal STBCs that are designed independently of the number of receive antennas.

10.7 Problems

1 Design an LDSTBC similar to the code in Example 10.2.1 for $N = 3$ transmit antennas. How does this code relate to the corresponding V-BLAST?

2 Consider the following LDSTBC:

$$\mathbf{C} = \begin{pmatrix} s_1 & s_2 & 0 \\ -s_2^* & s_1^* & 0 \\ 0 & s_3 & s_4 \\ 0 & -s_4^* & s_3^* \\ s_5 & 0 & s_6 \\ -s_6^* & 0 & s_5^* \end{pmatrix}.$$

(a) Find the dispersion matrices of this code.

(b) Compare this code with the code in Example 10.2.2.

3 Consider the code in Example 10.3.1 using QPSK and the optimal value of $\lambda = 0.5$.

(a) Calculate all possible transmitted symbols and draw the corresponding constellation.

(b) What is the PAPR of the code and how does it compare with the PAPR of the corresponding OSTBC?

4 Calculate the CGD for the DAST block code in Example 10.4.1 using QPSK.

5 Calculate the CGD for the TASTBC in Example 10.4.3 using QPSK and $\phi = e^{j\pi/6}$.

6 Design a TASTBC for $N = 4$ transmit and $M = 4$ receive antennas with the highest possible rate.

11

Additional topics in space-time coding

11.1 MIMO-OFDM

OFDM is an efficient technique for transmitting data over frequency selective channels. The main idea behind OFDM is to divide a broadband frequency channel into a few narrowband sub-channels. Then, each sub-channel is a flat fading channel despite the frequency selective nature of the broadband channel. To generate these parallel sub-carriers in OFDM, an inverse fast Fourier transform (IFFT) is applied to a block of L data symbols. To avoid ISI due to the channel delay spread, a few "cyclic prefix" (CP) symbols are inserted in the block. The cyclic prefix samples are also called guard intervals. Basically, the last g samples of the block are duplicated in front of the block as the cyclic prefix. The number of these cyclic prefix samples, g, should be bigger than the length of the channel impulse response. The effects of the cyclic prefix samples eliminate ISI and convert the convolution between the transmit symbols and the channel to a circular convolution. These cyclic prefix samples are removed at the receiver. Then, a fast Fourier transform (FFT) is utilized at the receiver to recover the block of L received symbols. Figure 11.1 shows the block diagram of a wireless communication system using OFDM over a SISO channel. In this section, we study how to use OFDM in a system designed for MIMO channels. This is usually called MIMO-OFDM.

11.1.1 Design criteria for MIMO-OFDM

A frequency selective channel provides an additional degree of diversity that may be called "frequency diversity." As discussed in Chapter 1, a frequency selective channel can be modeled by the faded sum of a few delta functions. As presented in (1.17), we assume that the length of the channel impulse response, the number of independent delta functions, is J. Then, as we show later, the maximum possible diversity gain that can be achieved for such a model is NMJ [17, 91]. To achieve

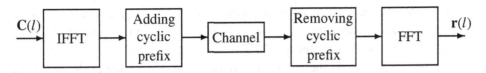

Fig. 11.1. Block diagram of an OFDM system.

such a gain, the code could operate over space, time, and frequency. In fact, the frequency selective model provides extra sources of independent paths that create additional diversity gain. Note that as long as the channel has a full rank correlation matrix, a diversity gain of NMJ is possible and the independence condition is not necessary. In fact, it may not be realistic to assume that the coefficients of the impulse response are independent. Also, a good assumption is usually to choose taps with an exponential power decay.

The goal of a space-time-frequency code is to achieve the maximum diversity by transmitting symbols through different antennas and different frequencies. Replacing the time with OFDM frequencies in the structure of space-time codes results in space-frequency codes [1]. If we treat the independent "frequency tones" as virtual antennas, space-time codes that we have studied so far can be used to achieve frequency diversity as well. Of course, the equivalent channel matrix will not be fully connected as different frequency tones will not affect each other. In what follows, we derive the design criteria for space-time-frequency codes.

For a channel with an impulse response length of J, let us denote the impulse response coefficients by $\mathbf{h}(0), \mathbf{h}(1), \ldots, \mathbf{h}(J-1)$. Note that $\mathbf{h}(i)$ is an $N \times M$ matrix including the impulse response coefficients for all path gains. We assume that a cyclic prefix of at least J samples is added at the transmitter after IFFT and is removed at the receiver before the FFT. Then, the $N \times M$ channel matrix, $\mathbf{H}(l)$, over the lth sub-carrier, frequency tone, is

$$\mathbf{H}(l) = \sum_{i=0}^{J-1} \mathbf{h}(i)\, e^{-j2\pi i l/L}, \quad l = 0, 1, \ldots, L-1. \tag{11.1}$$

Also, let us define $\mathbf{C}(l)$ as the $T \times N$ transmitted matrix over the lth sub-carrier and $\mathbf{r}(l)$ as the $T \times M$ received matrix. Then, the input–output relationship for the MIMO-OFDM system is

$$\mathbf{r}(l) = \mathbf{C}(l) \cdot \mathbf{H}(l) + \mathcal{N}(l), \quad l = 0, 1, \ldots, L-1, \tag{11.2}$$

where $\mathcal{N}(l)$ is the $T \times M$ noise matrix for the lth sub-carrier [90]. Note that the codeword $\mathbf{C}(l)$ is the collection of symbols before the IFFT and not the symbols transmitted over the channel. An IFFT is applied and cyclic prefix symbols are added before transmitting the codeword. Therefore, the codeword $\mathbf{C}(l)$ can be considered as the codeword in the frequency domain while its IFFT, transmitted over the channel, can be considered as the codeword in the time domain. This explains the use of IFFT in the transmitter and FFT in the receiver as shown in Figure 11.1. Similarly, $\mathbf{r}(l)$ represents the received signals after removing the cyclic prefix and performing FFT. Therefore, $\mathbf{H}(l)$ models the equivalent channel in the frequency domain.

Let us define $\mathbf{C}$ as the collection of L codewords $\mathbf{C}(l)$, $l = 0, 1, \ldots, L-1$, that is

$$\mathbf{C} = [\mathbf{C}(0)\,\mathbf{C}(1) \,\cdots\, \mathbf{C}(L-1)]. \tag{11.3}$$

Then, following the arguments in Section 3.2, the ML decoding for the above MIMO–OFDM system amounts to

$$\hat{\mathbf{C}} = \underset{\mathbf{C}}{\operatorname{argmin}} \sum_{l=0}^{L-1} ||\mathbf{r}(l) - \mathbf{C}(l) \cdot \mathbf{H}(l)||_F^2, \tag{11.4}$$

where $||A||_F$ is the Frobenius norm of matrix A as defined in (3.5). To find a design criterion for MIMO-OFDM systems, one may follow the same steps that we carried out in Section 3.2. The first step is to calculate the conditional pairwise error probability of transmitting $\mathbf{C}^1$ and decoding it as $\mathbf{C}^2$. This is calculated in [17, 90] as

$$P(\mathbf{C}^1 \to \mathbf{C}^2|\mathbf{H}(0), \mathbf{H}(1), \ldots, \mathbf{H}(L-1))$$
$$= Q\left(\sqrt{\frac{\gamma}{2} \sum_{l=0}^{L-1} ||[\mathbf{C}^2(l) - \mathbf{C}^1(l)] \cdot \mathbf{H}(l)||_F^2}\right). \tag{11.5}$$

Then, as in Section 3.2, using the upper bound $Q(x) \le \frac{1}{2} e^{\frac{-x^2}{2}}$ on the Q function results in

$$P\left(\mathbf{C}^1 \to \mathbf{C}^2|\mathbf{H}(0), \mathbf{H}(1), \ldots, \mathbf{H}(L-1)\right) \le \frac{1}{2} \exp\left(-\frac{\gamma}{4} d^2(\mathbf{C}^1, \mathbf{C}^2|\mathbf{H})\right), \tag{11.6}$$

where

$$d^2\left(\mathbf{C}^1, \mathbf{C}^2 \big| \mathbf{H}\right) = \sum_{l=0}^{L-1} \left|\left|[\mathbf{C}^2(l) - \mathbf{C}^1(l)] \cdot \mathbf{H}(l)\right|\right|_F^2. \tag{11.7}$$

The last step in finding an upper bound for the pairwise error probability is to calculate the expected value of the conditional bound in (11.6) over the channel matrices. Towards this goal, first, we define the following $NJ \times 1$ channel matrix in the time domain:

$$
\mathbf{h}_m = \begin{pmatrix} \mathbf{h}_{1,m}(0) \\ \mathbf{h}_{1,m}(1) \\ \vdots \\ \mathbf{h}_{1,m}(J-1) \\ \vdots \\ \mathbf{h}_{N,m}(0) \\ \mathbf{h}_{N,m}(1) \\ \vdots \\ \mathbf{h}_{N,m}(J-1) \end{pmatrix}.
\tag{11.8}
$$

We assume that $\mathbf{h}_m$ has an $N J \times N J$ full-rank correlation matrix $\mathbf{R}_h = E[\mathbf{h}_m \cdot \mathbf{h}_m^H]$. Note that $\mathbf{R}_h$ is independent of the receive antenna. We also assume that the matrices $\mathbf{h}_m$ are zero-mean complex Gaussian and independent from each other. Then, since $\mathbf{R}_h$ is positive definite and Hermetian, it can be decomposed to the product of its square root and the Hermetian of its square root, that is

$$
\mathbf{R}_h = \mathbf{R}_h^{1/2} \cdot \left(\mathbf{R}_h^{1/2} \right)^H.
\tag{11.9}
$$

Then, we define the following matrices from [90]:

$$
\underline{\omega}(l) = \begin{pmatrix} 1 \\ e^{-j2\pi l/L} \\ \vdots \\ e^{-j2\pi(J-1)l/L} \end{pmatrix},
\tag{11.10}
$$

$$
\mathbf{\Omega}(l) = I_N \otimes \underline{\omega}(l),
\tag{11.11}
$$

where $\underline{\omega}(l)$ is an $J \times 1$ vector and $\mathbf{\Omega}(l)$ is an $N J \times N$ matrix. Then, the upper bound for the pairwise error probability can be written in terms of an $N J \times N J$ matrix $\mathbf{\Lambda}$, which is defined as

$$
\mathbf{\Lambda} = \sum_{l=0}^{L-1} \mathbf{\Omega}(l) \cdot [\mathbf{C}^2(l) - \mathbf{C}^1(l)]^T \cdot [\mathbf{C}^2(l) - \mathbf{C}^1(l)]^* \cdot \mathbf{\Omega}(l)^H,
\tag{11.12}
$$

or its modified version

$$
\mathbf{\Lambda}' = \left(\mathbf{R}_h^{1/2} \right)^T \cdot \mathbf{\Lambda} \cdot \left(\mathbf{R}_h^{1/2} \right)^*.
\tag{11.13}
$$

Finally, an upper bound for the pairwise error probability is [90]:

$$
P\left(\mathbf{C}^1 \to \mathbf{C}^2 \right) \le \left(\frac{G_c \gamma}{4} \right)^{-G_d},
\tag{11.14}
$$

where

$$G_d = \text{rank}(\mathbf{\Lambda}')M, \tag{11.15}$$

$$G_c = \left[\prod_{i=1}^{\text{rank}(\mathbf{\Lambda}')} \frac{\lambda_i}{J} \right]^{1/\text{rank}(\mathbf{\Lambda}')}, \tag{11.16}$$

and λ_i is the ith non-zero eigenvalue of $\mathbf{\Lambda}'$. Similar to the case of flat fading with independent channel gains, G_d and G_c play the roles of diversity gain and coding gain, respectively. Note that if the channel correlation matrix $\mathbf{R}_h$ is full rank, then the maximum possible diversity gain is

$$\max G_d = NMJ. \tag{11.17}$$

Assuming a full rank $\mathbf{R}_h$, the matrix $\mathbf{\Lambda}'$ is full rank if and only $\mathbf{\Lambda}$ is full rank. One necessary condition, based on the definition of the matrix $\mathbf{\Lambda}$ is to have $L \geq J$. In other words, the number of sub-carriers, frequency tones, should be at least equal to the impulse response length of the channel. In the case of a full diversity space-time-frequency code, the coding gain in (11.16) can be simplified to

$$G_c = \frac{1}{J}[\det(\mathbf{\Lambda}')]^{1/NJ} = \frac{1}{J}[\det(\mathbf{R}_h)\det(\mathbf{\Lambda})]^{1/NJ}. \tag{11.18}$$

Therefore, the rank and determinant criteria in Chapter 3 translates to the rank and determinant of $\mathbf{\Lambda}'$. Equivalently, when the correlation matrix is full rank, a good design strategy is to make sure that $\mathbf{\Lambda}$ is full rank for all possible pairs of codewords and its minimum determinant is maximized.

11.1.2 Space-time-frequency coded OFDM

Space-time coding provides diversity gain by coding over spatial, that is multiple antennas, and time dimensions. Transmitting data over MIMO-OFDM systems is possible by applying space-time codes to each sub-carrier. While this allows the use of space-time codes that we have studied so far, it does not provide the maximum possible diversity gain. In fact, the frequency diversity and the correlation among different sub-carriers are ignored in such a system.

Another approach for transmission over MIMO channels using OFDM is to replace the "time" dimension with the "frequency" dimension [1]. In other words, different sub-carriers of OFDM can be used as a replacement for the time dimension of a space-time code. The result is called space-frequency coding and Figure 11.2 depicts its block diagram. The space-time encoder in Figure 11.2 generates $L \times N$ symbols at each OFDM time slot. The interleaver transmits the symbol (l, n) through the lth sub-carrier of the nth antenna. We describe a simple example of such

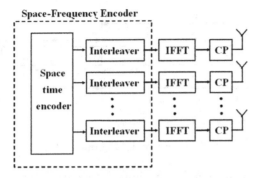

Fig. 11.2.   Block diagram of space-frequency coding.

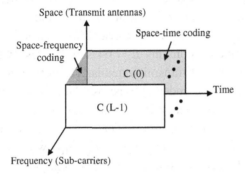

Fig. 11.3.   Space-time-frequency coded OFDM.

a code using the Alamouti code [17]. Let us assume L input symbols $s_1, s_2, \ldots, s_L$ at each time slot. The Alamouti code generates $\mathbf{C}(1) = (s_1\, s_2)$ and $\mathbf{C}(2) = (-s_2^* \, s_1^*)$ from the first two symbols, $\mathbf{C}(3) = (s_3\, s_4)$ and $\mathbf{C}(4) = (-s_4^* \, s_3^*)$ from the third and fourth symbols, and so on. Therefore, at each time slot, the first antenna transmits $s_1, -s_2^*, s_3, -s_4^*, \ldots, s_{L-1}, -s_L^*$ on sub-carriers $0, 1, \ldots, L-1$, respectively. Similarly, the second antenna transmits $s_2, s_1^*, s_4, s_3^*, \ldots, s_L, s_{L-1}^*$ on sub-carriers $0, 1, \ldots, L-1$, respectively. Now, let us pick two codewords $\mathbf{C}^1$ and $\mathbf{C}^2$ that differ only on the first symbols s_1^1 and s_1^2, that is $s_1^1 \neq s_1^2$ and $s_l^1 = s_l^2$, $l = 2, 3, \ldots, L$. Then, the rank of matrix $\mathbf{\Lambda}$, defined in (11.12), is two and the code provides a diversity of $2M$, which is less than the maximum possible NJM for $J > 1$.

To achieve the maximum possible diversity gain, one could code over three dimensions of space, time, and frequency as shown in Figure 11.3. It is clear that transmitting the codeword over different sub-carriers can provide additional frequency diversity. One approach to transmit a codeword over different sub-carriers is to use error correcting codes and interleaving [84, 52]. The role of interleaving is to make sure that the coding is over space, time, and frequency. Different error correcting codes can be utilized for this purpose. For example, convolutional codes, Reed–Solomon codes, or turbo codes can be viable candidates [84].

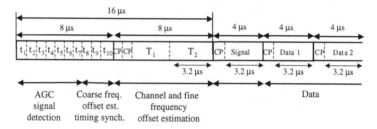

Fig. 11.4. IEEE 802.11a preamble structure and acquisition functions.

One approach to achieving the maximum possible diversity using the space-time codes designed for flat fading channels is to consider sub-carriers as additional transmit antennas [93]. For example, a space-time code designed for NL antennas can be used to transmit over N transmit antennas and L sub-carriers. This may not be practical as the number of sub-carriers, L, is relatively large. For example the IEEE 802.11 standard contains $L = 64$ sub-carriers. It is difficult to design codes for a large number of transmit antennas. Therefore, it is desirable to group the sub-carriers and design codes only across the sub-carriers in the same group [93, 90]. A space-time-frequency code over three dimensions of space, time, and frequency is designed for each group. This reduces the frequency dimension of the code and decreases the complexity. For groups with at least J sub-carriers, if designed carefully, such a grouping will not degrade the diversity gain of the space-time-frequency codes.

In this section, we have only discussed schemes to achieve the maximum diversity; however it is possible to design BLAST-OFDM systems based on spatial multiplexing. Usually, the impact of OFDM is very similar in the case of a BLAST system. In the next section, we discuss the implementation issues in MIMO-OFDM systems in a general way so that it covers both space-time-frequency coding and BLAST-OFDM systems.

11.2 Implementation issues in MIMO-OFDM

To implement an OFDM system, it is important to pay attention to practical issues like acquisition algorithms, synchronization, and channel estimation. These issues become more important in OFDM since the transmission is over different frequencies. Operating over MIMO channels makes acquisition and channel estimation even more difficult.

11.2.1 Acquisition and synchronization

Figure 11.4 shows the preamble structure and acquisition functions for IEEE 802.11a. In what follows, we define and explain the details of acquisition functions.

However, before that, first we demonstrate a typical block diagram of a receiver to provide the relationship between different blocks of the system. Figure 11.5 depicts the block diagram of an IEEE 802.11a receiver with two receive antennas. The system is designed for a SISO channel; however, the pair of receive antennas is utilized to provide diversity using selection combining as explained in Section 1.5.2. The main goal of this system is to take advantage of existing wireless standards. To increase the data rate of wireless communication systems or to improve the performance, equivalently to increase the coverage, it is desirable to benefit from the higher capacity and diversity gain of MIMO channels. As a result, the use of OFDM in MIMO channels has been the topic of extensive recent research [124]. Figure 11.6 demonstrates the block diagram of a MIMO-OFDM system including the acquisition functions. In this figure, the "Spatial processing" could be space-time-frequency coding or BLAST. Also, the inclusion and removal of cyclic prefix symbols are parts of the "Tx shaping" and "Filter" blocks, respectively.

The primary functions of a receiver acquisition subsystem are:

- automatic gain control (AGC);
- signal detection (preamble detection);
- automatic carrier frequency offset estimation and correction;
- timing synchronization

Note that the above acquisition subsystem includes synchronization as well. We explain each of these functions in the sequel. First, we discuss these functions for a system transmitting over SISO channels. At each case, we explain specific challenges due to the effects of MIMO channels (if any).

The function of the AGC is to make sure that the receiver can operate under a wide dynamic range. For example, an IEEE 802.11a receiver should work at SNRs of -30 dBm to -85 dBm. The analog signal must be automatically scaled prior to entering the analog to digital converter (ADC) block. This is for the purpose of maximizing the dynamic range while minimizing the required word width of the converter. In a MIMO channel, the various antenna outputs have different strengths. If we run independent AGC blocks on each path, the weaker signals are amplified more than the strong signals. As a result, the overall SNR will degrade when we combine all the streams. Generally, the measurements need to be made on each individual stream, while the optimum gains of the streams are decided jointly by a central processor.

The function of signal detection or preamble detection is to detect the presence of a known preamble pattern defined by wireless standards at the beginning of a packet. A typical approach for this operation is to perform an autocorrelation operation based on a known window size. The preamble patterns are usually defined in a manner to have strong autocorrelation properties.

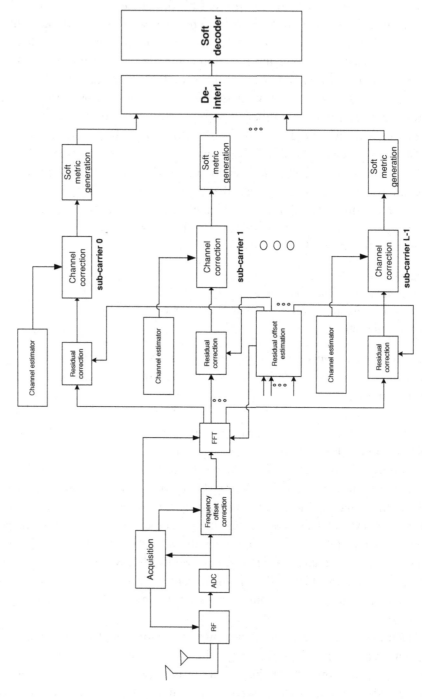

Fig. 11.5. An IEEE 802.11a receiver with selection combining using two receive antennas. (Courtesy of Hooman Honary.)

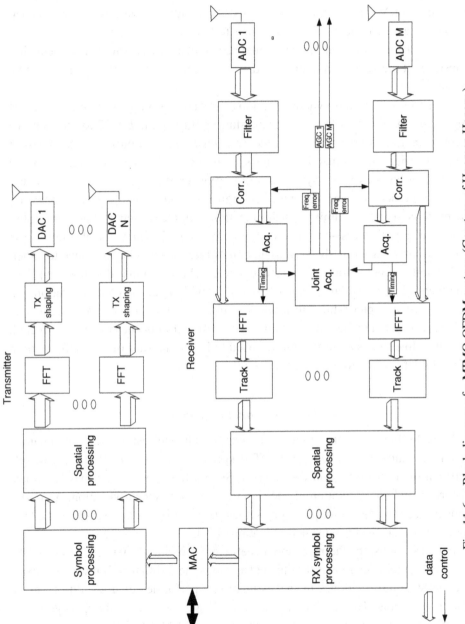

Fig. 11.6. Block diagram of a MIMO-OFDM system. (Courtesy of Hooman Honary.)

The function of automatic carrier frequency offset estimation and correction is to measure and correct the carrier frequency offset between the receiver and the transmitter. Typically this value is calculated using an autocorrelation operation. In the case of MIMO systems, the results of the autocorrelations applied to each antenna stream can be combined together to provide a more accurate estimate of the error. The idea of combining the antenna streams for frequency-offset detection is justified if the same oscillator is used at all the mixers of the transmit and receive antennas.

The function of timing synchronization is to determine the proper OFDM symbol alignment. In other words, it locates the starting time of the L FFT coefficients. To recover the transmitted symbols, a receiver must be synchronized to the received signal. Carrier synchronization is not specific to OFDM systems and is needed in other systems as well. Synchronization reduces the effects of frequency and phase differences between the received signal and the signal that is expected by the receiver. Typically timing synchronization is implemented as a matched filter. Since the signal is cross-correlated with the training sequence, the absolute phase is important. As a result, the various antenna streams cannot be combined together.

A detailed scheme for time and frequency synchronization is discussed in [124]. The main idea, as explained above, is to start with a coarse synchronization and detection of the signal while fine tuning it in different steps. The resulting frequency offset is estimated in the time domain and the residual frequency offset is corrected. Finally, a fine time synchronization is used to locate the start of the useful portion of the OFDM frame to within a few samples.

11.2.2 Channel estimation

A MIMO-OFDM system requires an accurate channel estimation. The channel parameters must be tracked closely. The system is sensitive to frequency offsets between the transmitter and the receiver. Such a frequency offset may introduce distortion and phase rotation in addition to the loss of synchronization. A continuous protection against frequency offset and phase offset is necessary even after a successful acquisition and synchronization [157].

In real systems, the fading channel coefficients are unknown to the receiver. The receiver explicitly estimates the channel and uses the estimate as if it were the exact channel state information. Channel estimation is performed by transmitting known training symbols. The receiver uses its knowledge about the training sequence to estimate the channel. To estimate the channel in a MIMO-OFDM system, the path gain for each pair of transmit-receive antennas and each carrier frequency needs to be estimated. The channel responses at different frequencies are correlated to each other. This can be used to simplify the channel estimation [81].

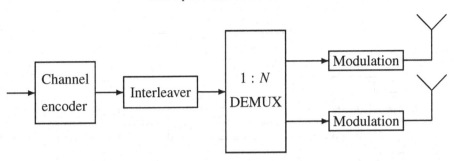

Fig. 11.7. Block diagram of a space-time turbo encoder using serial concatenation.

One approach to channel estimation for MIMO-OFDM systems is to multi-plex between different sub-carriers on different antennas at any given time. At the receiver, after the FFT, the channel can be estimated just like the channel estimation for a SISO channel. The role of multiplexing among different receive antennas is to guarantee that at any given time only one path exists between a transmit and a receive antenna. While this is a simple channel estimation approach that basically uses the channel estimation techniques used for SISO channels, it requires a long period of time to cover all the MIMO paths over all sub-carriers. Therefore, a long preamble should be used that results in a large overhead.

Another approach to channel estimation for MIMO-OFDM systems is to change the role of the codeword and channel during the channel estimation phase. A coher-ent detector uses the knowledge of channel $\mathbf{H}$ in equation $\mathbf{r} = \mathbf{C} \cdot \mathbf{H} + \mathcal{N}$ to decode the transmitted codeword $\mathbf{C}$. During the channel estimation period, the receiver knows the codeword $\mathbf{C}$ and can utilize it to estimate the channel matrix $\mathbf{H}$. For example, similar to a spatial multiplexing system, like V-BLAST, one can trans-mit a training sequence that spans all frequencies. Then, the receiver collects all the signals over a period of time and uses ZF or MMSE to estimate the channel. Another scheme is to use orthogonal designs as training sequences. This second approach uses a shorter and more efficient preamble, compared to the previous SISO approach. On the other hand, the corresponding estimation algorithm is more complicated.

11.3 Space-time turbo codes

One approach to provide spatial diversity is to use error correcting codes and time interleaving before a serial-to-parallel converter sends the symbols to different transmit antennas [16]. Note that this is a realization of bit-interleaved coded mod-ulation [19]. In this case, as in Figure 11.7, the spatial diversity is due to the fact that the signals transmitted through different antennas are dependent through the error correcting code in the system. It is shown that the spatial diversity gain for

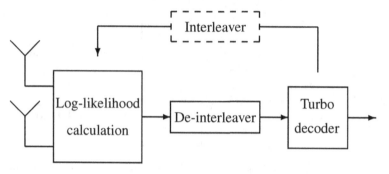

Fig. 11.8.   Block diagram of a space-time turbo decoder.

Rayleigh fading is determined by the minimum distance of the block codes and the free distance of the convolutional codes if soft-decision is used at the receiver. When hard-decision is used, the diversity gain is reduced by a factor of two. The diversity advantage provided through channel coding and interleaving motivates the use of more powerful codes like turbo codes. The performance of turbo codes [11] is very close to the Shannon's theoretical limits for AWGN channels. The main ideas behind turbo coding are serial or parallel concatenation of codes separated or connected by an interleaver at the transmitter and the use of iterative decoding at the receiver. While the interleaved concatenation resembles the performance of random codes, the iterative decoding makes the implementation practical. Since the introduction of turbo codes in 1993, different new turbo coding structures have been proposed in the literature. This is now the subject of a few books, for example [106, 147]. In this section, we show how similar principles can be utilized to design space-time turbo codes. The main goal is to achieve diversity gains of space-time codes in addition to coding gains of turbo codes. We do not discuss the detailed iterative decoding formulas for each scheme. Instead, we concentrate on the general structure of the codes and different possible configurations and refer the interested reader to the corresponding references for more details.

The main difference between the structure of different space-time turbo codes is in the choice of component codes and their concatenation with the interleaver. Trellis codes can be used as component codes instead of binary convolutional codes or block codes [88]. The block diagram of one such an example of space-time turbo code is depicted in Figure 11.7. The input bits are encoded by a binary turbo code and sent through an interleaver. Then, the resulting bits are converted to an N parallel set of bits to select symbols from the modulation constellation. The output of each parallel branch is transmitted from the corresponding transmit antenna. The turbo code may consist of the parallel concatenation of two systematic recursive convolutional codes via a pseudorandom interleaver [122]. Using QPSK, one may design a rate one 2 bits/(s Hz) code; however, maximum diversity is not guaranteed. Two possible decoders for the above code are depicted in Figure 11.8,

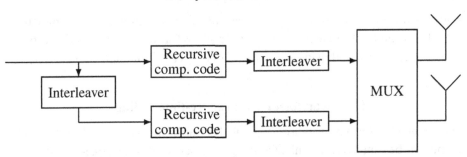

Fig. 11.9. Block diagram of a space-time turbo encoder using parallel concatenation.

where the dashed section of the diagram may be included or removed. The first decoder includes a log-likelihood calculation block, a de-interleaver, and a turbo decoder. In addition, the second decoder iterates between the turbo decoder and the log-likelihood calculation block by using the dashed connection through its interleaver [122]. This improves the performance while increasing the complexity. The main idea is to use the estimates of the probabilities of the transmitted symbols in the demodulation block. These probabilities are calculated from the log-likelihoods of the extrinsic bit information in the turbo decoder block. This is called an iterative demodulation-decoding algorithm [80] or a doubly-iterative decoder [15].

In the encoder block diagram of Figure 11.7, one may consider the combination of the serial to parallel converter and the modulation blocks as a "spatial multiplexing" block. In fact, the space-time processing is done by this spatial multiplexing block and the dependence of its input symbols. Another approach to space-time turbo codes is to replace such a spatial multiplexing block with a space-time coding block. One example of such a code is presented in [10] where a turbo encoder is concatenated with an orthogonal STBC. Another example is when the spatial multiplexing block is replaced with a recursive STTC and the channel code before interleaving is a convolutional code [95]. These are all examples of serially concatenated encoders. They provide maximum diversity gain but the rate is not one symbol per time slot. Rate one serially concatenated codes are possible by using an outer space-time trellis code, a set of interleavers, and a set of inner recursive convolutional codes [85, 89]. Each output of the space-time trellis code is interleaved and coded by the recursive convolutional code.

Parallel concatenation of component codes is also possible as shown in Figure 11.9. For two and three transmit antennas, it is possible to design full diversity codes for BPSK using recursive convolutional codes as component codes [125]. In [89], turbo codes are used as component codes. In this case, it is difficult to design the interleavers systematically to achieve full diversity; however, a randomly selected interleaver provides full diversity with high probability. It is possible to use recursive STTCs as component codes as well [95, 30]. However, the code construction procedure does not guarantee full diversity although the simulation results for a randomly generated interleaver show full diversity. A full diversity, full rate code is achieved using recursive SOSTTCs as component codes [104].

11.4 Beamforming and space-time coding

When channel information is available at the transmitter, a linear beamforming scheme can be employed to enhance the performance of a multiple-antenna system. When the transmitter knows the channel perfectly, space-time coding is not as advantageous as beamforming. Beamforming can be done in the analog domain by altering the radiation pattern of the arrays. Also, it is possible to "steer" the digital baseband signals by weighting them differently. This is named "digital beamforming" and we call it beamforming for the sake of brevity. The weighted sum of the baseband signals forms the desired signal. Then, to create the best "beam," one needs to find the optimal weights. The optimal weights are adaptively changed to accommodate the changing environments. This has been the subject of research for many years [86].

To perform beamforming, when perfect channel information is available at the transmitter, one needs to do singular value decomposition on the channel matrix **H**. This is also called "eigen-beamforming" since it uses eigenvectors to find the linear beamformer that optimizes the performance. The result of singular value decomposition can be found in (9.19), which is repeated here:

$$\mathbf{H} = U \cdot \Sigma \cdot V^H. \qquad (11.19)$$

Then, the beamforming is implemented by applying U^H to the data symbol matrices before transmission. Also, water-filling is used at the transmitter to optimally allocate the power on different directions. If there is only one receive antenna, the optimal allocation of the power results in transmitting only along the strongest direction. At the receiver, matrix V is applied to the receive vectors.

Normally, the channel parameters are estimated at the receiver and sent back to the transmitter. Channel information, if available at the transmitter, can be utilized to further improve the system performance. Due to the bandwidth limitation on the feedback channel, the CSI is usually partial and non-perfect [75]. When partial channel state information is available at the transmitter, the performance of the above eigen-beamforming deteriorates. On the other hand, utilizing space-time coding provides diversity gain. With partial channel information, the transmitted codeword can be tuned to better fit the channel. To maintain low implementation complexity, a simple linear beamforming scheme at the transmitter is preferred. To utilize a non-perfect CSI, one can combine beamforming and space-time coding as depicted in Figure 11.10. One example of such a system is the OSTBC beam former

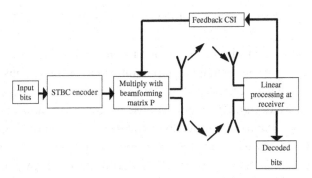

Fig. 11.10. Combined space-time block coding and beamforming in wireless communication systems.

from [76] and another example is the QOSTBC beamformer from [87]. Simulation results show that these beamformers significantly outperform the conventional one-directional beamformer when only partial CSI is available at the transmitter. Another similar method for combining orthogonal space-time block coding and beamforming was introduced in [160]. The power allocation scheme of [160] is similar to that of [76]. The optimality condition in [160] is equivalent to the optimality condition in [76] when OSTBCs are used. Moreover, a novel two-directional beamformer based on the Alamouti code was introduced in [160]. The two-directional beamformer includes the conventional one-directional beamformer as a special case and outperforms it uniformly. The two-directional beamformer also accomplishes full transmission rate for systems with more than two transmit antennas.

The simple linear beamforming scheme in Figure 11.10 utilizes a predetermined STBC. Before the codewords are sent out from the N transmit antennas, they are multiplied by a beamforming matrix $\mathbf{P}$. The actual transmitted codeword can be expressed as

$$\bar{\mathbf{C}} = \mathbf{C} \cdot \mathbf{P}, \tag{11.20}$$

where $\mathbf{C}$ is the predetermined un-beamformed codeword. Let us assume that $\mathbf{C}$ is the codeword of an $N \times N$ STBC with rate R that provides full diversity N. Note that by using an $N \times N'$ beamforming matrix $\mathbf{P}$, we can transmit over N' transmit antennas while providing an N-directional beamformer. For example, we may use the 2×2 OSTBC in (4.75) and a 2×2 beamforming matrix $\mathbf{P}$. The result is a rate one two-directional beamformer for two transmit antennas. Using the 4×4 OSTBC in (4.103) results in a rate 3/4 four-directional eigen-beamformer. If a QOSTBC is utilized instead of an OSTBC, a rate one four-directional beamformer is also possible. For example, we may use the 4×4 QOSTBC in (5.2) and any $4 \times N'$ beamforming matrix $\mathbf{P}$ to design a four-directional beamformer for any N' transmit antennas. Note that since the embedded STBC codeword **C** in (11.20) accomplishes full transmission rate, so does the beamformed codeword **C̄**.

Using the structure in Figure 11.10, one has to find the optimal beamforming matrix **P**. Using the union bound, the maximum value of the pairwise error probability dominates the performance. Minimizing this maximum value, the worst-case scenario, provides a design criterion to choose the optimal beamforming matrix **P**. Let us assume the channel mean feedback model in [76] that is a popular model for this case. Based on this model, the conditional mean of the path gains given the partial CSI is known at the transmitter. Also, the covariance between different path gains given the partial CSI is

$$\text{cov}(\alpha_{n,m}\alpha_{n',m'}|E[H]) = \begin{cases} \sigma^2 & \text{if } n = n' \text{ and } m = m' \\ 0 & \text{otherwise.} \end{cases} \quad (11.21)$$

Then, the design criterion results in minimizing the sum of two terms [76]:

$$\min\{f_1 + f_2\}. \quad (11.22)$$

The first term, f_1, is a function of the channel mean and becomes the dominating factor when a perfect CSI is available. The second term, f_2, is basically the same as the determinant criterion for space-time codes, presented in Chapter 3. Therefore, f_2 is a good indication of the degree of the spatial diversity of the system. When the quality of CSI is low, the second term becomes dominant and most of the performance gains come from the spatial diversity. Optimizing the beamforming matrix **P** based on the above design criterion results in a system that performs well in the case of partial CSI while providing optimal performance in other cases. When there is no CSI available, the transmit energy is evenly distributed on all beams. In other words, the system behaves like its STBC component in Figure 11.4. In this case, both the OSTBC beamformer and the QOSTBC beamformer fall back to their un-beamformed schemes. When the CSI at the transmitter is perfect, all the energy is allocated on the strongest beam. In this case, the beamformed STBCs fall back to the one-directional beamformer.

Figures 11.11 and 11.12 compare the performance of different schemes at rate 3 bits/(s Hz) assuming different channel feedback qualities. We consider a system with four transmit antennas and one receive antenna. We compare the performance of the QOSTBC beamformer [87], the conventional one-directional beamformer, the rate 3/4 OSTBC beamformer [76], and the full rate two-directional beamformer [160]. To make a fair comparison, a larger constellation has to be assigned to the rate 3/4 OSTBC beamformer to keep the transmission rate identical for all schemes. Therefore, the 16-QAM constellation is adopted for the rate 3/4 OSTBC beamformer; the 8-QAM constellation in [126] is used for the other three full rate beamformers. As it can be seen from the figures, at low SNRs, the QOSTBC

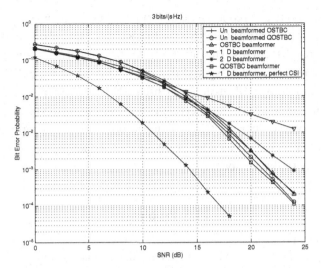

Fig. 11.11. Performance comparison for bad channel feedback quality at 3 bits/(s Hz) and the case of perfect CSI; four transmit antennas, one receive antenna.

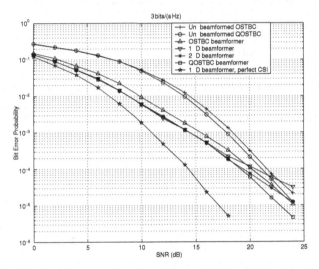

Fig. 11.12. Performance comparison for good channel feedback quality at 3 bits/(s Hz) and the case of perfect CSI; four transmit antennas, one receive antenna.

beamformer shows similar performance to the one-directional and two-directional beamformers of [160], they all outperform the OSTBC beamformer. This is because at low SNRs, all the above beamformers are equivalent to the conventional one-directional beamformer. While only the OSTBC beamformer suffers from a rate loss, the other three still maintain full rates. On the other hand, at high SNRs, both the one-directional and two-directional beamformers become inferior to the QOSTBC beamformer. This is because at high SNRs, the degree of spatial diversity dictates the slope of the BER-SNR curves. The one- and two-directional beamformers obviously suffer from a lower degree of spatial diversity, while the OSTBC and QOSTBC beamformers still enjoy full diversity at high SNRs. A better quality of channel feedback results in a larger gap between the un-beamformed and beamformed STBCs. In the case of perfect CSI a conventional 1-D beamformer is the best choice. Such results are provided as a reference of the lost performance due to the imperfect quality of the feedback.

11.5 Problems

1 Prove Equation (11.5).

2 Consider a space-frequency code for two transmit antennas that uses four tones and no cyclic prefix. Also, consider an Alamouti code as the space-time encoder building block in Figure 11.2. If the input bitstream is 01110010, using QPSK, what are the transmitted symbols from each antenna?

3 Provide an encoder block diagram of a space-time turbo code that uses recursive space-time trellis codes as component codes.

4 Consider a channel with N transmit antennas and one receive antenna. Let us assume that a data symbol s is transmitted by $\mathbf{C} = Ws$, where W is a $1 \times N$ vector. If both transmitter and receiver know the $N \times 1$ channel vector $\mathbf{H}$, the optimal beamforming maximizes the receive SNR for a given transmission power. Find the resulting optimal vector $W_{\text{beamforming}}$ if the data symbol s is from an equal energy constellation.

5 Consider a MIMO-OFDM system with two transmit antennas and one receive antenna that uses four sub-carriers. Also, assume that the channel has an impulse response length of two. Show that a space-time-frequency code defined as

$$\mathbf{C} = [\mathbf{C}(0) \ \mathbf{C}(1) \ \mathbf{C}(2) \ \mathbf{C}(3)] = \begin{pmatrix} s_1 & s_2 & s_1 & s_2 & s_3 & s_4 & s_3 & s_4 \\ -s_2^* & s_1^* & -s_2^* & s_1^* & -s_4^* & s_3^* & -s_4^* & s_3^* \end{pmatrix},$$

provides the maximum possible diversity gain.

References

[1] Agrawal, D., Tarokh, V., Naguib, A. and Seshadri, N. Space-time coded OFDM for high data-rate wireless communication over wideband channels. *IEEE Vehicular Technology Conference*, May 1998, 2232–6.

[2] Agrell, E., Eriksson, T., Vardy, A. and Zeger, K. Closest point search in lattices. *IEEE Trans. on Information Theory*, **48**(8): Aug. 2002, 2201–14.

[3] Aktas, D. and Fitz, M. P. Distance spectrum analysis of space-time trellis-coded modulations in quasi-static Rayleigh-fading channels. *IEEE Trans. on Information Theory*, **49**(12): Dec. 2003, 3335–44.

[4] Alamouti, S. M. A simple transmitter diversity scheme for wireless communications. *IEEE Journal on Selected Areas in Communications*, **16**(8): Oct. 1998, 1451–8.

[5] Alamouti, S., Tarokh, V. and Poon, P. Trellis-coded modulation and transmit diversity: design criteria and performance evaluation. *IEEE International Conference on Universal Personal Communication (ICUPC-98)*, **2**: 1998, 917–20.

[6] Babai, L. On Lovasz lattice reduction and the nearest lattice point problem. *Combinatorica*, **6**(1): 1986, 1–13.

[7] Baker, P. A. Phase-modulation data sets for serial transmission at 2000 and 2400 bits per second, part I. *IEEE Trans. Commun. Electron*, July 1962.

[8] Balaban, N. and Salz, J. Dual diversity combining and equalization in digital cellular mobile radio. *IEEE Trans. on Vehicular Technology*, **40**: May 1994, 342–54.

[9] Baro, S., Bauch, G. and Hansmann, A. Improved codes for space-time trellis-coded modulation. *IEEE Communications Letters*, **4**(1): Jan. 2000, 20–2.

[10] Bauch, G. Concatenation of space-time block codes and turbo-tcm. *IEEE International Conference on Communications (ICC)*, June 1999, 1202–6.

[11] Berrou, C., Glavieux, A. and Thitmajshima, P. Near Shannon limit error-correcting coding and decoding: turbo codes. *IEEE International Conference on Communications (ICC)*, May 1993, 1064–70.

[12] Bertsekas, D. P. *Nonlinear Programming*. Athena Scientific Publishing, 1999.

[13] Bertsekas, D. P. *Dynamic Programming and Optimal Control*. Athena Scientific, 2001.

[14] Biglieri, E., Divsalar, D., McLane, P. J. and Simon, M. K. *Introduction to Trellis Coded Modulation with Applications*. Prentice Hall, 1992.

[15] Biglieri, E., Nordio, A. and Taricco, G. Doubly-iterative decoding of space-time turbo codes with a large number of antennas. *IEEE International Conference on Communications (ICC)*, **1**: June 2004, 473–7.

[16] Bjerke, B. and Proakis, J. G. Multiple-antenna diversity techniques for transmission over fading channels. *IEEE Wireless Communications and Networking Conference (WCNC)*, **3**: Sept. 1999, 1038–42.

[17] Bolcskei, H. and Paulraj, A. Space-frequency coded broadband OFDM systems. *IEEE Wireless Communications and Networking Conference (WCNC)*, Sept. 2000, 1–6.

[18] Boutros, J. and Viterbo, E. Signal space diversity: A power- and bandwidth-efficient diversity technique for the Rayleigh fading channel. *IEEE Trans. on Information Theory*, **44**(4): July 1998, 1453–67.

[19] Caire, G., Taricco, G. and Biglieri, E. Bit-interleaved coded modulation. *IEEE Trans. on Information Theory*, **44**: May 1998, 927–46.

[20] Calderbank, A. R., Pottie, G. and Seshadri, N. Co-channel interference suppression through time/space diversity. *IEEE Trans. on Information Theory*, **46**(3): May 2000, 922–32.

[21] Chen, Z., Yuan, J. and Vucetic, B. An improved space-time trellis coded modulation scheme on slow Rayleigh fading channels. *IEEE International Conference on Communications (ICC)*, June 2001, 1110–16.

[22] Chen, Z., Yuan, J. and Vucetic, B. Improved space-time trellis coded modulation scheme on slow fading channels. *Electronics Letters*, **37**(7): Mar. 2001, 440–1.

[23] Chen, Z., Zhu, G., Shen, J. and Liu, Y. Differential space-time block codes from amicable orthogonal designs. *IEEE Wireless Communications and Networking Conference (WCNC)*, **2**: Mar. 2003, 768–72.

[24] Cimini, L. J. and Sollenberger, N. R. Peak-to-average power ratio reduction of an OFDM signal using partial transmit sequences. *IEEE Communications Letters*, **4**(3): Mar. 2000, 86–8.

[25] Clarkson, K. L., Sweldens, W. and Zheng, A. Fast multiple antenna differential decoding. *IEEE Trans. on Communications*, **49**: Feb. 2001, 253–61.

[26] Conway, J. H. and Sloane, N. J. A. *Sphere Packings, Lattices and Groups.* Springer-Verlag, 1999.

[27] Cover, T. M. and Thomas, J. A. *Elements of Information Theory.* Wiley, 1991.

[28] Coveyou, R. R. and MacPherson, R. D. Fourier analysis of uniform random number generators. *Journal Assoc. Comput. Mach.*, **14**: Jan. 1967, 100–19.

[29] Craig, J. W. A new, simple and exact result for calculating the probability of error for two-dimensional signal constellations. *IEEE MILCOM*, 1991, 25.5.1–25.5.5.

[30] Cui, D. and Haimovich, A. M. Performance of parallel concatenated space-time codes. *IEEE Communications Letters*, **5**(6): June 2001, 236–8.

[31] Dalton, L. A. and Georghiades, C. N. A four transmit antenna, quasi-orthogonal space-time block code with full diversity and rate. *Allerton Conference on Communication, Control and Computing*, Oct. 2002, 1104–5.

[32] Damen, M. O., Abed-Meraim, K. and Belfiore, J. C. Diagonal algebraic space-time block codes. *IEEE Trans. on Information Theory*, **48**: Mar. 2002, 628–36.

[33] Damen, M. O., El-Gamal, H. and Beaulieu, N. C. Linear threaded algebraic space-time constellations. *IEEE Trans. on Information Theory*, **49**(10): Oct. 2003, 2372–88.

[34] Damen, M. O., Tewfik, A. and Belfiore, J.-C. A construction of a space-time code based on number theory. *IEEE Trans. on Information Theory*, **48**(3): Mar. 2002, 753–60.

[35] Damen, O., Chkeif, A. and Belfiore, J.-C. Lattice code decoder for space-time codes. *IEEE Communications Letters*, **4**(5): May 2000, 161–3.

[36] Dieter, U. How to calculate shortest vectors in a lattice. *Math. of Comput.*, **29**: July 1975, 827–33.

[37] Divsalar, D. and Simon, M. K. Multiple trellis coded modulation (MTCM). *IEEE Trans. on Communications*, **36**: Apr. 1988, 410–19.

[38] EIA/TIA IS-54. Cellular system dual-mode mobile station–base station compatibility standard.

[39] El-Gamal, H. and Damen, M. O. Universal space-time coding. *IEEE Trans. on Information Theory*, **49**(5): May 2003, 1097–1119.

[40] El-Gamal, H. and Hammons, Jr., A. R. A new approach to layered space-time coding and signal processing. *IEEE Trans. on Information Theory*, **47**(6): Sept. 2001, 2321–34.

[41] Fincke, U. and Pohst, M. Improved methods for calculating vectors of short length in a lattice, including a complexity analysis. *Math. Comput.*, **44**: Apr. 1985, 463–71.

[42] Foschini, G. J. and Gans, M. On the limits of wireless communication in a fading environment when using multiple antennas. *Wireless Pers. Commun.*, **6**: Mar. 1998, 311–35.

[43] Foschini, Jr., G. J. Layered space-time architecture for wireless communication in a fading environment when using multi-element antennas. *Bell Labs Technical Journal*, Autumn 1996, 41–59.

[44] Foschini, Jr., G. J., Chizhik, D., Gans, M. J., Papadias, C. and Valenzuela, R. A. Analysis and performance of some basic space-time architectures. *IEEE Journal on Selected Areas in Communications*, **21**(3): Apr. 2003, 303–20.

[45] Foschini, Jr., G. J., Golden, G. D., Valenzuela, R. A. and Wolniansky, P. W. Simplified processing for high spectral efficiency wireless communications employing multi-element arrays. *IEEE Journal on Selected Areas in Communications*, **17**: Nov. 1999, 1841–52.

[46] Ganesan, G. and Stoica, P. Differential detection based on space-time block codes. *Wireless Personal Communications*, **21**: 2002, 163–80.

[47] Ganesan, G. and Stoica, P. Differential modulation using space-time block codes. *IEEE Signal Processing Letters*, **9**(2): 2002, 57–60.

[48] Gao, C., Haimovich, A. M. and Lao, D. Multiple-symbol differential detection for space-time block codes. *Conference on Information Sciences and Systems*, Mar. 2002.

[49] Geramita, A. V. and Seberry, J. *Orthogonal Designs, Quadratic Forms and Hadamard Matrices*. Lecture Notes in Pure and Applied Mathematics, 1979.

[50] Giraud, X., Boutillon, E. and Belfiore, J.-C. Algebraic tools to build modulation schemes for fading channels. *IEEE Trans. on Information Theory*, **43**: May 1997, 938–52.

[51] Guey, J.-C., Fitz, M. P., Bell, M. R. and Kuo, W.-Y. Signal design for transmitter diversity wireless communication systems over Rayleigh fading channels. *IEEE Trans. on Communications*, **47**(4): Apr. 1999, 527–37.

[52] Guo, Z., Zhu, W. and Letaief, K. B. Space-frequency trellis coding for MIMO-OFDM systems. *IEEE Vehicular Technology Conference*, **1**: Apr. 2003, 557–61.

[53] Hammons, Jr., A. R. and El-Gamal, H. On the theory of space-time codes for PSK modulation. *IEEE Trans. on Information Theory*, **46**: Mar. 2000, 524–42.

[54] Hassibi, B. A fast square-root implementation for BLAST. *Asilomar Conference*, **2**: Oct. 2000, 1255–9.

[55] Hassibi, B. and Hochwald, B. High-rate codes that are linear in space and time. *IEEE Trans. on Information Theory*, **48**(7): July 2002, 1804–24.

[56] Hassibi, B. and Hochwald, B. M. How much training is needed in multiple-antenna wireless links? *IEEE Trans. on Information Theory*, **49**: Apr. 2003, 951–63.

[57] Haykin, S. and Moher M. *Modern Wireless Communications*. Pearson Prentice Hall, 2005.

[58] Helfrich, B. Algorithms to construct Minkowski reduced and Hermite reduced lattice bases. *Theor. Comput. Sci.*, **41**(2-3): 1985, 125–39.

[59] Hochwald, B. M. and Marzetta, T. L. Unitary space-time modulation for multiple-antenna communications in Rayleigh flat fading. *IEEE Trans. on Information Theory*, **46**(2): Mar. 2000, 543–64.

[60] Hochwald, B. M., Marzetta, T. L., Richardson, T. J., Sweldens, W. and Urbanke, R. Systematic design of unitary space-time constellation. *IEEE Trans. on Information Theory*, **46**(6): Sept. 2000, 1962–73.

[61] Hochwald, B. M. and Sweldens, W. Differential unitary space-time modulation. *IEEE Trans. on Communications*, **48**(12): Dec. 2000, 2041–52.

[62] Horn, A. H. and Johnson, C. R. *Matrix Analysis*. Cambridge University Press, 1999.

[63] Hughes, B. L. Differential space-time modulation. *IEEE Trans. on Information Theory*, **46**(7): Nov. 2000, 2567–78.

[64] Hui, A. L. C. and Letaief, K. B. Successive interference cancellation for mutiuser asynchronouns DS/CDMA detectors in multipath fading links. *IEEE Trans. on Communications*, **46**(3): Mar. 1998, 1067–75.

[65] Ionescu, D. M. New results on space-time code design criteria. *IEEE Wireless Communications and Networking Conference (WCNC)*, 1999, 684–7.

[66] Ionescu, D. M., Mukkavilli, K. K., Yan, Z. and Lilleberg, J. Improved 8-state and 16-state space time codes for 4-PSK with two transmit antennas. *IEEE Communications Letters*, **5**(7): July 2001, 301–3.

[67] Jafarkhani, H. A quasi-orthogonal space-time block code. *IEEE Wireless Communications and Networking Conference (WCNC)*, **1**: Sept. 2000, 23–8.

[68] Jafarkhani, H. A quasi-orthogonal space-time block code. *IEEE Trans. on Communications*, **49**(1): Jan. 2001, 1–4.

[69] Jafarkhani, H. A Noncoherent detection scheme for space-time block codes. In V. Bhargava, H. V. Poor, V. Tarokh, and S. Yoon, eds., *Communications, Information, and Network Security*. Kluwer Academic Publishers, 2003 pp. 768–72.

[70] Jafarkhani, H. and Hassanpour, N. Super-quasi-orthogonal space-time trellis codes. *IEEE Trans. on Wireless Communications*, **4**(1), Jan. 2005, 215–27.

[71] Jafarkhani, H. and Seshadri, N. Super-orthogonal space-time trellis codes. *IEEE Trans. on Information Theory*, **49**(4): Apr. 2003, 937–50.

[72] Jafarkhani, H. and Taherkhani, F. Pseudo orthogonal designs as space-time block codes. *IEEE International Symposium on Advances in Wireless Communications (ISWC'02)*, Sept. 2002.

[73] Jafarkhani, H. and Tarokh, V. Multiple transmit antenna differential detection from generalized orthogonal designs. *IEEE Trans. on Information Theory*, **47**(6): Sept. 2001, 2626–31.

[74] Jakes, W. C. *Microwave Mobile Communication*. Wiley, 1974.

[75] Jongren, G. and Skoglund, M. Quantized feedback information in orthogonal space-time block coding. *IEEE Trans. on Information Theory*, **50**(10): Oct. 2004, 2473–86.

[76] Jongren, G., Skoglund, M. and Ottersten, B. Combining beamforming and orthogonal space-time block coding. *IEEE Trans. on Information Theory*, **48**(3): Mar. 2002, 611–27.

[77] Kannan, R. Minkowski's convex body theorem and integer programming. *Math. Oper. Res.*, **12**: Aug. 1987, 415–40.

[78] Kim, I.-M. and Tarokh, V. Variable-rate spacetime block codes in M-ary PSK systems. *IEEE Journal on Selected Areas in Communications*, **21**(3): Apr. 2003, 362–73.

[79] Lebrun, G., Faulkner, M., Shafi, M. and Smith, P. J. MIMO Ricean channel capacity. *IEEE International Conference on Communications (ICC)*, **5**: June 2004, 2939–43.

[80] Li, X. and Ritcey, J. A. Trellis-coded modulation with bit interleaving and iterative decoding. *IEEE Journal on Selected Areas in Communications*, **17**: Apr. 1999, 715–24.

[81] Li, Y., Seshadri, N. and Ariyavisitakul, S. Channel estimation for OFDM systems with transmitter diversity in mobile wireless channels. *IEEE Journal on Selected Areas in Communications*, **17**: Mar. 1999, 461–71.

[82] Liang, X.-B. A high-rate orthogonal space-time block code. *IEEE Communications Letters*, **7**(5): May 2003, 222–3.

[83] Liang, X.-B. and Xia, X.-G. On the nonexistence of rate-one generalized complex orthogonal designs. *IEEE Trans. on Information Theory*, **49**(11): Nov. 2003, 2984–9.

[84] Lin, L., Cimini, L. J. and Chuang, J. C.-I. Comparison of convolutional and turbo codes for OFDM with antenna diversity in high-bit-rate wireless applications. *IEEE Communications Letters*, **4**(9): Sept. 2000, 277–9.

[85] Lin, X. and Blum, R. S. Improved space-time codes using serial concatenation. *IEEE Communications Letters*, **4**: July 2000, 221–3.

[86] Litva, J. and Lo, T. K.-Y. *Digital Beamforming in Wireless Communications*. Artech House Publishers, 1996.

[87] Liu, L. and Jafarkhani, H. Application of quasi-orthogonal space-time block codes in beamforming. *IEEE Trans. on Signal Processing*, Jan. 2005.

[88] Liu, Y. and Fitz, M. P. Space-time turbo codes. *Allerton Conference*, Sept. 1999.

[89] Liu, Y., Fitz, M. P. and Takeshita, O. Y. Full rate space-time turbo codes. *IEEE Journal on Selected Areas in Communications*, **19**(5): May 2001, 969–80.

[90] Liu, Z., Xin, Y. and Giannakis, G. B. Space-time-frequency coded OFDM over frequency-selective fading channels. *IEEE Trans. on Signal Processing*, **50**(10): Oct. 2002, 2465–76.

[91] Lu, B. and Wang, X. Space-time code design in OFDM systems. *IEEE Globecom*, **2**: Nov. 2000, 1000–4.

[92] Lu, H.-F., Kumar, P. V. and Chung, H. On orthogonal designs and space-time codes. *IEEE International Symposium on Information Theory (ISIT)*, June 2002, 418.

[93] Molisch, A. F., Win, M. Z. and Winters, J. H. Space-time-frequency (STF) coding for MIMO-OFDM systems. *IEEE Communications Letters*, **6**(9): Sept. 2002, 370–2.

[94] Naguib, A. F., Seshadri, N. and Calderbank, A. R. Applications of space-time block codes and interference suppression for high capacity and high data rate wireless systems. *Asilomar Conference on Signals, Systems and Computers*, **2**: Nov. 1998, 1803–10.

[95] Narayanan, K. R. Turbo decoding of concatenated space-time codes. *Allerton Conference*, Sept. 1999.

[96] Narula, A., Trott, M. and Wornell, G. Performance limits of coded diversity methods for transmitter antenna arrays. *IEEE Trans. on Information Theory*, **45**: Nov. 1999, 2418–33.

[97] Paterson, K. G. On codes with low peak-to-average power ratio for multicode CDMA. *IEEE Trans. on Information Theory*, **50**(3): Mar. 2004, 550–9.

[98] Pohst, M. On the computation of lattice vectors of minimal length, successive minima and reduced bases with applications. *ACM SIGSAM Bull.*, **15**: Feb. 1981, 37–44.

[99] Pottie, G. J. System design choices in personal communications. *IEEE Personal Commun. Mag.*, **2**: Oct. 1995, 50–67.

[100] Proakis, J. G. *Digital Communications*. McGraw-Hill Inc., 1989.

[101] Radon, J. Lineare scharen orthogonaler matrizen. *Abhandlungen aus dem Mathematischen Seminar der Hamburgishen Universitat*, **I**: 1922, 1–14.

[102] Raleigh, G. and Cioffi, J. M. Spatio-temporal coding for wireless communications. *IEEE Globecom*, 1996, 1809–14.

[103] Rappaport, T. S. *Wireless Communications: Principles and Practice*. Prentice Hall, 1996.

[104] Rathinakumar, A. and Motwani, R. Full rate space-time turbo codes for general constellations. *IEEE Workshop on Signal Processing Advances in Wireless Communications*, 2003, 264–8.

[105] Sandhu, S. and Paulraj, A. Space-time block codes: A capacity perspective. *IEEE Communications Letters*, **4**: Dec. 2000, 384–6.

[106] Schlegel, C. *Trellis and Turbo Coding*. Wiley-IEEE Press, 2003.

[107] Schnorr, C. P. and Euchner, M. Lattice basis reduction: improved practical algorithms and solving subset sum problems. *Math. Programming*, **66**: 1994, 181–91.

[108] Sellathurai, M. and Haykin, S. TURBO-BLAST for wireless communications: theory and experiments. *IEEE Trans. on Signal Processing*, **50**(10): Oct. 2002, 2538–46.

[109] Sellathurai, M. and Haykin, S. TURBO-BLAST for wireless communications: first experimental results. *IEEE Trans. on Vehicular Technology*, **52**(3): May 2003, 530–5.

[110] Seshadri, N. and Winters, J. H. Two signaling schemes for improving the error performance of frequency-division-duplex (FDD) transmission systems using transmitter antenna diversity. *IEEE Vehicular Technology Conference*, May 1993, 508–11.

[111] Shankar, P. M. *Introduction to Wireless Systems*. John Wiley & Sons, 2002.

[112] Sharma, N. and Papadias, C. B. Improved quasi-orthogonal codes through constellation rotation. *IEEE Trans. on Communications*, **51**(3): Mar. 2003, 332–5.

[113] Simon, M. K. Evaluation of average bit error probability for space-time coding based on a simpler exact evaluation of pairwise error probability. *Int. Jour. Commun. and Networks*, **3**(3): Sept. 2001, 257–64.

[114] Simon, M. K. A moment generating function (MGF)-based approach for performance evaluation of space-time coded communication systems. *Wireless Communication and Mobile Computing*, **2**(7): Nov. 2002, 667–92.

[115] Simon, M. K. and Alouini, M. *Digital Communication Over Fading Channels: A Unified Approach to Performance Analysis*. John Wiley & Sons, 2000.

[116] Simon, M. K. and Alouini, M. *Digital Communication Over Fading Channels*. John Wiley & Sons, 2004.

[117] Simon, M. K. and Jafarkhani, H. Performance evaluation of super-orthogonal space-time trellis codes using a moment generating function-based approach. *IEEE Trans. on Signal Processing*, **51**(11): Nov. 2003, 2739–51.

[118] Siwamogsatham, S. and Fitz, M. P. Improved high rate space-time codes via expanded STBC-MTCM constructions. *IEEE International Symposium on Information Theory (ISIT)*, June 2002, 106.

[119] Siwamogsatham, S. and Fitz, M. P. Improved high rate space-time codes via orthogonality and set partitioning. *IEEE Wireless Communications and Networking Conference (WCNC)*, **1**: Mar. 2002, 264–70.

[120] Siwamogsatham, S. and Fitz, M. P. Robust space-time coding for correlated Rayleigh fading channels. *IEEE Trans. on Signal Processing*, **50**(10): Oct. 2002, 2408–16.

[121] Smith, P. J. and Shafi, M. On a Gaussian approximation to the capacity of wireless MIMO systems. *IEEE International Conference on Communications (ICC)*, 2002, 406–10.

[122] Stefanov, A. and Duman, T. M. Turbo-coded modulation for systems with transmit and receive antenna diversity over block fading channels: system model, decoding approaches, and practical considerations. *IEEE Journal on Selected Areas in Communications*, **19**(5): May 2001, 958–68.

[123] Stuber, G. L. *Principles of Mobile Communication*. Kluwer Academic Publishers Group, 2001.

[124] Stuber, G. L., Barry, J. R., McLaughlin, S. W., Li, Y., Ingram, M. A. and Pratt, T. G. Broadband MIMO-OFDM wireless communications. *Proceedings of the IEEE*, **92**(2): Feb. 2004, 271–94.

[125] Su, H. and Geraniotis, E. Space-time turbo codes with full antenna diversity. *IEEE Trans. on Communications*, **49**(1): Jan. 2001, 47–57.

[126] Su, W. and Xia, X. Signal constellations for quasi-orthogonal space-time block codes with full diversity. *IEEE Trans. on Information Theory*, **50**(10): Oct. 2004, 2331–47.

[127] Su, W. and Xia, X.-G. Two generalized complex orthogonal space-time block codes of rates 7/11 and 3/5 for 5 and 6 transmit antennas. *IEEE Trans. on Information Theory*, **49**(1): Jan. 2003, 313–16.

[128] Su, W., Xia, X.-G. and Liu, K. J. R. A systematic design of high-rate complex orthogonal space-time block codes. *IEEE Communications Letters*, **8**: June 2004, 380–2.

[129] Sundberg, C.-E. W. and Seshadri, N. Digital cellular systems for North America. *IEEE Globecom*, Dec. 1990, 533–7.

[130] Tao, M. and Cheng, R. S. Differential space-time block codes. *IEEE Globecom*, **2**: Nov. 2001, 1098–102.

[131] Taricco, G. and Biglieri, E. Exact pairwise error probability of space-time codes. *IEEE Trans. on Information Theory*, **48**(2): Feb. 2002, 510–13.

[132] Tarokh, V., Alamouti, S. M. and Poon, P. New detection schemes for transmit diversity with no channel estimation. *IEEE International Conference on Universal Personal Communications*, **2**: Oct. 1998, 917–20.

[133] Tarokh, V. and Jafarkhani, H. A differential detection scheme for transmit diversity. *IEEE Journal on Selected Areas in Communications*, **18**(7): July 2000, 1169–74.

[134] Tarokh, V. and Jafarkhani, H. On the computation and reduction of the peak to average power ratio in multicarrier communications. *IEEE Trans. on Communications*, **48**(1): Jan. 2000, 37–44.

[135] Tarokh, V., Jafarkhani, H. and Calderbank, A. R. Space-time block codes for high data rate wireless communication: performance results. *IEEE Journal on Selected Areas in Communications*, **17**(3): Mar. 1999, 451–60.

[136] Tarokh, V., Jafarkhani, H. and Calderbank, A. R. Space-time block codes from orthogonal designs. *IEEE Trans. on Information Theory*, **45**(5): July 1999, 1456–67.

[137] Tarokh, V., Naguib, A., Seshadri, N. and Calderbank, A. R. Combined array processing and space-time coding. *IEEE Trans. on Information Theory*, **45**(4): May 1999, 1121–8.

[138] Tarokh, V., Naguib, A., Seshadri, N. and Calderbank, A. R. Space-time codes for high data rates wireless communications: performance criteria in the presence of channel estimation errors, mobility and multiple paths. *IEEE Trans. on Communications*, **47**(2): Feb. 1999, 199–207.

[139] Tarokh, V., Seshadri, N. and Calderbank, A. R. Space-time codes for high data rate wireless communication: performance analysis and code construction. *IEEE Trans. on Information Theory*, **44**: Mar. 1998, 744–65.

[140] Telatar, E. Capacity of multi-antenna Gaussian channels. *European Transactions on Telecommunications*, **10**: Nov./Dec. 1999, 585–95.

[141] Tirkkonen, O. Optimizing space-time block codes by constellation rotations. *Finnish Wireless Communications Workshop (FWWC)*, Oct. 2001.

[142] Tirkkonen, O., Boariu, A. and Hottinen, A. Minimal non-orthogonality rate 1 space-time block code for 3+ tx antennas. *IEEE 6th Int. Symp. on Spread-Spectrum Tech. and Appl. (ISSSTA2000)*, Sept. 2000, 429–32.

[143] Turin, G. L. The characteristic function of Hermitian quadratic forms in complex normal random variables. *Biometrika*, June 1960, 199–201.

[144] Ungerboeck, G. Channel coding for multilevel/phase signals. *IEEE Trans. on Information Theory*, **28**(1): Jan. 1982, 55–67.

[145] Uysal, M. and Georghiades, C. N. Error performance analysis of space-time codes over Rayleigh fading channels. *Int. Jour. Commun. and Networks*, **2**(4): Dec. 2000, 351–5.

[146] Viterbo, E., and Boutros, J. A universal lattice code decoder for fading channels. *IEEE Trans. on Information Theory*, **45**(5): July 1999, 1639–42.

[147] Vucetic, B. & Yuan, J. S. *Turbo Codes: Principles and Applications*. Kluwer Academic Publishers, 2000.

[148] Wang, H. and Xia, X.-G. Upper bounds of rates of space-time block codes from complex orthogonal designs. *IEEE Trans. on Information Theory*, **49**(10): Oct. 2003, 2788–96.

[149] Wicker, S. B. *Error Control Systems for Digital Communication and Storage*. Prentice Hall, 1995.

[150] Win, M. Z. and Winters, J. H. Analysis of hybrid selection/maximal-ratio combining in Rayleigh fading. *IEEE Trans. on Communications*, **47**(12): Dec. 1999, 1773–6.

[151] Winters, J. H. The diversity gain of transmit diversity in wireless systems with Rayleigh fading. *IEEE International Conference on Communications (ICC)*, **2**: 1994, 1121–5.

[152] Winters, J., Salz, J. and Gitlin, R. D. The impact of antenna diversity on the capacity of wireless communication systems. *IEEE Trans. on Communications*, **42**: Feb./Mar./Apr. 1994, 1740–51.

[153] Wittneben, A. Base station modulation diversity for digital SIMULCAST. *IEEE Vehicular Technology Conference*, May 1991, 848–53.

[154] Wittneben, A. A new bandwidth efficient transmit antenna modulation diversity scheme for linear digital modulation. *IEEE International Conference on Communications (ICC)*, **3**: May 1993, 1630–4.

[155] Wolniansky, P. W., Foschini, Jr., G. J., Golden, G. D. and Valenzuela, R. A. V-BLAST: an architecture for realizing very high data rates over the rich-scattering wireless channel. *International Symposium on Signals, Systems and Electronics (ISSSE)*, Sept. 1998, 295–300.

[156] Yan, Q. and Blum, R. S. Improved space-time convolutional codes for quasi-static fading channels. *IEEE Trans. on Wireless Communications*, **1**(4): Oct. 2002, 563–71.

[157] Yang, B., Letaief, K. B., Cheng, R. S. and Cao, Z. Timing recovery for OFDM transmission. *IEEE Journal on Selected Areas in Communications*, **18**: Nov. 2000, 2278–91.

[158] Yoon, C. Y., Kohno, R. and Imai, H. A spread-spectrum multiaccess system with cochannel interference cancellation for mutipath fading channels. *IEEE Journal on Selected Areas in Communications*, **11**: Sept. 1993, 1067–75.

[159] Zheng, L. and Tse, D. N. C. Diversity and multiplexing: a fundamental trade-off in multiple-antenna channels. *IEEE Trans. on Information Theory*, **49**(5): May 2003, 1073–96.

[160] Zhou, S. and Giannakis, G. B. Optimal transmitter eigen-beamforming and space-time block coding based on channel mean feedback. *IEEE Trans. on Signal Processing*, **50**(10): Oct. 2002, 2599–613.

[161] Zhu, Y. and Jafarkhani, H. Differential modulation based on quasi-orthogonal codes. *IEEE Wireless Communications and Networking Conference (WCNC)*, Mar. 2004.

Index